I0755494

CASSERLY'S TRANSLATION TO JACOB'S GREEK READER; for the use of Schools, Colleges, and private lessons, with copious notes, and a complete Parsing Index. 12mo.

LEUSDEN'S GREEK AND LATIN TESTAMENT. 12mo.

GRÆCA MINORA; with extensive English Notes and a Lexicon.

VALPY'S GREEK GRAMMAR; greatly enlarged and improved, by Charles Anthon, LL. D. 12mo.

BECK'S CHEMISTRY; a new and improved edition.

THE SCHOOL FRIEND. By Miss Robbins. 18mo.

LEVIZAC'S FRENCH GRAMMAR; revised and corrected by Mr. Stephen Pasquier, M. A. With the Voltarian Orthography, according to the Dictionary of the French Academy. 12mo.

CHRESTOMATHE DE LA LITTERATURE FRANCAISE, &c. By C. Ladreyt. 12mo.

RECUEIL CHOISI de Traits Historiques et de Contes Moraux; with the signification of Words in English at the bottom of each page; for the use of Young Persons of both Sexes, by N. Wanostrocht. Corrected and enlarged, with the Voltarian Orthography, according to the Dictionary of the French Academy, by Paul Moules. 12mo.

HISTORY OF CHARLES XII., in French, by Voltaire. 18mo.

LE BRETHON'S FRENCH GRAMMAR; especially designed for persons who wish to study the elements of that language. First American from the seventh London edition, corrected, enlarged and improved; by P. Bekeart. 1 vol. 12mo.

SIMPLE AND EASY GUIDE TO THE STUDY OF THE FRENCH GRAMMAR. By Wm. P. Wilson. 12mo.

FRENCH COMPANION, consisting of familiar conversations on every topic that can be useful: together with models of letters, notes and cards. The whole exhibiting the true pronunciation of the French Language, the silent letters being printed in Italic throughout the work. By Mr. De Rouillon. Second American, from the tenth London edition. By Prof. Mouls. 1 vol. 18mo.

BLACKSTONE'S COMMENTARIES on the Laws of England; with Notes by Christian, Chitty, Lee, Hovendon, and Ryland. Also, a life of the Author, and References to American Cases. By a member of the New York bar. 2 vols. 8vo.

DUBLIN PRACTICE OF MIDWIFERY, with Notes and Additions. By Dr. Gilman. 12mo.

BLAIR'S LECTURES ON RHETORIC; abridged, with questions for the use of Schools. 18mo.

ENGLISH HISTORY; adapted to the use of Schools, and young persons. Illustrated by a map and engravings, by Miss Robbins. Third edition. 1 vol. 12mo.

ENGLISH EXERCISES; adapted to Murray's English Grammar, consisting of Exercises in Parsing, instances of False Orthography, violations of the Rules of Syntax, Defects in Punctuation; and violations of the Rules respecting Perspicuous and Accurate Writing. Designed for the benefit of private learners, as well as for the use of Schools. By Lindley Murray. 18mo.

RYAN'S ASTRONOMY on an improved plan, in three Books; systematically arranged and scientifically illustrated with several cuts and engravings, and adapted to the instruction of youth, in Schools and Academies. 18mo.

MYTHOLOGICAL FABLES; translated by Dryden, Pope, Congreve, Addison, and others; prepared expressly for the use of Youth. 12mo.

YOUTH'S PLUTARCH, or Select Lives of Greeks and Romans By Miss Robbins. 18mo.

Dean's Stereotype Edition.

BONNYCASTLE'S INTRODUCTION TO ALGEBRA;

CONTAINING THE

INDETERMINATE AND DIOPHANTINE ANALYSIS,

AND THE

APPLICATION OF ALGEBRA TO GEOMETRY.

REVISED, CORRECTED AND ENLARGED

By JAMES RYAN

AUTHOR OF "AN ELEMENTARY TREATISE ON ALGEBRA THEORETICAL AND PRACTICAL," &c.

TO WHICH IS ADDED

A LARGE COLLECTION OF PROBLEMS FOR EXERCISE,

ORIGINAL AND SELECTED,

By JOHN F. JENKINS, A.M.,

PRINCIPAL OF THE MECHANICS' SOCIETY SCHOOL.

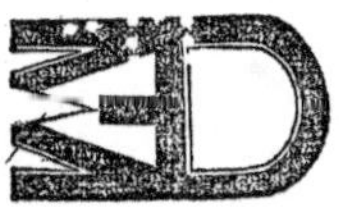

NEW YORK:

W. E. DEAN, PRINTER & PUBLISHER,

2 ANN STREET.

1851.

ADVERTISEMENT

TO

THE SECOND NEW YORK EDITION.

It would be superfluous to advance any thing in commendation of "Bonnycastle's Introduction to Algebra," as the number of European editions, and the increase of demand for it since its publication in this country, are sufficient proofs of its great utility.

But to make it universally useful both to the tutor and scholar, I have given, in this edition, the answers that were omitted by the author in the original.

In the course of the work, particularly in Addition, Subtraction, Multiplication, Division, Fractions, Simple Equations, and Quadratics, I have added a great variety of practical examples, as being essentially necessary to exercise young students in the elementary principles.

Several new rules are introduced, those of principal note are the following: Case 12. Surds, containing two rules for finding any root of a Binomial Surd, the Solution of Cubics by Converging Series, the Solution of Biquadratics by Simpson's and Euler's methods: all these rules are investigated in the plainest manner possible, with notes and remarks, interspersed throughout the work, containing some very useful matter.

There is also given all the Diophantine Analysis, contained in Bonnycastle's Algebra, Vol. I. 8vo. 1820, being a methodical abstract of this part of the science, which comprehends most of the methods hitherto known for resolving problems of this kind, and will be found a ready compendium for such readers as may acquire some knowledge of the Analytic Art.

JAMES RYAN.

New York, January 1, 1822.

PREFACE.

THE powers of the mind, like those of the body, are increased by frequent exertion; application and industry supply the place of genius and invention; and even the creative faculty itself may be strengthened and improved by use and perseverance. Uncultivated nature is uniformly rude and imbecile, it being by imitation alone that we at first acquire knowledge, and the means of extending its bounds. A just and perfect acquaintance with the simple elements of science, is a necessary step towards our future progress and advancement; and this assisted by laborious investigation and habitual inquiry, will constantly lead to eminence and perfection.

Books of rudiments, therefore, concisely written, well digested, and methodically arranged, are treasures of inestimable value; and too many attempts cannot be made to render them perfect and complete. When the first principles of any art or science are firmly fixed and rooted in the mind, their application soon becomes easy, pleasant and obvious ; the understanding is delighted and enlarged; we conceive clearly, reason distinctly, and form just and satisfactory conclusions. But, on the contrary, when the mind, instead of reposing on the stability of truth and received principles, is wandering in doubt and uncertainty, our ideas will necessarily be confused and obscure; and every step we take must be attended with fresh difficulties and endless perplexity.

That the grounds, or fundamental parts of every science, are dull and unentertaining, is a complaint universally made, and a truth not to be denied; but then, what is obtained with difficulty is usually remembered with ease; and what is purchased with pain is often possessed with pleasure. The seeds of knowledge are sown in every soil, but it is by proper culture alone that they are cherished and brought to maturity. A few years of early and assiduous application never fails to procure us the reward of our industry; and who is there, who knows the pleasures and advantages which the sciences afford, that would think his time, in this case, misspent, or his labours useless? Riches and honours are the gift of fortune, casually bestowed, or hereditarily received, and are frequently abused by their possessors; but the superiority of wisdom and knowledge is a pre-eminence of merit, which originates with the man, and is the noblest of all distinctions.

Nature, bountiful and wise in all things, has provided us with an infinite variety of scenes, both for our instruction and entertainment; and, like a kind and indulgent parent, admits all her children to an equal participation of her blessings. But as the modes, situations, and circumstances of life are various, so accident, habit, and education, have each their predominating influence, and give to every mind its particular bias. Where examples of excellence are wanting, the attempts to attain it are but few; but eminence excites attention, and produces imitation. To raise the curiosity, and to awaken the listless and dormant powers of younger minds, we have only to point out to them a valuable acquisition, and the means of obtaining it; the active principles are immediately put into motion, and the certainty of the conquest is ensured from a determination to conquer.

But of all the sciences which serve to call forth this spirit of enterprise and inquiry, there are none more eminently useful than Mathematics. By an early attachment to these elegant and sublime studies, we acquire a habit of reasoning, and an elevation of thought, which fixes the mind, and prepares it for every other pursuit. From a few axioms, and evident principles, we proceed gradually to the most general propositions and remote analogies; deducing one truth from another, in a chain of argument well connected and logically pursued; which brings us at last, in the most satisfactory manner, to the conclusion, and serves as a general direction in all our inquiries after truth.

And it is not only in this respect that mathematical learning is so highly valuable; it is, likewise, equally estimable for its practical utility. Almost all the works of art and devices of man, have a dependance upon its principles, and are indebted to it for their origin and perfection. The cultivation of these admirable sciences is, therefore, a thing of the utmost importance, and ought to be considered as a principal part of every liberal and well-regulated plan of education. They are the guide of our youth, the perfection of our reason, and the foundation of every great and noble undertaking.

From these considerations, I have been induced to compose an introductory course of mathematical science; and from the kind encouragement which I have hitherto received, am not without hopes of a continuance of the same candour and approbation. Considerable practice as a teacher, and a long attention to the difficulties and obstructions which retard the progress of learners in general, have enabled me to accommodate myself the more easily to their capacities and understandings. And as an earnest desire of promoting and diffusing useful knowledge is the chief motive for this undertaking, so no pains or attention shall be wanting to make it as complete and perfect as possible.

The subject of the present performance is ALGEBRA; which is one of the most important and useful branches of those sciences, and may be justly considered as the key to all the rest. Geometry delights us by the simplicity of its principles, and the elegance of its demonstrations; Arithmetic is confined in its object, and partial in its application; but Algebra, or the Analytic Art, is general and comprehensive, and may be applied with success in all cases where truth is to be obtained and proper data can be established.

To trace this science to its birth, and to point out the various alterations and improvements it has undergone in its progress, would far exceed the limits of a preface.* It will be sufficient to observe that the invention is of the highest antiquity, and has challenged the praise and adoration of all ages. *Diophantus*, a Greek mathematician, of Alexandria in Egypt, who flourished in or about the third century after Christ, appears to have been the first, among the ancients, who applied it to the solution of indeterminate or unlimited problems; but it is to the moderns that we are principally indebted for the most curious refinements of the art, and its great and extensive usefulness in every abstruse and difficult inquiry. *Newton*, *Maclaurin*, *Sanderson*, *Simpson*,

* Those who are desirous of a knowledge of this kind, may consult the Introduction to my *Treatise on Algebra*; where they will find a regular historical detail of the rise and progress of the science, from its first rude beginnings to the present times.

and *Emerson*, among our own countrymen, and *Clairaut*, *Euler*, *Lagrange*, and *Lacroix*, on the continent, are those who have particularly excelled in this respect; and it is to their works that I would refer the young student, as the patterns of elegance and perfection.

The following compendium is formed entirely upon the model of those writers, and is intended as a useful and necessary introduction to them. Almost every subject, which belongs to pure Algebra, is concisely and distinctly treated of; and no pains have been spared to make the whole as easy and intelligible as possible. A great number of elementary books have already been written upon this subject; but there are none, which I have yet seen, but what appear to me to be extremely defective. Besides being totally unfit for the purpose of teaching, they are generally calculated to vitiate the taste, and mislead the judgment. A tedious and inelegant method prevails through the whole, so that the beauty of the science is generally destroyed by the clumsy and awkward manner in which it is treated; and the learner, when he is afterwards introduced to some of our best writers, is obliged, in a great measure, to unlearn and forget every thing which he has been at so much pains in acquiring.

There is a certain taste and elegance in the sciences, as well as in every branch of polite literature, which is only to be obtained from the best authors, and a judicious use of their instructions. To direct the student in his choice of books, and to prepare him properly for the advantages he may receive from them, is therefore, the business of every writer who engages in the humble but useful task of a preliminary tutor. This information I have been careful to give, in every part of the present performance, where it appeared to be in the least necessary; and, though the nature and confined limits of my plan admitted not of diffused observations or a formal enumeration of particulars, it is presumed nothing of real use and importance has been omitted. My principal object was to consult the ease, satisfaction, and accommodation of the learner; and the favourable reception the work has met with from the public, has afforded me the gratification of believing that my labours have not been unsuccessfully employed.

CONTENTS.

ALGEBRA.

Algebra is the science which treats of a general method of performing calculations, and resolving mathematical problems, by means of the letters of the alphabet.

Its leading rules are the same as those of arithmetic; and the operations to be performed are denoted by the following characters:—

$+$ *plus*, or *more*, the sign of addition; signifying that the quantities between which it is placed are to be added together.

Thus, $a + b$ shows that the number, or quantity, represented by b, is to be added to that represented by a; and is read a plus b.

$-$ *minus*, or *less*, the sign of substraction; signifying that the latter of the two quantities between which it is placed is to be taken from the former.

Thus $a - b$ shows that the quantity represented by b is to be taken from that represented by a: and is read a minus b.

Also, $a \sim b$ represents the difference of the two quantities a and b, when it is not known which of them is the greater.

$\times$ *into*, the sign of multiplication; signifying that the quantities between which it is placed are to be multiplied together.

Thus, $a \times b$ shows that the quantity represented by a is to be multipled by that represented by b; and is read a into b.

The multiplication of simple quantities is also frequently denoted by a point, or by joining the letters together in the form of a word.

Thus, $a \times b$, $a . b$, and ab, all signify the product of a and b; also, $3 \times a$, or $3a$, is the product of 3 and a; and is read 3 times a.

$\div$ *by*, the sign of division; signifying that the former of the two quantities between which it is placed is to be divided by the latter.

Thus, $a \div b$, shows that the quantity represented by a is to be divided by that represented by b; and is read a by b. or a divided by b.

Division is also frequently denoted by placing one of the two quantities over the other, in the form of a fraction

Thus, $b \div a$ and $\frac{b}{a}$ both signify the quotient of b divided by a; and $\frac{a-b}{a+c}$ signifies that $a - b$ is to be divided by $a + c$.

$=$ *equal to*, the sign of equality; signifying that the quantities between which it is placed are equal to each other.

Thus, $x = a + b$ shows that the quantity denoted by x is equal to the sum of the quantities a and b; and is read x equal to a plus b.

Any two algebraic expressions are said to be *identical*, when they are of the same value, for all the values of the letters of which they are composed.

* Thus $(x + a) \times (x - a) = x^2 - a^2$, whatever numeral values may be given to the quantities represented by x and a.

$>$ *greater than*, the sign of majority; signifying that the former of the two quantities between which it is placed is greater than the latter.

Thus $a > b$ shows that the quantity represented by a is greater than that represented by b; and is read a greater than b.

$<$ *less than*, the sign of minority; signifying that the former of the two quantities between which it is placed is less than the latter.

Thus, $a < b$ shows that the quantity represented by a is less than that represented by b; and is read a less than b.

: *as*, or *to*, and : : *so is*, the signs of an equality of ratios; signifying that the quantities between which they are placed are proportional.

Thus, $a : b :: c : d$ denotes that a has the same ratio to b that c has to d, or that a, b, c, d, are proportionals; and is read, as a is to b so is c to d, or, a is to b as c is to d.

$\surd$ *the radical sign*, signifying that the quantity before which is placed is to have some root of it extracted.

---

* Woodhouse, in his Principles of Analytical Calculation, says that $x^2 - a^2$ is not *generally* $= (x-a) \,.\, (x+a)$: for instance, the particular case of $x = a$ is to be excluded; the proof essentially demanding this circumstance, to wit, that $x - a$ be a quantity, or that x be greater than a. Euler calls $x - 1 = x - 1$ an *identical equation*; and shows that x is indeterminate, or that any number whatever may be substituted for it. See Euler's Algebra, page 289, Vol. I.—Ed.

Thus, $\sqrt{a}$ is the square root of a; $\sqrt[3]{a}$ is the cube root of a; and $\sqrt[4]{a}$ is the fourth root of a; &c.

The roots of quantities are also represented by figures placed at the righthand corner of them, in the form of a fraction.

Thus, $a^{\frac{1}{2}}$ is the square root of a; $a^{\frac{1}{3}}$ is the cube root of a; and $a^{\frac{1}{n}}$ is the nth root of a, or a root denoted by any number n.

In like manner, a^2 is the square of a; a^3 is the cube of a; and a^m is the mth power of a, or any power denoted by the number m.

∞ is the sign of infinity, signifying that the quantity standing before it is of an unlimited value, or greater than any quantity that can be assigned.

The coefficient of a quantity is the number or letter which is prefixed to it.

Thus, in the quantities $3b$, $-\frac{2}{3}b$, 3 and $-\frac{2}{3}$ are the coefficients of b; and a is the coefficient of x in the quantity ax.

A quantity without any coefficient prefixed to it is supposed to have 1 or unity; and when a quantity has no sign before it, $+$ is always understood.

Thus, a is the same as $+a$, or $+1a$; and $-a$ is the same as $-1a$.

A term is any part or member of a compound quantity, which is separated from the rest by the signs $+$ or $-$

Thus a and b are the terms of $a+b$; and $3a$, $-2b$, and $+5cd$, are the terms of $3a-2b+5cd$.

In like manner, the terms of a product, fraction, or proportion, are the several parts or quantities of which they are composed.

Thus, a and b are the terms of ab, or of $\frac{a}{b}$; and a, b, c, d, are the terms of the proportion $a:b::c:d$.

A factor is one of the terms, or multipliers which form the product of two or more quantities.

Thus, a and b are the factors of ab; also, 2, a, and b^2, are the factors of $2ab^2$; and $a-x$ and $b-x$ are the factors of the product $(a-x)\times(b-x)$.

A composite number, or quantity, is that which is produced by the multiplication of two or more terms or factors.

Thus, 6 is a composite number, formed of the factors 2 and 3, or 2×3; and $3abc$ is a composite quantity, the factors of which are 3, a, b, c.

Like quantities, are those which consist of the same letters or combinations of letters; as a and $3a$, or $5ab$ and $7ab$, or $2a^2b$ and $9a^2b$.

Unlike quantities, are those which consist of different letters, or combinations of letters; as a and b, or $3a$, and a^2, or $5ab^2$ and $7a^2b$.

Given quantities, are such as have known values, and are generally represented by some of the first letters of the alphabet; as a, b, c, d, &c.

Unknown quantities, are such as have no fixed values, and are usually represented by some of the final letters of the alphabet; as x, y, z.

Simple quantities are those which consist of one term only; as $3a$, $5ab$, $-8a^2b$, &c.

Compound quantities, are those which consist of several terms; as $2a + b$, or $3a - 2c$, or $a + 2b - 3c$, &c.

Positive, or affirmative quantities, are those which are to be added; as a, or $+a$, or $+3ab$, &c.

Negative quantities are those which are to be subtracted: as $-a$, or $-3ab$, or $-7ab^2$, &c.

Like signs, are such as are all positive, or all negative; as $+$ and $+$, or $-$ and $-$

Unlike signs, are when some are positive and others negative; as $+$ and $-$, or $-$ and $+$.

A monomial, is a quantity consisting of one term only; as a, $2b$, $-3a^2b$, &c.

A binomial, is a quantity consisting of two terms, as $a + b$, or $a - b$; the latter of which is, also, sometimes called a residual quantity.

A trinomial, is a quantity consisting of three terms, as $a + 2b - 3c$; a quadrinomial of four, as $a - 2b + 3c - d$, and a polynomial, or multinomial, is that which has many terms.

The power of a quantity, is its square, cube, biquadrate, &c.; called also its second, third, fourth power, &c.; as a^2, a^3, a^4, &c.

The index, or exponent of a quantity, is the number which denotes its power or root.

Thus, -1 is the index of a^{-1}, 2 is the index of a^2, and $\frac{1}{2}$ of $a^{\frac{1}{2}}$ or $\sqrt{a}$.

When a quantity appears without any index, or exponent, it is always understood to have unity, or 1.

Thus, a is the same as a^1, and $2x$ is the same as $2x^1$; the 1, in such cases, being usually omitted.

A rational quantity, is that which can be expressed in finite terms, or without any radical sign, or fractional index; as a, or $\frac{2}{3}a$, or $5a$, &c.

An irrational Quantity, or Surd, is that of which the value cannot be accurately expressed in numbers, as the square roots of 2, 3, 5. Surds are commonly expressed by means of the radical sign $\sqrt{\ }$; as $\sqrt{2}$, $\sqrt{a}$, $\sqrt[4]{a^3}$, or a fractional index; as $2^{\frac{1}{2}}$, $a^{\frac{1}{2}}$, $a^{\frac{3}{4}}$, &c

A square or cube number, &c., is that which has an exact square or cube root, &c.

Thus, 4 and $\frac{9}{16}a^2$ are square numbers; and 64 and $\frac{8}{27}a^3$ are cube numbers, &c.

A measure of any quantity, is that by which it can be divided without leaving a remainder.

Thus, 3 is a measure of 6, $7a$ is a measure of $35a$, and $9ab$ of $27\ a^2b^2$.

Commensurable quantities, are such as can be each divided by the same quantity, without leaving a remainder.

Thus, 6 and 8, $2\sqrt{2}$ and $3\sqrt{2}$, $5a^2b$ and $7ab^2$, are commensurable quantities; the common divisors being 2, $\sqrt{2}$, and ab.

Incommensurable quantities, are such as have no common measure, or divisor, except unity.

Thus, 15 and 16, $\sqrt{2}$ and $\sqrt{3}$, and $a+b$ and a^2+b^2 are incommensurable quantities.

A multiple of any quantity, is that which is some exact number of times that quantity.

Thus, 12 is a multiple of 4, $15a$ is a multiple of $3a$, and $20a^2b^2$ of $5ab$.

The reciprocal of any quantity, is that quantity inverted, or unity divided by it.

Thus, the reciprocal of a, or $\frac{a}{1}$ is $\frac{1}{a}$; and the reciprocal of $\frac{a}{b}$ is $\frac{b}{a}$.

A function of one or more quantities, is an expression into which those quantities enter, in any manner whatever, either combined, or not, with known quantities.

---

* This definition of a Surd, or irrational Quantity, is due to Robert Adrian, LL. D., Professor of Mathematics and Natural Philosophy in the University of Pennsylvania, who had first published it in *his* edition of Hutton's Course of Mathematics.—Ed.

Thus, $a - 2x$, $ax + 3x^3$, $2x - a\,(a^2 - x^2)^{\frac{1}{2}}$, ax^m, a^x, &c., are functions of x; and $axy + bx^2$, $ay + x\,(ax^2 - by^2)^{\frac{1}{2}}$, &c., are functions of x and y.

A vinculum, is a bar ——, or parentheses ( ), made use of to collect several quantities into one.

Thus $\overline{a + b} \times c$, or $(a + b)\,c$, denotes that the compound quantity $a + b$ is to be multiplied by the simple quantity c; and $\sqrt{ab + c^2}$, or $(ab + c^2)^{\frac{1}{2}}$, is the square root of the compound quantity $ab + c^2$.

Practical Examples for computing the numeral Values of various Algebraic Expressions, or Combinations of Letters.

Supposing $a = 6$, $b = 5$, $c = 4$, $d = 1$, and $e = 0$.

Then

1. $a^2 + 2ab - c + d = 36 + 60 - 4 + 1 = 93$.
2. $2a^3 - 3a^2b + c^3 = 432 - 540 + 64 = -44$.
3. $a^2 \times (a + b) - 2abc = 36 \times 11 - 240 = 156$.
4. $2a\sqrt{(b^2 - ac)} + \sqrt{(2ac + c^2)} = 12 \times 1 + 8 = 20$.
5. $3a\sqrt{(2ac + c^2)}$, or $3a\,(2ac + c^2)^{\frac{1}{2}} = 18\sqrt{64} = 144$.
6. $\sqrt{[2a^2 - \sqrt{(2ac + c^2)}]} = \sqrt{(72 - \sqrt{64})} = \sqrt{(72 - 8)} = \sqrt{64} = 8$.
7. $\dfrac{2a + 3c}{6d + 4e} + \dfrac{4bc}{\sqrt{(2ac + c^2)}} = \dfrac{12 + 12}{6 + 0} + \dfrac{80}{\sqrt{(48 + 16)}} = \dfrac{24}{6} + \dfrac{80}{8} = 14$.

Required the numeral values of the following quantities; supposing a, b, c, d, e, to be 6, 5, 4, 1, and 0, respectively, as above.

1. $2a^2 + 3bc - 5d = 127$
2. $5a^2b - 10ab^2 + 2e = -600$
3. $7a^2 + b - c \times d + e = 253$
4. $5\sqrt{ab + b^2} - 2ab - e^2 = -7.613875$
5. $\dfrac{a}{c} \times d - \dfrac{a - b}{d} + 2a^2e = \frac{1}{2}$.
6. $3\sqrt{c} + 2\sqrt{(2a + b - d)} = 14$
7. $a\sqrt{(a^2 + b^2)} + 3bc\sqrt{(a^2 - b^2)} = 245.8589862$
8. $3a^2b + \sqrt[3]{[c^2 + \sqrt{(2ac + c^2)}]} = 542.8844991$
9. $\dfrac{2b + c}{3a - c} - \dfrac{\sqrt{5b} + 3\sqrt{c + d}}{2a + c} = \frac{1}{4}$.

ADDITION.

ADDITION is the connecting of quantities together by means of their proper signs, and incorporating such as are like, or that can be united into one sum; the rule for performing which is commonly divided into the three following cases.*

CASE I.

When the Quantities are like, and have like signs.

RULE.—Add all the coefficients of the several quantities together, and to their sum annex the letter or letters belonging to each term, prefixing, when necessary, the common sign.

EXAMPLES.

$3a$	$-3ax$	$2b+3y$
$5a$	$-6ax$	$5b+7y$
a	$-ax$	$b+2y$
$7a$	$-2ax$	$8b+y$
$12a$	$-7ax$	$4b+4y$
$28a$	$-19ax$	$20b+17y$

$2ay$	$-2by^2$	$a-2x^2$
$5ay$	$-6by^2$	$a-6x^2$
$4ay$	$-by^2$	$4a-x^2$
$7ay$	$-8by^2$	$3a-5x^2$
$16ay$	$-by^2$	$7a-x^2$
$34ay$	$-18by^2$	$16a-15x^2$

$3ax^2$	$7x-4y$	$2a+x^2$
$2ax^2$	$x-8y$	$3a+x^2$
$12ax^2$	$3x-y$	$a+2x^2$
$9ax^2$	$x-3y$	$9a+3x^2$
$10ax^2$	$4x-y$	$4a+x^2$
$36ax^2$	$16x-17y$	$19a+8x^2$

* The term Addition, which is generally used to denote this rule, is too scanty to express the nature of the operations that are to be performed in it; which are sometimes those of addition, and sometimes subtraction, according as the quantities are negative or positive. It should, therefore, be called by some name signifying incorporation, or striking a balance; in which case, the incongruity here mentioned would be removed.

CASE II.

When the Quantities are like, but have unlike signs.

Rule.—Add all the affirmative coefficients into one sum, and those that are negative into another, when there are several of the same kind: then subtract the less of these sums from the greater, and to the difference prefix the sign of the greater, annexing the common letter or letters as before.

EXAMPLES.

$-3a$	$2a-3x^2$	$6x+5ay$
$+7a$	$-7a+5x^2$	$-3x+2ay$
$+8a$	$-3a+x^2$	$x-6ay$
$-a$	$+a-3x^2$	$2x+ay$
$+11a$	$-7a$	$6x+2ay$

$-2a^2$	$3ay-7$	$-3ab+7x$
$-3a^2$	$-ay+8$	$+3ab-10x$
$-8a^2$	$+2ay-9$	$+3ab-6x$
$+10a^2$	$-3ay-11$	$-ab-2x$
$+13a^2$	$+12ay+13$	$+2ab+7x$
$+10a^2$	$+13ay-6$	$+4ab-4x$

$-2a\sqrt{x}$	$-6a^2+2b$	$6ax^2+5x^{\frac{1}{2}}$
$+a\sqrt{x}$	$+2a^2-3b$	$-2ax^2-6x^{\frac{1}{2}}$
$-3a\sqrt{x}$	$-5a^2-8b$	$+3ax^2-10x^{\frac{1}{2}}$
$+7a\sqrt{x}$	$+4a^2-2b$	$-7ax^2+3x^{\frac{1}{2}}$
$-4a\sqrt{x}$	$-3a^2+9b$	$+ax^2+11x^{\frac{1}{2}}$
$-a\sqrt{x}$	$-8a^2-2b$	$+ax^2+3x^{\frac{1}{2}}$

CASE III.

When the Quantities are unlike; or some like and others unlike.

Rule.—Collect all the like quantities together, by taking their sums or differences, as in the foregoing cases, and set down those that are unlike, one after another, with their proper signs.

EXAMPLES.

$5xy$	$2xy - 2x^2$	$2ax - 30$
$4ax$	$3x^2 + xy$	$3x^2 - 2ax$
$-xy$	$x^2 + xy$	$5x^2 - 3x^{\frac{1}{2}}$
$-4ax$	$4x^2 - 3xy$	$3\sqrt{x} + 10$
$4xy$	$6x^2 + xy$	$8x^2 - 20$

$+ax^{\frac{1}{2}}$	$8a^2x^2 - 3ax$	$10b^2 - 3a^2x$
$-ax^2$	$7ax - 5xy$	$-b^2 + 2a^2x^2$
$+3ax^2$	$9xy - 5ax$	$50 + 2a^2x$
$-ax^{\frac{1}{2}}$	$2a^2x^2 + xy$	$a^2x^2 + 120$
$+2ax^2$	$10a^2x^2 + 5xy - ax$	$9b^2 + 3a^2x^2 - a^2x + 170$

$+3a^2y$	$2\sqrt{x} - 17y$	$2a^2 - 3a\sqrt{x}$
$-2xy^2$	$3\sqrt{xy} + 10x$	$x^2 - 2a^{\frac{1}{2}}x^{\frac{1}{2}}$
$-3y^2x$	$2x^2y + 25y$	$3a^2 - 13xy$
$-8x^2y$	$12xy - \sqrt{xy}$	$xy + 32a^2$
$+2xy^2$	$-9y + 18x^{\frac{1}{2}}$	$20 - 65x^2$
$3a^2y - 3y^2x - 8x^2y$	$\left\{ \begin{array}{l} 19\sqrt{x} + 12xy \\ +2x^2y + 2\sqrt{xy} \\ +10x - y \end{array} \right\}$	$37a^2 - 3a\sqrt{x} - 12xy$ $- 64x^2 - 2a^{\frac{1}{2}}x^{\frac{1}{2}} +$ 20

EXAMPLES FOR PRACTICE.

1. Required the sum of $\frac{1}{2}(a+b)$ and $\frac{1}{2}(a-b)$. Ans. a.
2. Add $5x - 3a + b + 7$ and $-4a - 3x + 2b - 9$ together. Ans. $2x - 7a + 3b - 2$.
3. Add $2a + 3b - 4c - 9$ and $5a - 3b + 2c - 10$ together. Ans. $7a - 2c - 19$.
4. Add $3a + 2b - 5$, $a + 5b - c$, and $6a - 2c + 3$ together. Ans. $10a + 7b - 3c - 2$.
5. Add $x^3 + ax^2 + bx + 2$ and $x^3 + cx^2 + dx - 1$ together. Ans. $2x^3 + (a+c)x^2 + (b+d)x + 1$.
6. Add $6xy - 12x^2$, $-4x^2 + 3xy$, $4x^2 - 2xy$, and $-3xy + 4x^2$ together. Ans. $4xy - 8x^2$

7. Add $4ax - 130 + 3x^{\frac{1}{2}}$, $5x^2 + 3ax + 9x^2$, $7xy - 4x^{\frac{1}{2}} + 90$, and $\sqrt{x} + 40 - 6x^2$ together. Ans. $7ax + 8x^2 + 7xy$.

8. Add $2a^2 - 3ab + 2b^3 - 3a^2$, $3b^3 - 2a^2 + a^3 - 5c^3$, $4c^3 - 2b^3 + 5ab + 100$, and $20ab + 16a^2 - bc - 80$ together.

Ans. $13a^2 + 22ab + 3b^3 + a^3 - c^3 + 20 - bc$.

9. Add $\frac{5a}{b} - \frac{3c^2}{a} + \frac{7\sqrt{bc}}{x} - 9(\frac{ab+x}{d})$ and $\frac{8a}{b} + \frac{7c^2}{a} - 12\frac{\sqrt{bc}}{x} + 6(\frac{ab+x}{d})$ together.

Ans. $\frac{13a}{b} + \frac{4c^2}{a} - 5\frac{\sqrt{(bc)}}{x} - 3(\frac{ab+x}{d})$.

10. Add $3a^2 + 4bc - e^2 + 10$, $-5a^2 + 6bc + 2e^2 - 15$, and $-4a^2 - 9bc - 10e^2 + 21$ together.

Ans. $bc - 6a^2 - 9e^2 + 16$.

SUBTRACTION.

Subtraction is the taking of one quantity from another; or the method of finding the difference between any two quantities of the same kind; which is performed as follows:—*

Rule.—Change all the signs (+ and −) of the lower line, or quantities that are to be subtracted, into the contrary signs, or rather conceive them to be so changed, and then collect the terms together, as in the several cases of addition.

EXAMPLES.

$5a^2 - 2b$	$x^2 - 2y + 3$	$5xy + 8x - 2$
$2a^2 + 5b$	$4x^2 + 9y - 5$	$3xy - 8x - 7$
$3a^2 - 7b$	$-3x^2 - 11y + 8$	$2xy + 16x + 5$

* This rule being the reverse of addition, the method of operation must be so likewise. It depends upon this principle, that to subtract an affirmative quantity from an affirmative, is the same as to add a negative quantity to an affirmative.

Thus, according to Laplace, we can write

$$a = a + b - b \ldots\ldots\ldots (1),$$
$$a - c = a - c + b - b \ldots (2);$$

so that if from a we are to subtract $+b$, or $-b$; or what amounts to the same thing, if in a we suppress $+b$, or $-b$; the remainder from transformation (1), must be $a - b$ in the first case, and $a + b$ in the second. Also, if from $a - c$ we take away $+b$, or $-b$, the remainder, from (2), will be $a - c - b$, or $a - c + b$.—Ed.

$5xy - 18$	$8y^2 - 2y - 5$	$10 - 8x - 3xy$
$-xy + 12$	$-y^2 + 3y + 2$	$-x + 3 - xy$
$6xy - 30$	$9y^2 - 5y - 7$	$7 - 7x - 2xy$

$-5x^2y - 8a$	$4\sqrt{ax} - 2x^2y$	$5x^2 + \sqrt{x} - 4y$
$+3x^2y - 7b$	$3\sqrt{ax} - 5xy^2$	$6x^2 - 8x - x^{\frac{1}{2}}$
$-8x^2y - 8a + 7b$	$\sqrt{ax} - 2x^2y + 5xy^2$	$-x^2 + 8x + 2\sqrt{x} - 4y$

EXAMPLES FOR PRACTICE.

1. Find the difference of $\frac{1}{2}(a + b)$ and $\frac{1}{2}(a - b)$. Ans. b.
2. From $3x - 2a - b + 7$, take $8 - 3b + a + 4x$.
   Ans. $2b - x - 3a - 1$.
3. From $3a + b + c - 2d$, take $b - 8c + 2d - 8$.
   Ans. $3a + 9c - 4d + 8$.
4. From $13x^2 - 2ax + 9b^2$, take $5x^2 - 7ax - b^2$.
   Ans. $8x^2 + 5ax + 10b^2$.
5. From $20ax - 5\sqrt{x} + 3a$, take $4ax + 5x^{\frac{1}{2}} - a$.
   Ans. $16ax - 10\sqrt{x} + 4a$.
6. From $5ab + 2b^2 - c + bc - b$, take $b^2 - 2ab + bc$.
   Ans. $7ab + b^2 - c - b$.
7. From $ax^3 - bx^2 + cx - d$, take $bx^2 + ex - 2d$.
   Ans. $ax^3 - 2bx + (c - e)x + d$.
8. From $-6a - 4b - 12c + 13x$, take $4x - 9a + 4b - 5c$.
   Ans. $3a + 9x - 8b - 7c$.
9. From $6x^2y - 3\sqrt{(xy)} - 6ay$, take $3x^2y + 3(xy)^{\frac{1}{2}} - 4ay$.
   Ans. $3x^2y - 6\sqrt{(xy)} - 2ay$.
10. From the sum of $4ax - 150 + 4x^{\frac{1}{2}}$, $5x^2 + 3ax + 10x^{\frac{1}{2}}$, and $90 - 2ax - 12\sqrt{(x)}$; take the sum of $2ax - 80 + 7x^2$, $7x^{\frac{1}{2}} - 8ax - 70$, and $30 - 4\sqrt{(x)} - 2x^2 + 4a^2x^2$.
    Ans. $11ax + 60 - x^{\frac{1}{2}} - 4a^2x^2$.

MULTIPLICATION.

Multiplication, or the finding of the product of two or more quantities, is performed in the same manner as in arithmetic; except that it is usual, in this case, to begin the operation at the lefthand, and to proceed towards the right, or contrary to the way of multiplying numbers.

The rule is commonly divided into three cases; in each of which it is necessary to observe, that like signs, in multiplying, produce +, and unlike signs, —.

It is likewise to be remarked, that powers, or roots of the same quantity, are multiplied together by adding their indices: thus,

$$a \times a^2, \text{ or } a^1 \times a^2 = a^3;\ a^2 \times a^3 = a^5;\ a^{\frac{1}{2}} \times a^{\frac{1}{3}} = a^{\frac{5}{6}}; \text{ and } a^m \times a^n = a^{m+n}.$$

The multiplication of compound quantities, is also, sometimes, barely denoted by writing them down, with their proper signs, under a vinculum, without performing the whole operation, as

$$3ab\ (a - b), \text{ or } 2a\sqrt{(a^2 + b^2)}.$$

Which method is often preferable to that of executing the entire process, particularly when the product of two or more factors is to be divided by some other quantity, because, in this case, any quantity that is common to both the divisor and dividend, may be more readily suppressed; as will be evident from various instances in the following part of the work.*

CASE I.

When the factors are both simple quantities.

RULE.—Multiply the coefficients of the two terms together, and to the product annex all the letters, or their powers, belonging to each, after the manner of a word; and the result, with the proper sign prefixed, will be the product required.†

---

* The above rule for the signs may be proved thus: If B, b, be any two quantities, of which B is the greater, and B — b is to be multiplied by a, it is plain that the product, in this case, must be less than aB, because B — b is less than B; and, consequently, when each of the terms of the former are multiplied by a, as above, the result will be

$$(B - b) \times a = aB - ab.$$

For if it were aB + ab, the product would be greater than aB, which is absurd.

Also, if B be greater than b, and A greater than a, and it is required to multiply B — b by A — a, the result will be

$$(B - b) \times (A - a) = AB - aB - bA + ab.$$

For the product of B — b by A is A (B — b), or AB — Ab, and that of B — b by — a, which is to be taken from the former, is — a (B — b) as has been already shown; whence B — b being less than B, it is evident that the part, which is to be taken away must be less than aB; and consequently since the first part of this product is — aB, the second part must be + ab; for if it were — ab, a greater part than aB would be to be taken from A (B — b), which is absurd.

† When any number of quantities are to be multiplied together, it is the same thing in whatever order they are placed: thus, if ab is to be

EXAMPLES.

$12a$	$-2a$	$+5a$	$-9x^2$
$3b$	$+4b$	$-6x$	$-5bx$
$36ab$	$-8ab$	$-30ax$	$+45bx^3$
$7ab$	$-6a^2x$	$-2xy^2$	$-7axy$
$-5ac$	$+5x$	$-xy$	$+6ay$
$-35a^2bc$	$-30a^2x^2$	$+2x^2y^3$	$-42a^2xy^2$
$3a^2b$	$12a^2x$	$-6xyz$	$-a^2xy$
$2ba^2$	$-2x^2y$	$+ay^2z$	$+2xy^2$
$6a^4b^2$	$-24a^2x^3y$	$-6axy^3z^2$	$-2a^2x^2y^3$

CASE II.

When one of the factors is a compound quantity.

RULE.—Multiply every term of the compound factor, considered as a multiplicand, separately, by the multiplier, as in the former case; then these products, placed one after another with their proper signs, will be the whole product required.

EXAMPLES.

$3a-2b$	$6xy-8$	a^2-2x+1
$4a$	$3x$	$4x$
$12a^2-8ab$	$18x^2y-24x$	$4a^2x-8x^2+4x$
$12x-ab$	$35x-7a$	$3y^2+y-2$
$5a$	$-2x$	xy
$60ax-5a^2b$	$-70x^2+14ax$	$3xy^3+xy^2-2xy$
$13x^2-a^2b$	$25xy+3a^2$	$3x^2-xy+2y^2$
$-2a$	$13x^2$	$5x^2$
$-26ax^2+2a^3b$	$325x^3y+39a^2x^2$	$15x^4-5x^3y-10x^2y^2$

multiplied by c, the product is either abc, acb, or bca, &c.; though it is usual, in this case, as well as in addition and subtraction, to put them according to their rank in the alphabet. It may here also be observed in conformity to the rule given above for the signs, that $(+a)\times(+b)$, or $(-a)\times(-b)=+ab$; and $(+a)\times(-b)$, or $(-a)\times(+b)$ $=-ab$.

CASE III.

When both the factors are compound quantities.

RULE.—Multiply every term of the multiplicand separately, by each term of the multiplier, setting down the products one after another, with their proper signs; then add the several lines of products together, and their sum will be the whole product required.

EXAMPLES.

$$\begin{array}{l} x+y \\ x+y \\ \hline x^2+xy \\ \quad +xy+y^2 \\ \hline x^2+2xy+y^2 \end{array} \qquad \begin{array}{l} 5x+4y \\ 3x-2y \\ \hline 15x^2+12xy \\ \quad -10xy-8y^2 \\ \hline 15x^2+2xy-8y^2 \end{array} \qquad \begin{array}{l} x^2+xy-y^2 \\ x-y \\ \hline x^3+x^2y-xy^2 \\ \quad -x^2y-xy^2+y^3 \\ \hline x^3 \quad * \quad -2xy^2+y^3 \end{array}$$

$$\begin{array}{l} x+y \\ x-y \\ \hline x^2+xy \\ \quad -xy-y^2 \\ \hline x^2 \; * \; -y^2 \end{array} \qquad \begin{array}{l} x^2+y \\ x^2+y \\ \hline x^4+x^2y \\ \quad +x^2y+y^2 \\ \hline x^4+2x^2y+y^2 \end{array} \qquad \begin{array}{l} x^2+xy+y^2 \\ x-y \\ \hline x^3+x^2y+xy^2 \\ -x^2y-xy^2-y^3 \\ \hline x^3 \quad ** \quad -y^3 \end{array}$$

EXAMPLES FOR PRACTICE.

1. Required the product of $x^2 - xy + y^2$ and $x + y$.
Ans. $x^3 + y^3$.

2. Required the product of $x^3 + x^2y + xy^2 + y^3$ and $x - y$.
Ans $x^4 - y^4$.

3. Required the product of $x^2 + xy + y^2$ and $x^2 - xy + y^2$.
Ans. $x^4 + x^2y^2 + y^4$.

4. Required the product of $3x^2 - 2xy + 5$, and $x^2 + 2xy - 3$.
Ans. $3x^4 + 4x^3y - 4x^2y^2 - 4x^2 + 16xy - 15$.

5. Required the product of $2a^2 - 3ax + 4x^2$ and $5a^2 - 6ax - 2x^2$. Ans. $10a^4 - 27a^3x + 34a^2x^2 - 18ax^3 - 8x^4$.

6. Required the product of $5x^3 + 4ax^2 + 3a^2x + a^3$, and $2x^2 - 3ax + a^2$. Ans. $10x^5 - 7ax^4 - a^2x^3 - 3a^3x^2 + a^5$.

7. Required the product of $3x^3 + 2x^2y^2 + 3y^3$ and $2x^3 - 3x^2y^2 + 5y^3$.
Ans. $6x^6 - 5x^5y^2 - 6x^4y^4 + 21x^3y^3 + x^2y^5 + 15y^6$.

8. Required the product of $x^3 - ax^2 + bx - c$ and $x^2 - dx + e$. Ans. $x^5 - ax^4 - dx^4 + (b + ad + e)\,x^3 - (c + bd + ae)\,x^2 + (cd + eb)\,x - ce$.

9. * Required the product of the four following factors, viz.

I. II. III. IV.

$(a+b)$, (a^2+ab+b^2), $(a-b)$, and (a^2-ab+b^2).

Ans. a^6-b^6.

10. Required the product of $a^3+3a^2x+3ax^2+x^3$ and $a^3-3a^2x+3ax^2-x^3$. Ans. $a^6-3a^4x^2+3a^2x^4-x^6$.

11. Required the product of $a^4+a^2c^2+c^4$ and a^2-c^2.

Ans. a^6-c^6.

12. Required the product of $a^2+b^2+c^2-ab-ac-bc$ and $a+b+c$. Ans. $a^3-3abc+b^3+c^3$.

DIVISION.

Division is the converse of mulplication, and is performed like that of numbers; the rule being usually divided into three cases; in each of which like signs give + in the quotient, and unlike signs −, as in finding their products.†

It is here also to be observed, that powers and roots of the same quantity, are divided by subtracting the index of the divisor from that of the dividend.

Thus, $a^3 \div a^2$, or $\dfrac{a^3}{a^2}=a$; $a^{\frac{1}{2}} \div a^{\frac{1}{3}}$, or $\dfrac{a^{\frac{1}{2}}}{a^{\frac{1}{3}}}=a^{\frac{1}{6}}$;

$a^{\frac{3}{4}} \div a^{\frac{2}{3}}$, or $\dfrac{a^{\frac{3}{4}}}{a^{\frac{2}{3}}}=a^{\frac{1}{12}}$; and $a^m \div a^n$, or $\dfrac{a^m}{a^n}=a^{m-n}$.

CASE I.

When the divisor and dividend are both simple quantities.

Rule.—Set the dividend over the divisor, in the manner of a fraction, and reduce it to its simplest form, by cancelling the letters and figures that are common to each term.

---

* I would advise the learner to perform the calculation of this example several ways; viz. First, by multiplying the product of the factors I. and II. by the product of the factors III. and IV. Secondly, by multiplying the product of the factors I. and III. by the product of the factors II. and IV. Thirdly, by multiplying the product of the factors I. and IV. by the product of the factors II. and III. The last method is the most concise. See Euler's Algebra, page 119, Vol. I.—Ed.

† According to the rule here given for signs, it follows that

$$\frac{+ab}{+b}=+a,\ \frac{-ab}{-b}=+a,\ \frac{-ab}{+b}=-a,\ \frac{+ab}{-b}=-a,$$

as will readily appear by multiplying the quotient by the divisor; the signs of the product being then the same as would take place in the former rule.

EXAMPLES.

$6ab \div 2a$, or $\frac{6ab}{2a} = 3b$; and $12ax^2 \div 3x$, or $\frac{12ax^2}{3x} = 4ax$,

$a \div a$, or $\frac{a}{a} = 1$; and $a \div -a$, or $\frac{a}{-a} = -1$.

Also $-2a \div 3a$, or $\frac{-2a}{3a} = -\frac{2}{3}$; and $9x^{\frac{1}{2}} \div 3x^{\frac{1}{4}} = 3x^{\frac{1}{4}}$.

1. Divide $16x^2$ by $8x$, and $12a^2x^2$ by $-8a^2x$.

Ans. $2x$, and $-\frac{3x}{2}$.

2. Divide $-15ay^2$ by $3ay$ and $-18ax^2y$ by $-8ax$.

Ans. $-5y$, and $\frac{9xy}{4}$.

3. Divide $-\frac{2}{3}a^{\frac{1}{2}}$ by $\frac{1}{5}a^{\frac{1}{2}}$, and $ax^{\frac{1}{3}}$ by $-\frac{3}{5}a^{\frac{1}{2}}a^{\frac{1}{4}}$.

Ans. $-3\frac{1}{3}$, and $-\frac{5}{3}a^{\frac{1}{2}}x^{\frac{1}{12}}$.

4. Divide $12a^2b^2$ by $-3a^2b$, and $-15ay^{\frac{2}{3}}$ by $-3ay^{\frac{1}{2}}$.

Ans. $-4b$, and $5y^{\frac{1}{6}}$.

5. Divide $-15a^2x^2$ by $5ax^2$, and $21a^2c^2x^{\frac{1}{2}}$ by $-7ac^2x^{\frac{1}{4}}$.

Ans. $-3a$, and $-3ax^{\frac{1}{4}}$.

6. Divide $-17x^{\frac{1}{2}}a^3c$ by $-5x^{\frac{1}{3}}a^2c^{\frac{1}{2}}$, and $24xy$ by $8\sqrt{(xy)}$.

Ans. $\frac{17x^{\frac{1}{6}}ac^{\frac{1}{2}}}{5}$, and $3\sqrt{(xy)}$

CASE II.

When the divisor is a simple quantity, and the dividend a compound one.

Rule.—Divide each term of the dividend by the divisor, as in the former case; setting down such as will not divide in the simplest form they will admit of.

EXAMPLES.

$(ab + b^2) \div 2b$, or $\frac{ab + b^2}{2b} = \frac{1}{2}a + \frac{1}{2}b = \frac{a + b}{2}$.

$(10ab - 15ax) \div 5a$, or $\frac{10ab - 15ax}{5a} = 2b - 3x$.

$(30ax - 48x^2) \div 6x$, or $\frac{30ax - 48x^2}{6x} = 5a - 8x$.

1. Let $3x^3 + 6x^2 + 3ax - 15x$ be divided by $3x$.
Ans. $x^2 + 2x + a - 5$.

2. Let $3abc + 12abx - 9a^2b$ be divided by $3ab$.
Ans. $c + 4x - 3a$.

3. Let $40a^3b^3 + 60a^2b^2 - 17ab$ be divided by $-ab$.
Ans. $-40a^2b^2 - 60ab + 17$.

4. Let $15a^2bc - 12acx^2 + 5ad^2$ be divided by $-5ac$.
Ans. $-3ab + \frac{12x^2}{5} - \frac{d^2}{c}$.

5. Let $20ax^3 + 15ax^2 + 10ax + 5a$ be divided by $5a$.
Ans. $4x^3 + 3x^2 + 2x + 1$.

6. Let $6bcdz + 4bzd^2 - 2b^2z^2$ be divided by $2bz$.
Ans. $3cd + 2d^2 - bz$.

7. Let $14a^2 - 7ab + 21ax - 28a$ be divided by $7a$.
Ans. $2a - b + 3x - 4$

8. Let $-20ab + 60ab^3 - 12a^2b^2$ be divided by $-4ab$.
Ans. $5 - 15b^2 + 3ab$.

CASE III.

When the divisor and dividend are both compound qualities.

RULE.—Set them down in the same manner as in division of numbers, ranging the terms of each of them so, that the higher power of one of the letters may stand before the lower.

Then divide the first term of the dividend by the first term of the divisor, and set the result in the quotient, with its proper sign, or simply by itself, if it be affirmative.

This being done, multiply the whole divisor by the term thus found; and, having subtracted the result from the dividend, bring down as many terms to the remainder as are requisite for the next operation, which perform as before; and so on, till the work is finished, as in common arithmetic.

EXAMPLES.

$$x + y)\ x^2 + 2xy + y^2\ (x + y$$
$$x^2 + xy$$
$$\overline{\quad xy + y^2}$$
$$xy + y^2$$
$$\overline{\qquad\qquad}$$

$$
\begin{array}{l}
a+x)\,a^3+5a^2x+5ax^2+x^3\,(a^2+4ax+x^2\\
\quad a^3+a^2x\\
\hline
\quad 4a^2x+5ax^2\\
\quad 4a^2x+4ax^2\\
\hline
\quad ax^2+x^3\\
\quad ax^2+x^3\\
\hline
\quad *
\end{array}
$$

$$
\begin{array}{l}
x-3)\,x^3-9x^2+27x-27\,(x^2-6x+9\\
\quad x^3-3x^2\\
\hline
\quad -6x^2+27x\\
\quad -6x^2+18x\\
\hline
\quad 9x-27\\
\quad 9x-27\\
\hline
\quad *
\end{array}
$$

$$
\begin{array}{l}
2x^2-3ax+a^2)\,4x^4-9a^2x^2+6a^3x-a^4\,(2x^2+3ax-a^2\\
\quad 4x^4-6ax^3+2a^2x^2\\
\hline
\quad 6ax^3-11a^2x^2+6a^3x\\
\quad 6ax^3-9a^2x^2+3a^3x\\
\hline
\quad -2a^2x^2+3a^3x-a^4\\
\quad -2a^2x^2+3a^3x-a^4\\
\hline
\quad *
\end{array}
$$

NOTE 1. If the divisor be not exactly contained in the dividend, the quantity that remains after the division is finished, must be placed over the divisor, at the end of the quotient, in the form of a fraction; thus,*

$$
\begin{array}{l}
a+x)\,a^3-x^3\,(a^2-ax+x^2-\dfrac{2x^3}{a+x}\\
\quad a^3+a^2x\\
\hline
\quad -a^2x-x^3\\
\quad -a^2x-ax^2\\
\hline
\quad ax^2-x^3\\
\quad ax^2+x^3\\
\hline
\quad -2x^3
\end{array}
$$

* In the case here given, the operation of division may be considered

$$x + y)x^4 + y^4(x^3 - x^2y + xy^2 - y^3 + \frac{2y^4}{x+y}$$
$$x^4 + x^3y$$
$$- x^3y + y^4$$
$$- x^3y - x^2y^2$$
$$x^2y^2 + y^4$$
$$x^2y^2 + xy^3$$
$$- xy^3 + y^4$$
$$- xy^3 - y^4$$
$$2y^4$$

2. The division of quantities may also be sometimes carried on, *ad infinitum*, like a decimal fraction; in which case a few of the leading terms of the quotient will generally be sufficient to indicate the rest, without its being necessary to continue the operation; thus,

$$a + x)a \ldots . (1 - \frac{x}{a} + \frac{x^2}{a^2} - \frac{x^3}{a^3} + \frac{x^4}{a^4}, \text{ \&c.*}$$
$$a + a$$
$$- x$$
$$- x - \frac{x^2}{a}$$

as terminated, when the highest power of the letter, in the first or leading term of the remainder, by which the process is regulated, is less than the power of the first term of the divisor; or when the first term of the divisor is not contained in the first term of the remainder; as the succeeding part of the quotient, after this, instead of being integral, as it ought to be, would necessarily become fractional.

* Now, it is easy to perceive that the next or 6th term of the quotient will be $-\frac{x^5}{a^5}$, and the seventh term $\frac{x^6}{a^6}$ and so on, alternately *plus* and *minus*; this is called the law of *continuation* of the series. And the sum of all the terms when infinitely continued is said to be equal to the fraction $\frac{a}{a+x}$. Thus we say the *vulgar fraction* $\frac{2}{9}$, when reduced to a decimal, is $= .22222$, &c., infinitely continued. The terms in the quotient are found by dividing the remainder by a, the first term of the divisor; thus, the first remainder $-x$ divided by a, gives $-\frac{x}{a}$ the second term in the quotient; and the second remainder $+\frac{x^2}{a}$ divided by a gives $+\frac{x^2}{a^2}$ the third term, &c.

$$\frac{x^2}{a}$$

$$\frac{x^2}{a}+\frac{x^3}{a^2}$$

$$\frac{x^3}{a^2}$$

$$-\frac{x^3}{a^2}-\frac{x^4}{a^3}$$

And by a process similar to the above, it may be shown that

$$\frac{a}{a-x}=1+\frac{x}{a}+\frac{x^2}{a^2}+\frac{x^3}{a^3}+\frac{x^4}{a^4}+\frac{x^5}{a^5}+,\ \&c.$$

Where the law, by which either of these series may be continued at pleasure, is obvious.*

* In this example, if x be less than a, the series is convergent, or the value of the terms continually diminish; but when x is greater than a, it is said to diverge.

To explain this by numbers: suppose $a=3$, and $x=2$.

Then, $1+\frac{x}{a}+\frac{x^2}{a^2}+\frac{x^3}{a^3}$, &c.

The corresponding values are,

$1+\frac{2}{3}+\frac{4}{9}+\frac{8}{27}$, &c.

where the fractions or terms of the series grow less and less, and the farther they are extended, the more they converge or approximate to 0, which is supposed to be the last term or limit.

But if $a=2$, and $x=3$,

Then $1+\frac{x}{a}+\frac{x^2}{a^2}+\frac{x^3}{a^3}$, &c.

The corresponding values are,

$1+\frac{3}{2}+\frac{9}{4}+\frac{27}{8}$, &c.

in which the terms become larger and larger. This is called a diverging series.

If $x=1$, and $a=1$ in the preceding example:

Then, $\frac{a}{a+x}=1-\frac{x}{a}+\frac{x^2}{a^2}-\frac{x^3}{a^3}$, &c., will be $\frac{1}{1+1}=1-1+1-1$, &c. Now, because, $\frac{1}{1+1}=\frac{1}{2}$, it has been said that $1-1+1-1$, &c., infinitely continued, is $=\frac{1}{2}$; a singular conclusion, when it is perceived from the terms themselves, that their sum must necessarily be either 0 or $+1$, to whatever extent the division is supposed to be continued. The real question, however, results from the fractional parts, which

EXAMPLES FOR PRACTICE.

1. Let $a^2 - 2ax + x^2$ be divided by $a - x$. Ans. $a - x$.

2. Let $x^3 - 3ax^2 + 3a^2x - a^3$ be divided by $x - a$.
Ans. $x^2 - 2ax + a^2$

3. Let $a^3 + 5a^2x + 5ax^2 + x^3$ be divided by $a + x$.
Ans. $a^2 + 4ax + x^2$.

4. Let $2y^3 - 19y^2 + 26y - 16$ be divided by $y - 8$.
Ans $2y^2 - 3y + 2$.

5. Let $x^5 + 1$ be divided by $x + 1$, and $x^6 - 1$ by $x - 1$.
Ans. $x^4 - x^3 + x^2 - x + 1$, and $x^5 + x^4 + x^3 + x^2 + x + 1$.

6. Let $48x^3 - 76ax^2 - 64a^2x + 105a^3$ be divided by $2x - 3a$.
Ans. $24x^2 - 2ax - 35a^2$

7. Let $4x^4 - 9x^2 + 6x - 1$ be divided by $2x^2 + 3x - 1$.
Ans. $2x^2 - 3x + 1$.

8. Let $x^4 - a^2x^2 + 2a^3x - a^4$ be divided by $x^2 - ax + a^2$.
Ans. $x^2 + ax - a^2$.

9. Let $6x^4 - 96$ be divided by $3x - 6$, and $a^5 + x^5$ by $a + x$.
Ans. $2x^3 + 4x^2 + 8x + 16$, and $a^4 - a^3x + a^2x^2 - ax^3 + x^4$.

10. Let $32x^5 + 243$ be divided by $2x + 3$, and $x^6 - a^6$ by $x - a$.
Ans. $16x^4 - 24x^3 + 36x^2 - 54x + 81$, and $x^5 + ax^4 + a^2x^3 + a^3x^2 + a^4x + a^5$.

11. Let $b^4 - 3y^4$ be divided by $b - y$, and $a^4 + 4a^2b + 8b^4$ by $a + 2b$.

Ans. $b^3 + b^2y + by^2 + y^3 - \dfrac{2y^4}{b - y}$, and $a^3 - 2a^2b + 4ab + 4ab^2 - 8b^2 - 8b^3 + \dfrac{16b^3 + 24b^4}{a + 2b}$,

12. Let $x^2 + px + q$ be divided by $x + a$, and $x^3 - px^2 + qx - r$ by $x - a$.

Ans. $x + p - a + \dfrac{q - pa + a^2}{x + a}$, and $x^2 + (a - p)x - ap + a^2 + q + \dfrac{a^3 - a^2p + aq - r}{x - a}$,

13. Let $1 - 5x + 10x^2 - 10x^3 + 5x^4 - x^5$ be divided by $1 - 2x + x^2$. Ans. $1 - 3x + 3x^2 - x^3$.

14. Let $a^4 + 4b^4$ be divided by $a^2 - 2ab + 2b^2$.
Ans. $a^2 + 2ab + 2b^2$.

15. $a^5 - 5a^4x + 10a^3x^2 - 10a^2x^3 + 5ax^4 - x^5$ be divided by $a^2 - 2ax + x^2$. Ans. $a^3 - 3a^2x + 3ax^2 - x^3$

16. Let $a^4 + b^4$ be divided by $a^2 + ab\sqrt{2} + b^2$.
Ans. $a^2 - ab\sqrt{2} + b^2$

(by the division) is always $+\frac{1}{2}$ when the sum of the terms is 0, and $-\frac{1}{2}$ when the sum is $+1$: consequently $\frac{1}{2}$ is the true quotient in the former case, and $1 - \frac{1}{2}$ in the other.—Ed.

OF ALGEBRAIC FRACTIONS.

Algebraic fractions have the same names and rules of operation as numeral fractions in common arithmetic; and the methods of reducing them, in either of these branches, to their most convenient forms, are as follows:—

CASE I.

To find the greatest common measure of the terms of a fraction.

Rule.—1. Arrange the two quantities according to the order of their powers, and divide that which is of the highest dimensions by the other, having first expunged any factor, that may be contained in all the terms of the divisor, without being common to those of the dividend.

2. Divide this divisor by the remainder, simplified, if necessary, as before; and so on, for each successive remainder and its preceding divisor, till nothing remains, when the divisor last used will be the greatest common measure required; and if such a divisor cannot be found, the terms of the fraction have no common measure.*

Note. If any of the divisors, in the course of the operation, become negative, they may have their signs changed, or be taken affirmatively, without altering the truth of the result; and if the first term of a divisor should not be exactly contained in the first term of the dividend, the several terms of the latter may be multiplied by any number or quantity, that will render the division complete.†

---

* If, by proceeding in this manner, no compound divisor can be found, that is, if the last remainder be only a simple quantity, we may conclude the case proposed does not admit of any, but is already in its lowest terms. Thus, for instance, if the fraction proposed were to be

$$\frac{a^3+2a^2x+3ax^2+4x^3}{a^2+ax+x^2};$$

it is plain by inspection, that it is not reducible by any simple divisor; but to know whether it may not, by a compound one, I proceed as above, and find the last remainder to be the simple quantity $7x^2$: whence I conclude that the fraction is already in its lowest terms.

† In finding the greatest common measure of two quantities, either of them may be multiplied, or divided, by any quantity, which is not a divisor of the other, or that contains no factor which is common to them both, without in any respect changing the result.

It may here also be farther added, that the common measure, or

EXAMPLES.

1. Required the greatest common measure of the fraction $\frac{x^4 - 1}{x^5 + x^3}$.

$$x^4 - 1)\ x^5 + x^3\ (x$$
$$x^5 - x$$
$$\overline{\qquad x^3 + x\qquad}$$
$$\text{or } x^2 + 1 \mid x^4 - 1\ (x^2 - 1$$
$$x^4 + x^2$$
$$\overline{\qquad - x^2 - 1\qquad}$$
$$- x^2 - 1$$
$$\overline{\qquad * \qquad}$$

Whence $x^2 + 1$ is the greatest common measure required.

2. Required the greatest common measure of the fraction $\frac{x^3 - b^2x}{x^2 + 2bx + b^2}$.

$$x^2 + 2bx + b^2)\ x^3 - b^2x\ (x$$
$$x^3 + 2bx^2 + b^2x$$
$$\overline{\qquad * - 2bx^2 - 2b^2x\qquad}$$
$$\text{or } x + b \mid x^2 + 2bx + b^2 (x + b$$
$$x^2 + bx$$
$$\overline{\qquad bx + b^2\qquad}$$
$$bx + b^2$$
$$\overline{\qquad * \qquad}$$

Where $x + b$ is the greatest common measure required.

3. Required the greatest common measure of the fraction $\frac{3a^2 - 2a - 1}{4a^3 - 2a^3 - 3a + 1}$.

---

divisor, of any number of quantities, may be determined in a similar manner to that given above, by first finding the common measure of two of them, and then of that common measure and a third; and so on to the last.

* Here, I divide the remainder $-2bx^2 - 2b^2x$ by $-2xb$, (its greatest simple divisor) and the quotient is $x + b$; and then I divide the last divisor by $x + b$, &c.—Ed.

$$
\begin{array}{l}
3a^2-2a-1)\,4a^3-2a^2-3a+1 \\
\qquad\qquad\qquad 3 \\
\hline
\qquad\qquad 12a^3-6a^2-9a+3\,(4a \\
\qquad\qquad 12a^3-8a^2-4a \\
\hline
\qquad\qquad\qquad 2a^2-5a+3)\,3a^2-2a-1 \\
\qquad\qquad\qquad\qquad\qquad 2 \\
\hline
\qquad\qquad\qquad\qquad\qquad 6a^2-4a-2\,(3 \\
\qquad\qquad\qquad\qquad\qquad 6a^2-15a+9 \\
\hline
\qquad\qquad\qquad\qquad\qquad 11a-11 \text{ or } a-1
\end{array}
$$

Where, since $a-1\,)\,2a^2-5a+3\,(3a-3$, it follows that the last divisor $a-1$ is the common measure required.

4. It is required to find the greatest common measure of $\dfrac{x^3-a^3}{x^4-a^4}$. Ans. $x-a$.

5. Required the greatest common measure of the fraction $\dfrac{a^4-x^4}{a^3-a^2x-ax^2+x^3}$. Ans. a^2-x^2.

6. Required the greatest common measure of the fraction $\dfrac{x^4+a^2x^2+a^4}{x^4+ax^3-a^3x-a^4}$. Ans. x^2+ax+a^2.

7. Required the greatest common measure of the fraction $\dfrac{7a^2-23ab+6b^2}{5a^3-18a^2b+11ab^2-6b^3}$. Ans. $a-3b$.

8. *Required the greatest common measure of the fraction $\dfrac{x^3+ax^2+bx^2-2a^2x+bax-2ba^2}{x^2-bx+2ax-2ab}$ Ans. $x+2a$.

---

* This fraction can be reduced by Simpson's rule (page 48) thus:—Fractions that have in them more than two different letters, and one of the letters rises only to a single dimension, either in the numerator or denominator, it will be best to divide the said numerator or denominator (whichever it is) into two parts, so that the said letter may be found in every term of the one part, and be totally excluded out of the other; this being done, let the greatest common divisor of these two parts be found, which will evidently be a divisor to the whole, and by which the division of the quantity is to be tried; as in the following example, where the fraction given is

$$\frac{x^3+ax^2+bx^2-2a^2x+bax-2ba^2}{x^2-bx+2ax-2ab}.$$

Here the denominator being the least compounded, and b rising therein

9. Required the greatest common measure of the fraction $\frac{x^4 - 3ax^3 - 8a^2x^2 + 18a^3x - 8a^4}{x^3 - ax^2 - 8a^2x + 6a^3}$. Ans. $x^2 + 2ax - 2a^2$.

10. Required the greatest common measure of the fraction $\frac{5a^5 + 10a^4b + 5a^3b^2}{a^3b + 2a^2b^2 + 2ab^3 + b^4}$. Ans. $a + b$.

11. Required the greatest common measure of the fraction $\frac{6a^5 + 15a^4b - 4a^3c^2 - 10a^2bc^2}{9a^3b - 27a^2bc - 6abc^2 + 18bc^3}$, Ans. $3a^2 - 2c^2$.

CASE II.

To reduce fractions to their lowest or most simple terms.

Rule.—Divide the terms of the fraction by any number, or quantity, that will divide each of them without leaving a remainder; or find their greatest common measure, as in the last rule, by which divide both the numerator and denomina tor, and it will give the fraction required.

EXAMPLES.

1. Reduce $\frac{a^2bc}{5a^2b^2}$ and $\frac{x^2}{ax + x^2}$ to their lowest terms.

Here $\frac{a^2bc}{5a^2b^2} = \frac{c}{5b}$ Ans. And $\frac{x^2}{ax + x^2} = \frac{x}{a + x}$. Ans.

2. It is required to reduce $\frac{cx + x^2}{a^2c + a^2x}$ to its lowest terms.

$$\begin{array}{r|l} \text{Here } cx + x^2 & a^2c + a^2x \\ \text{or } \quad c + x & a^2c + a^2x \; (a^2 \\ & a^2c + a^2x \\ & \overline{\qquad\qquad} \\ & \quad * \end{array}$$

Whence $c + x$ is the greatest common measure;

and $c + x)\frac{cx + x^2}{a^2c + a^2x} = \frac{x}{a^2}$ the fraction required.

---

to a single dimension only, I divide the same into the parts $x^2 + 2ax$, and $-bx - 2ab$; which, by inspection, appear to be equal to $(x + 2a) \times x$, and $(x + 2a) \times -b$. Therefore $x + 2a$ is a divisor to both the parts, and likewise to the whole, expressed by $(x + 2a) \times (x - b)$; so that one of these two factors, if the fraction given can be reduced to lower terms, must also measure the numerator; but the former will be found to succeed, the quotient coming out $x^2 - ax + bx - ab$, exactly; whence the fraction itself is reduced to $\frac{x^2 - ax + bx - ab}{x - b}$, which is not reducible farther by $x - b$, since the division does not terminate without a remainder, as upon trial will be found.

This rule is sometimes of great utility, because it spares great labour, and is very expeditious in reducing several fractions.—Ed.

3. It is required to reduce $\frac{x^3 - b^2x}{x^2 + 2bx + b^2}$ to its lowest terms.

$$\begin{array}{l} x^2 + 2bx + b^2)\, x^3 - b^2x\, (x \\ \qquad\qquad\quad x^3 + 2bx^2 + b^2x \\ \hline \qquad\qquad\quad -2bx^2 - 2b^2x \\ \qquad\qquad\qquad \text{or } x + b \;\big|\; x^2 + 2bx + b^2\, (x + b \\ \qquad\qquad\qquad\qquad\qquad x^2 + bx \\ \hline \qquad\qquad\qquad\qquad\qquad\quad bx + b^2 \\ \qquad\qquad\qquad\qquad\qquad\quad bx + b^2 \\ \hline \qquad\qquad\qquad\qquad\qquad\qquad * \end{array}$$

Whence $x + b$ is the greatest common measure; and $x + b)$ $\frac{x^3 - b^2x}{x^2 + 2bx + b^2} = \frac{x^2 - bx}{x + b}$ the fraction required.

And the same answer would have been found, if $x^3 - b^2x$ had been made the divisor instead of $x^2 + 2bx + b^2$.

4. It is required to reduce $\frac{x^4 - a^4}{x^5 - a^2x^3}$ to its lowest terms.

Ans. $\frac{x^2 + a^2}{x^3}$.

5. It is required to reduce $\frac{6a^2 + 7ax - 3x^2}{6a^2 + 11ax + 3x^2}$ to its lowest terms.

Ans. $\frac{3a - x}{3a + x}$.

6. It is required to reduce $\frac{2x^3 - 16x - 6}{3x^3 - 24x - 9}$ to its lowest terms.

Ans. $\frac{2}{3}$.

7. It is required to reduce $\frac{9x^5 + 2x^3 + 4x^2 - x + 1}{15x^4 - 2x^3 + 10x^2 - x + 2}$ to its lowest terms.

Ans. $\frac{3x^3 + x^2 + 1}{5x^2 + x + 2}$

8. It is required to reduce $\frac{a^2d^2 - c^2d^2 - a^2c^2 + c^4}{4a^2d - 4acd - 2ac^2 + 2c^3}$ to its lowest terms.

Ans. $\frac{ad^2 + cd^2 - ac^2 - c^3}{4ad - 2c^2}$

CASE III.

To reduce a mixed quantity to an improper fraction.

Rule.—Multiply the integral part by the denominator of the fraction, and to the product add the numerator, when it is affirmative, or subtract it when negative; then the result, placed over the denominator, will give the improper fraction required.

EXAMPLES.

1. Reduce $3\frac{2}{5}$ and $a-\frac{b}{c}$ to improper fractions.

Here $3\frac{2}{5}=\frac{3\times5+2}{5}=\frac{15+2}{5}=\frac{17}{5}$. Ans.

And $a-\frac{b}{c}=\frac{a\times c-b}{c}=\frac{ac-b}{c}$. Ans.

2. Reduce $x+\frac{a}{x}$ and $x-\frac{a^2-x^2}{x}$ to improper fractions.

Here $x+\frac{a}{x}=\frac{x\times x+a}{x}=\frac{x^2+a}{x}$. Ans.

And x* $-\frac{a^2-x^2}{x}=\frac{x^2-a^2+x^2}{x}=\frac{2x^2-a^2}{x}$. Ans.

3. Let $1-\frac{2x}{a}$ be reduced to an improper fraction.

Ans. $\frac{a-2x}{a}$.

4. Let $5a-\frac{3x-b}{a}$ be reduced to an improper fraction.

Ans. $\frac{5a^2-3x+b}{a}$.

5. Let $x-\frac{a+x^2}{2a}$ be reduced to an improper fraction.

Ans. $\frac{2ax-a-x^2}{2a}$.

6. Let $5+\frac{2x-7}{3x}$ be reduced to an improper fraction.

Ans. $\frac{17x-7}{3x}$.

7. Let $1-\frac{x-a-1}{a}$ be reduced to an improper fraction.

Ans. $\frac{2a-x+1}{a}$

* $x\times x=x^2$. In adding the numerator a^2-x^2, the sign $-$ affixed to the fraction $\frac{a^2-x^2}{x}$, denotes that the whole of that fraction is to be subtracted, and consequently that the signs of each term of the numerator must be changed when it is combined with x^2; hence the improper fraction is $\frac{x^2-a^2+x^2}{x}$, or $\frac{2x^2-a^2}{x}$. Ed.

8. Let $1 + 2x - \frac{x-3}{5x}$ be reduced to an improper fraction.

Ans. $\frac{10x^2 + 4x + 3}{5x}$

CASE IV.

To reduce an improper fraction to a whole or mixed quantity.

RULE.—Divide the numerator by the denominator, for the integral part, and place the remainder, if any, over the denominator, for the fractional part; then the two, joined together, with the proper sign between them, will give the mixed quantity required.

EXAMPLES.

1. Reduce $\frac{27}{5}$ and $\frac{ax + a^2}{x}$ to mixed quantities.

Here $\frac{27}{5} = 27 \div 5 = 5\frac{2}{5}$. Ans.

And $\frac{ax + a^2}{x} = (ax + a^2) \div x = a + \frac{a^2}{x}$. Ans.

2. It is required to reduce the fraction $\frac{ax - x^3}{x}$ to a whole quantity.

Ans. $a - x^2$.

3. It is required to reduce the fraction $\frac{ab - 2a^2}{ab}$ to a mixed quantity.

Ans. $1 - \frac{2a}{b}$

4. It is required to reduce the fraction $\frac{a^2 + x^2}{a - x}$ to a mixed quantity.

Ans. $a + x + \frac{2x^2}{a - x}$.

5. It is required to reduce the fraction $\frac{x^3 - y^3}{x - y}$ to a whole quantity.

Ans. $x^2 + xy + y^2$

6. It is required to reduce the fraction $\frac{10x^2 - 5x + 3}{5x}$ to a mixed quantity.

Ans. $2x - 1 + \frac{3}{5x}$.

CASE V.

To reduce fractions to other equivalent ones, that shall have a common denominator.

RULE.—Multiply each of the numerators, separately, into

all the denominators, except its own, for the new numerators, and all the denominators together for a common denominator.*

EXAMPLES.

1. Reduce $\frac{a}{b}$ and $\frac{b}{c}$ to fractions that shall have a common denominator.

Here $\left.\begin{array}{l} a \times c = ac \\ b \times b = b^2 \end{array}\right\}$ the new numerators.

$b \times c = bc$ the common denominator.

Whence $\frac{a}{b}$ and $\frac{b}{c} = \frac{ac}{bc}$ and $\frac{b^2}{bc}$, the fractions required.

2. Reduce $\frac{2x}{a}$ and $\frac{b}{c}$ to equivalent fractions having a common denominator. Ans. $\frac{2cx}{ac}$ and $\frac{ab}{ac}$.

3. Reduce $\frac{a}{b}$ and $\frac{a+b}{c}$ to equivalent fractions having a common denominator. Ans. $\frac{ac}{bc}$ and $\frac{ab+b^2}{bc}$.

4. Reduce $\frac{3x}{2a}$, $\frac{2b}{3c}$, and d, to equivalent fractions having a common denominator. Ans. $\frac{9cx}{6ac}$, $\frac{4ab}{6ac}$, and $\frac{6acd}{6ac}$.

5. Reduce $\frac{3}{4}$, $\frac{2x}{3}$, and $a+\frac{4x}{5}$, to fractions having a common denominator. Ans. $\frac{45}{60}$, $\frac{40x}{60}$, and $\frac{60a+48x}{60}$.

6. Reduce $\frac{a}{2}$, $\frac{3x}{7}$, and $\frac{a+x}{a-x}$, to fractions having a common denominator. Ans. $\frac{7a^2-7ax}{14a-14x}$, $\frac{6ax-6x^2}{14a-14x}$, and $\frac{14a+14x}{14a-14x}$.

---

* It may here be remarked, that if the numerator and denominator of a fraction be either both multiplied, or both divided, by the same number or quantity, its value will not be altered; thus

$\frac{2}{3}=\frac{2\times 3}{3\times 3}=\frac{6}{9}$, and $\frac{3}{12}=\frac{3\div 3}{12\div 3}=\frac{1}{4}$; or $\frac{a}{b}=\frac{ac}{bc}$, and $\frac{ab}{bc}=\frac{a}{c}$.

which method is often of great use in reducing fractions more readily to a common denominator.

CASE VI.

To add fractional quantities together.

RULE.—Reduce the fractions, if necessary, to a common denominator; then add all the numerators together, and under their sum put the common denominator, and it will give the fractions required.*

EXAMPLES.

1. It is required to find the sum of $\frac{x}{2}$ and $\frac{x}{3}$.

Here $\left.\begin{array}{l} x \times 3 = 3x \\ x \times 2 = 2x \end{array}\right\}$ the numerators.

And $2 \times 3 = 6$ the common denominator.

Whence $\frac{3x}{6} + \frac{2x}{6} = \frac{5x}{6}$, the sum required.

2. It is required to find the sum of $\frac{a}{b}$, $\frac{c}{d}$, and $\frac{e}{f}$.

Here $\left.\begin{array}{l} a \times d \times f = adf \\ c \times b \times f = cbf \\ e \times b \times d = ebd \end{array}\right\}$ the numerators.

And $b \times d \times f = bdf$ the common denominator.

Whence $\frac{adf}{bdf} + \frac{cbf}{bdf} + \frac{ebd}{bdf} = \frac{adf + cbf + ebd}{bdf}$ the sum.

3. It is required to find the sum of $a - \frac{3x^2}{b}$ and $b + \frac{2ax}{c}$.

Here, taking only the fractional parts,

we shall have $\left\{\begin{array}{l} 3x^2 \times c = 3cx^2 \\ 2ax \times b = 2abx \end{array}\right\}$ the numerators.

And $b \times c = bc$ the common denominators.

Whence $a - \frac{3cx^2}{bc} + b + \frac{2abx}{bc} = a + b + \frac{2abx - 3cx^2}{bc}$ the sum.

4. It is required to find the sum of $\frac{2x}{5}$ and $\frac{5x}{7}$.

Ans. $\frac{39x}{35}$

5. It is required to find the sum of $\frac{3x}{2a}$ and $\frac{x}{5}$.

Ans. $\frac{15x + 2ax}{10a}$

* In the adding or subtracting of mixed quantities, it is best to bring the fractional parts only to a common denominator, and then to affix their sum or difference to the sum or difference of the integral parts, interposing the proper sign.

6. It is required to find the sum of $\frac{x}{2}$, $\frac{x}{3}$, and $\frac{x}{4}$.

Ans. $\frac{13x}{12}$.

7. It is required to find the sum of $\frac{4x}{7}$ and $\frac{x-2}{5}$.

Ans. $\frac{27x-14}{35}$.

8. Required the sum of $2a$, $3a+\frac{2x}{5}$, and $a-\frac{8x}{9}$.

Ans. $6a-\frac{22x}{45}$.

9. Required the sum of $2a+\frac{3x}{5}$, $\frac{a}{a-x}$, and $\frac{a-x}{a}$

Ans. $2a+2+\frac{3a^2x-3ax^2+5x^3}{5a^2-5ax}$.

10. Required the sum of $5x+\frac{x-2}{3}$ and $4x-\frac{2x-3}{5x}$.

Ans. $9x+\frac{5x^2-16x+9}{15x}$.

11. It is required to find the sum of $5x$, $\frac{2a}{3x^2}$, and $\frac{a+2x}{4x}$.

Ans. $5x+\frac{8a+3ax+6x^2}{12x^2}$.

CASE VII.

To subtract one fractional quantity from another.

RULE.—Reduce the fractions to a common denominator, if necessary, as in addition; then subtract the less numerator from the greater, and under the difference write the common denominator, and it will give the difference of the fractions required.

EXAMPLES.

1. It is required to find the difference of $\frac{2x}{3}$ and $\frac{3x}{5}$

Here $\left\{\begin{array}{l} 2x \times 5 = 10x \\ 3x \times 3 = 9x \end{array}\right\}$ the numerators.

And $3 \times 5 = 15$ the common denominator.

Whence $\frac{10x}{15} - \frac{9x}{15} = \frac{x}{15}$, the difference required.

2. It is required to find the difference of $\frac{x-a}{2b}$ and $\frac{2a-4x}{3c}$.

Here $\left\{ \begin{array}{l} (x-a) \times 3c = 3cx - 3ac \\ (2a-4x) \times 2b = 4ab - 8bx \end{array} \right\}$ the numerators.

And $2b \times 3c = 6bc$ the common denominator.

Whence $\frac{3cx - 3ac}{6bc} - \frac{4ab - 8bx}{6bc} = \frac{3cx - 3ac - 4ab + 8bx}{6bc}$ the difference required.

3. Required the difference of $\frac{12x}{7}$ and $\frac{3x}{5}$. Ans. $x + \frac{4x}{35}$.

4. Required the difference of $15y$ and $\frac{1+2y}{8}$.

Ans. $\frac{118y-1}{8}$

5. Required the difference of $\frac{ax}{b-c}$ and $\frac{ax}{b+c}$.

Ans. $\frac{2acx}{b^2-c^2}$.

6. Required the difference of $x - \frac{x-a}{c}$ and $x + \frac{x}{2b}$.

Ans. $\frac{2ba - 2bx - cx}{2bc}$.

7. Required the difference of $a + \frac{a-x}{a+x}$ and $a - \frac{a+x}{a-x}$.

Ans. $\frac{2a^2 + 2x^2}{a^2 - x^2}$.

8. Required the difference of $ax + \frac{2x+7}{8}$ and $x - \frac{5x-6}{21}$.

Ans. $ax - \frac{86x-99}{168}$.

9 Required the difference of $2x + \frac{3x-5}{7}$, and $3x + \frac{11x-10}{15}$. Ans. $x + \frac{32x+5}{105}$.

10. Required the difference of $a + \frac{a - x}{a(a + x)}$ and $\frac{a + x}{a(a - x)}$.

Ans. $a - \frac{4x}{a^2 - x^2}$.

CASE VIII.

To multiply fractional quantities together.

RULE.—Multiply the numerators together for a new numerator and the denominators for a new denominator; and the former of these being placed over the latter, will give the product of the fractions, as required.*

EXAMPLES.

1. It is required to find the product of $\frac{x}{6}$ and $\frac{2x}{9}$.

Here $\frac{x \times 2x}{6 \times 9} = \frac{2x^2}{54} = \frac{x^2}{27}$ the product required.

2. It is required to find the continued product of $\frac{x}{2}$, $\frac{4x}{5}$, and $\frac{10x}{21}$.

Here $\frac{x \times 4x \times 10x}{2 \times 5 \times 21} = \frac{40x^3}{210} = \frac{4x^3}{21}$ the product.

3. It is required to find the product of $\frac{x}{a}$ and $\frac{a+x}{a-x}$.

Here $\frac{x \times (a + x)}{a \times (a - x)} = \frac{x^2 + ax}{a^2 - ax}$ the product.

4. It is required to find the product of $\frac{3x}{2}$ and $\frac{5x}{3b}$.

Ans. $\frac{5x^2}{2b}$.

5. It is required to find the product of $\frac{2x}{5}$ and $\frac{3x^2}{2a}$.

Ans. $\frac{3x^3}{5a}$.

---

* When the numerator of one of the fractions to be multiplied, and the denominator of the other, can be divided by some quantity, which is common to each of them, the quotients may be used instead of the fractions themselves.

Also, when a fraction is to be multiplied by an integer, it is the same thing whether the numerator be multiplied by it, or the denominator divided by it. Or if an integer is to be multiplied by a fraction, or a fraction by an integer, the integer may be considered as having unity for its denominator, and the two be then multiplied together as usual.

6. It is required to find the continued product of $\frac{2x}{3}$, $\frac{4x^2}{7}$, and $\frac{a}{a+x}$. Ans. $\frac{8ax^3}{21a+21x}$.

7. It is required to find the continued product of $\frac{2x}{a}$, $\frac{3ab}{c}$, and $\frac{5ac}{2b}$. Ans. $15ax$.

8. It is required to find the product of $2a+\frac{bx}{a}$ and $3a-\frac{b}{ax}$.

Ans. $6a^2+3bx-\frac{2b}{x}-\frac{b^2}{a^2}$.

9. It is required to find the continued product of $3x$, $\frac{x+1}{2a}$, and $\frac{x-1}{a+b}$. Ans. $\frac{3x^3-3x}{2a^2+2ab}$.

10. It is required to find the continued product of $\frac{a^2-x^2}{a+b}$, $\frac{a^2-b^2}{ax+x^2}$, and $a+\frac{ax}{a-x}$. Ans. $\frac{a^3-a^2b}{x}$.

CASE IX.

To divide one fractional quantity by another.

RULE.—Multiply the denominator of the divisor by the numerator of the dividend, for the numerator; and the numerator of the divisor by the denominator of the dividend, for the denominator. Or, which is more convenient in practice, multiply the dividend by the reciprocal of the divisor, and the product will be the quotient required.*

EXAMPLES.

1. It is required to divide $\frac{x}{3}$ by $\frac{2x}{9}$.

Here $\frac{x}{3}\div\frac{2x}{9}=\frac{x}{3}\times\frac{9}{2x}=\frac{9x}{6x}=\frac{3}{2}=1\frac{1}{2}$ Ans.

2. It is required to divide $\frac{2a}{b}$ by $\frac{4c}{d}$.

---

* When a fraction is to be divided by an integer, it is the same thing whether the numerator be divided by it, or the denominator multiplied by it.

Also, when the two numerators, or the two denominators, can be divided by some common quantity, that quantity may be thrown out of each, and the quotients used instead of the fractions first proposed.

Here $\frac{2a}{b} \times \frac{d}{4c} = \frac{2ad}{4bc} = \frac{ad}{2bc}$ Ans.

3. It is required to divide $\frac{x+a}{x-b}$ by $\frac{x+b}{5x+a}$.

Here $\frac{x+a}{x-b} \times \frac{5x+a}{x+b} = \frac{5x^2+6ax+a^2}{x^2-b^2}$ Ans.

4. It is required to divide $\frac{2x^2}{a^3+x^3}$ by $\frac{x}{x+a}$.

Here $\frac{2x^2}{a^3+x^3} \times \frac{x+a}{x} = \frac{2x^2(x+a)}{x(a^3+x^3)} = \frac{2x}{x^2-ax+a^2}$.

5. It is required to divide $\frac{7x}{5}$ by $\frac{3}{x}$ Ans. $\frac{7x^2}{15}$.

6. It is required to divide $\frac{4x^2}{7}$ by $5x$. Ans. $\frac{4x}{35}$.

7. It is required to divide $\frac{x+1}{6}$ by $\frac{2x}{3}$ Ans. $\frac{x+1}{4x}$.

8. It is required to divide $\frac{x}{1-x}$ by $\frac{x}{5}$. Ans. $\frac{5}{1-x}$.

9. It is required to divide $\frac{2ax+x^2}{c^3-x^3}$ by $\frac{x}{c-x}$.

Ans. $\frac{2a+x}{c^2+cx+x^2}$

10. It is required to divide $\frac{x^4-b^4}{x^2-2bx+b^2}$ by $\frac{x^2+bx}{x-b}$.

Ans. $\frac{x^2+b^2}{x}$.

INVOLUTION.

Involution is the raising of powers from any proposed root; or the method of finding the square, cube, biquadrate, &c., of any given quantity.

Rule 1.—Multiply the index of the quantity by the index of the power to which it is to be raised, and the result will be the power required.

Or multiply the quantity into itself as many times less one as is denoted by the index of the power, and the last product will be the answer.

Note. When the sign of the root is +, all the powers of it will be +; and when the sign is −, all the even powers will be +, and the odd powers −: as is evident from multiplication.*

EXAMPLES.

a, the root. $a^2 =$ square. $a^3 =$ cube. $a^4 =$ 4th power. $a^5 =$ 5th power. &c.	a^2 the root. $a^4 =$ square. $a^6 =$ cube. $a^8 =$ 4th power. $a^{10} =$ 5th power. &c.
$-3a$ the root. $+9a^2 =$ square. $-27a^3 =$ cube. $+81a^4 =$ 4th power. &c.	$-2ax^2$ the root. $+4a^2x^4 =$ square. $-8a^3x^6 =$ cube. $+16a^4x^8 =$ 4th power. &c.
$\frac{x}{a}$ the root. $\frac{x^2}{a^2} =$ square. $\frac{x^3}{a^3} =$ cube. $\frac{x^4}{a^4} =$ 4th power. &c.	$-\frac{2ax^2}{3b}$ the root. $+\frac{4a^2x^4}{9b^2} =$ square. $-\frac{8a^3x^6}{27b^3} =$ cube. $+\frac{16a^4x^8}{81b^4} =$ 4th power. &c.
$x - a$ the root. $x - a$ $x^2 - ax$ $\quad - ax + a^2$ $x^2 - 2ax + a^2$ square. $x - a$ $x^3 - 2ax^2 + a^2x$ $\quad - ax^2 + 2a^2x - a^3$ $x^3 - 3ax^2 + 3a^2x - a^3$ cube.	$x + a$ the root. $x + a$ $x^2 + ax$ $\quad + ax + a^2$ $x^2 + 2ax + a^2$ square. $x + a$ $x^3 + 3ax^2 + a^2$ $\quad + ax^2 + 2a^2x + a^3$ $x^3 + 3ax^2 + 3a^2x + a^3$ cube.

* Any power of the product of two or more quantities is equal to the same power of each of the factors multiplied together. And any power of a fraction is equal to the same power of the numerator divided by the like power of the denominator.

Also, a^m raised to the nth power is a^{mn}; and $-a^m$ raised to the nth power is $\pm a^{mn}$, according as n is an even or an odd number.

EXAMPLES FOR PRACTICE.

1. Required the cube, or third power, of $2a^2$. Ans. $8a^6$.
2. Required the biquadrate, or fourth power, of $2a^2x$.
Ans. $16a^3x^4$.
3. Required the cube, or third power, of $-\frac{2}{3}x^2y^3$.
Ans. $-\frac{8}{27}x^6y^9$.
4. Required the biquadrate, or fourth power of $\frac{3a^2x}{5b^2}$.
Ans. $\frac{81a^8x^4}{625b^8}$,
5. Required the fourth power of '$a+x$;' and the fifth power of $a-y$. Ans. $a^4+4a^3x+6a^2x^2+4ax^3+x^4$, and $a^5-5a^4y+10a^3y^2-10a^2y^3+5ay^4-y^5$.

RULE 2.—A binomial or residual quantity may also be readily raised to any power whatever, as follows:—

1. Find the terms without the coefficients, by observing that the index of the first, or leading quantity, begins with that of the given power, and decreases continually by 1, in every term to the last; and that in the following quantity, the indices of the terms are 1, 2, 3, 4, &c.

2. To find the coefficients, observe that those of the first and last terms are always 1; and that the coefficient of the second term is the index of the power of the first: and for the rest, if the coefficient of any term be multiplied by the index of the leading quantity in it, and the product be divided by the number of terms to that place, it will give the coefficient of the term next following.

Note. The whole number of terms will be one more than the index of the given power; and when both terms of the root are +, all the terms of the power will be +; but if the second term be —, all the odd terms will be +, and the even terms —; or, which is the same thing, the terms will be + and — alternately.*

* The rule here given, which is the same in the cases of integral powers as in the binomial theorem of Newton, may be expressed in general terms, as follows:—

$$(a+b)^m=a^m+ma^{m-1}b+m\cdot\frac{m-1}{2}a^{m-2}b^2+m\cdot\frac{m-1}{2}\cdot\frac{m-2}{3}a^{m-3}b^3,$$
&c.

$$(a-b)^m=a^m-ma^{m-1}b+m\cdot\frac{m-1}{2}a^{m-2}b^2-m\cdot\frac{m-1}{2}\cdot\frac{m-2}{3}a^{m-3}b^3,$$

EXAMPLES.

1. Let $a+x$ be involved, or raised to the fifth power.

Here the terms, without the coefficients, are,

$$a^5,\ a^4x,\ a^3x^2,\ a^2x^3,\ ax^4,\ x^5.$$

And the coefficients, according to the rule, will be

$$1,\ 5,\ \frac{5\times4}{2},\ \frac{10\times3}{3},\ \frac{10\times2}{4},\ \frac{5\times1}{1};$$

$$\text{or } 1,\ 5,\ 10,\ 10,\ 5,\ 1,$$

Whence the entire fifth power of $a+x$ is

$$a^5+5a^4x+10a^3x^2+10a^2x^3+5ax^4+x^5.$$

3. Let $a-x$ be involved, or raised, to the sixth power.

Here the terms, without their coefficients, are,

$$a^6,\ a^5x,\ a^4x^2,\ a^3x^3,\ a^2x^4,\ ax^5,\ x^6.$$

And the coefficients, found as before, are,

$$1,\ 6,\ \frac{6\times5}{2},\ \frac{15\times4}{3},\ \frac{20\times3}{4},\ \frac{15\times2}{5},\ \frac{6\times1}{6};$$

$$\text{or } 1,\ 6,\ 15,\ 20,\ 15,\ 6,\ 1.$$

Whence the entire sixth power of $a-x$ is

$$a^6-6a^5x+15a^4x^2-20a^3x^3+15a^2x^4-6ax^5+x^6.$$

3. Required the fourth power of $a+x$, and the fifth power of $a-x$. Ans. $a^4+4a^3x+6a^2x^2+4ax^3+x^4$, and $a^5-5a^4x+10a^3x^2-10a^2x^3+5ax^4-x^5$.

5. Required the sixth power of $a+b$, and the seventh power of $a-y$. Ans. $a^6+6a^5b+15a^4b^2+20a^3b^3+15a^2b^4+6ab^5+b^6$, and $a^7-7a^6y+21a^5y^2-35a^4y^3+35a^3y^4-21a^2y^5+7ay^6-y^7$.

6. Required the fifth power of $2+x$, and the cube of $a-bx+c$. Ans. $32+80x+80x^2+40x^3+10x^4+x^5$, and $a^3+3a^2c+3ac^2+c^3-3a^2bx-6acbx-3c^2bx+3ab^2x^2+3cb^2x^2-b^3x^3$.

EVOLUTION.

Evolution, or the extraction of roots, is the reverse of involution, or the raising powers; being the method of finding the square root, cube root, &c., of any given quantity.

CASE I.

To find any root of a simple quantity.

Rule.—Extract the root of the coefficient for the numeral

&c. which formulæ will also equally hold when m is a fraction, as will be more fully explained hereafter.

It may also be farther observed, that the sum of the coefficients in every power, is equal to the number 2 raised to that power. Thus $1+1=2$, for the first power; $1+2+1=4=2^2$, for the square; $1+3+3+1=8=2^3$, for the cube, or third power; and so on.

part, and the root of the quantity subjoined to it for the literal part; then these joined together, will be the root required.

And if the quantity proposed be a fraction, its root will be found by taking the root both of its numerator and denominator.

Note. The square root, the fourth root, or any other even root, of an affirmative quantity, may be either + or —. Thus $\sqrt{a^2} = +a$ or $-a$, and $\sqrt[4]{b^4} = +b$ or $-b$, &c. But the cube root, or any other odd root, of a quantity, will have the same sign as the quantity itself. Thus,

$\sqrt[3]{a^3} = a$; $\sqrt[3]{-a^3} = -a$; and $\sqrt[5]{-a^5} = -a$, &c.*

It may here, also, be farther remarked, that any even root of of a negative quantity is unassignable.

Thus, $\sqrt{-a^2}$ cannot be determined, as there is no quantity, either positive or negative, (+ or —), that, when multiplied by itself, will produce $-a^2$.

EXAMPLES.

1. Find the square root of $9x^2$; and the cube root of $8x^3$.
   Here $\sqrt{9x^2} = \sqrt{9} \times \sqrt{x^2} = 3 \times x = 3x$. Ans.
   And $\sqrt[3]{8x^3} = \sqrt[3]{8} \times \sqrt[3]{x^3} = 2 \times x = 2x$. Ans.

2. It is required to find the square root of $\frac{a^2x^2}{4c^2}$ and the cube root of $-\frac{8a^3x^3}{27c^3}$.

Here $\sqrt{\frac{a^2x^2}{4c^2}} = \frac{\sqrt{a^2x^2}}{\sqrt{4c^2}} = \frac{ax}{2c}$; and $\sqrt[3]{-\frac{8a^3x^3}{27c^3}} = -\frac{2ax}{3c}$.

3. It is required to find the square root of $4a^2x^6$. Ans. $2ax^3$.

4. It is required to find the cube root of $-125a^3x^6$. Ans. $-5ax^2$.

5. It is required to find the 4th root of $256a^4x^8$. Ans. $4ax^2$.

6. It is required to find the square root of $\frac{4a^4}{9x^2y^2}$. Ans. $\frac{2a^2}{3xy}$.

7. It is required to find the cube root of $\frac{8a^3}{125x^6}$. Ans. $\frac{2a}{5x^2}$.

---

* The reason why $+a$ and $-a$ are each the square root of a^2 is obvious, since, by the rule of multiplication, $(+a) \times (+a)$ and $(-a) \times (-a)$ are both equal to a^2.

And for the cube root, fifth root, &c., of a negative quantity, it is plain, from the same rule, that

$(-a) \times (-a) \times (-a) = -a^3$; and $(-a^3) \times (+a^2) = -a^5$.

And consequently $\sqrt[3]{a^3} = -a$, and $\sqrt[5]{-a^5} = -a$.

8. It is required to find the 5th root of $-\frac{32a^5x^{10}}{243}$.

Ans. $\frac{-2ax^2}{3}$.

CASE II.

To extract the square root of a compound quantity.

RULE 1.—Range the terms, of which the quantity is composed, according to the dimensions of some letter in them, beginning with the highest, and set the root of the first term in the quotient.

2. Subtract the square of the root thus found, from the first term, and bring down the two next terms to the remainder for a dividend.

3. Divide the dividend, thus found, by double that part of the root already determined, and set the result both in the quotient and divisor.

4. Multiply the divisor, so increased, by the term of the root last placed in the quotient, and subtract the product from the dividend; and so on, as in common arithmetic.

EXAMPLES.

1. Extract the square root of $x^4 - 4x^3 + 6x^2 - 4x + 1$.

$$x^4 - 4x^3 + 6x^2 - 4x + 1\ (x^2 - 2x + 1$$
$$x^4$$

$$2x^2 - 2x)\ -4a^3 + 6x^2$$
$$-4x^3 + 4x^2$$

$$2x^2 - 4x + 1)\ 2x^2 - 4x + 1$$
$$2x^2 - 4x + 1$$

Ans. $x^2 - 2x + 1$, the root required.

2. Extract the square root of $4a^4 + 12a^3x + 13a^2x^2 + ax^36 + x^4$.

$$4a^4 + 12a^3x + 13a^2x^2 + 6ax^3 + x^4\ (2a^2 + 3ax + x^2$$
$$4a^4$$

$$4a^2 + 3ax)\ 12a^3x + 13a^2x^2$$
$$12a^3x + 9a^2x^2$$

$$4a^2 + 6ax + x^2)\ 4a^2x^2 + 6ax^3 + x^4$$
$$4a^2x^2 + 6ax^3 + x^4$$

*

Note. When the quantity to be extracted has no exact root, the operation may be carried on as far as is thought necessary, or till the regularity of the terms show the law by which the series would be continued.

EXAMPLE.

1. It is required to extract the square root of $1 + x$.

$$1 + x)\, 1 + \frac{x}{2} - \frac{x^2}{8} + \frac{x^3}{16} - \frac{5x^4}{128}, \text{ \&c.}$$

$$1$$

$$2 + \frac{x}{2})\; x$$

$$x + \frac{x^2}{4}$$

$$2 + x - \frac{x^2}{8})\; -\frac{x^2}{4}$$

$$-\frac{x^2}{4} - \frac{x^3}{8} + \frac{x^4}{64}$$

$$2 + x - \frac{x^2}{4} + \frac{x^3}{16})\; \frac{x^3}{8} - \frac{x^4}{64}$$

$$\frac{x^3}{8} + \frac{x^4}{16} - \frac{x^5}{64} + \frac{x^6}{256}$$

$$-\frac{5x^4}{64} + \frac{x^5}{64} - \frac{x^6}{256}.$$

Here, if the numerators and denominators of the two last terms be each multiplied by 3, which will not alter their values, the root will become

$$1 + \frac{x}{2} - \frac{x^2}{2.4} + \frac{3x^3}{2.4.6} - \frac{3.5x^4}{2.4.6.8} + \frac{3.5.7x^5}{2.4.6.8.10}, \text{ \&c.}$$

where the law of the series is manifest.

EXAMPLES FOR PRACTICE.

2. It is required to find the square root of $a^4 + 4a^3x + 6a^2x^2 + 4ax^3 + x^4$. Ans. $a^2 + 2ax + x^2$.

3. It is required to find the square root of $x^4 - 2x^3 + \frac{3}{2}x^2 - \frac{1}{2}x + \frac{1}{16}$. Ans. $x^2 - x + \frac{1}{4}$.

4. It is required to find the square root of $4x^6 - 4x^4 + 12x^3 + x^2 - 6x + 9$. Ans. $2x^3 - x + 3$.

5. Required the square root of $x^6 + 4x^5 + 10x^4 + 20x^3 + 25x^2 + 24x + 16$. Ans. $x^3 + 2x^2 + 3x + 4$.

6. It is required to extract the square root of $a^2 + b$.

Ans. $a + \frac{b}{2a} - \frac{b^2}{8a^3} + \frac{b^3}{16a^5} - \frac{5b^4}{128a^7}$, &c.

7. It is required to extract the square root of 2, or of $1 + 1$.

Ans. $1 + \frac{1}{2} - \frac{1}{8} + \frac{1}{16} - \frac{1}{32} + \frac{1}{64}$, &c.

CASE III.

To find any root of a compound quantity.

Rule.—Find the root of the first term, which place in the quotient; and having subtracted its corresponding power from that term, bring down the second term for a dividend.

Divide this by twice the part of the root above determined, for the square root; by three times the square of it, for the cube root, and so on; and the quotient will be the next term of the root.

Involve the whole of the root, thus found, to its proper power, which subtract from the given quantity, and divide the first term of the remainder by the same divisor as before; and proceed in this manner till the whole is finished.*

EXAMPLES.

1. Required the square root of $a^4 - 2a^3x + 3a^2x^2 - 2ax^3 + x^4$.

$$\begin{array}{l}
a^4 - 2a^3x + 3a^2x^2 - 2ax^3 + x^4 \;(a^2 - ax + x^2 \\
a^4 \\
\hline
2a^2)\; - 2a^3x \\
\hline
a^4 - 2a^3x + a^2x^2 \\
\hline
\qquad 2a^2)\; 2a^2x^2 \\
\hline
a^4 - 2a^3x + 3a^2x^2 - 2ax^3 + x^4 \\
\hline
\qquad\qquad *
\end{array}$$

* As this rule, in high powers, is often found to be very laborious, it may be proper to observe, that the roots of various compound quantities may sometimes be easily discovered, as follows:—

Extract the roots of all the simple terms, and connect them together by the signs + or —, as may be judged most suitable for the purpose; then involve the compound root, thus found, to its proper power, and if it be the same with the given quantity, it is the root required. But if it

2. Required the cube root of $x^6+6x^5-40x^3+96x-64$.

$x^6+6x^5-40x^3+96x-64\ (x^2+2x-4$

x^6

$3x^4)\ 6x^5$

$x^6+6x^5+12x^4+8x^3$

$3x^4)-12x^4$

$x^6+6x^5-40x^3+96x-64$

*

3. Required the square root of $4a^2-12ax+9x^2$. Ans. $2a-3x$.

4. Required the square root of $a^2+2ab+2ac+b^2+2bc+c^2$. Ans. $a+b+c$.

5. Required the cube root of $x^6-6x^5+15x^4-20x^3+15x^2-6x+1$. Ans. x^3-2x+1.

6. Required the 4th root of $16a^4-96a^3x+216a^2x^2-216ax^3+81x^4$. Ans. $2a-3x$.

7. Required the 5th root of $32x^5-80x^4+80x^3-40x^2+10x-1$. Ans. $2x-1$.

OF IRRATIONAL QUANTITIES, OR SURDS.

IRRATIONAL *Quantities* or *Surds*, are those of which the values cannot be accurately expressed in numbers; and are usually expressed by means of the radical sign $\surd$, or by fractional indices; in which latter case, the numerator shows the power the quantity is to be raised to, and the denominator its root.

Thus, $\sqrt{2}$, or $2^{\frac{1}{2}}$, denotes the square root of 2; $\sqrt[3]{a^2}$, or $a^{\frac{2}{3}}$ the cube root of the square of a, &c.*

be found to differ only in some of the signs, change them from + to —, or from — to +, till its power agrees with the given one throughout.

Thus, in the third example next following, the root is $2a-3x$, which is the difference of the roots of the first and last terms; and in the fourth example, the root is $a+b+c$, which is the sum of the roots of the first, fourth, and sixth terms. The same may also be observed of the sixth example, where the root is found from the first and last terms.

* A quantity of the kind here mentioned, as for instance $\sqrt{2}$, is called an irrational number, or a surd, because no number, either whole or fractional, can be found, which, when multiplied by itself, will produce 2. But its approximate value may be determined to any degree of exactness, by the common rule for extracting the square root, being 1 and certain non-periodic decimals, which never terminate.

CASE I.

To reduce a rational quantity to the form of a surd.

RULE.—Raise the quantity to a power corresponding with that denoted by the index of the surd; and over this new quantity place the radical sign, or proper index, and it will be of the form required.

EXAMPLES.

1. Let 3 be reduced to the form of the square root.
   Here $3 \times 3 = 3^2 = 9$; whence $\sqrt{9}$. Ans.
2. Reduce $2x^2$ to the form of the cube root.
   Here $(2x^2)^3 = 8x^6$; whence $\sqrt[3]{8x^6}$, or $(8x^6)^{\frac{1}{3}}$.
3. Let 5 be reduced to the form of the square root.
   Ans. $\sqrt{(25)}$.
4. Let $-3x$ be reduced to the form of the cube root.
   Ans. $\sqrt[3]{-(27x^3)}$.
5. Let $-2a$ be reduced to the form of the fourth root.
   Ans. $-\sqrt[4]{(16a^4)}$.
6. Let a^2 be reduced to the form of the fifth root, and $\sqrt{a} + \sqrt{b}$, $\dfrac{\sqrt{a}}{2a}$ and $\dfrac{a}{b\sqrt{a}}$ to the form of the square root.

   Ans. $\sqrt[5]{a^{10}}$, $\sqrt{(a + 2\sqrt{ab} + b)}$, $\sqrt{(\frac{1}{4a})}$, and $\sqrt{\frac{a}{b^2}}$.

Note. Any rational quantity may be reduced by the above rule, to the form of the surd to which it is joined, and their product be then placed under the same index or radical sign.

EXAMPLES.

Thus $2\sqrt{2} = \sqrt{4} \times \sqrt{2} = \sqrt{(4 \times 2)} = \sqrt{8}$
And $2\sqrt[3]{4} = \sqrt[3]{8} \times \sqrt[3]{4} = \sqrt[3]{(8 \times 4)} = \sqrt[3]{32}$
Also $3\sqrt{a} = \sqrt{9} \times \sqrt{a} = \sqrt{(9 \times a)} = \sqrt{9a}$
And $\frac{1}{2}\sqrt[3]{4a} = \sqrt[3]{\frac{1}{8}} \times \sqrt[3]{4a} = \sqrt[3]{(\frac{1}{8} \times 4a)} = \sqrt[3]{\frac{a}{2}}$

1. Let $5\sqrt{6}$ be reduced to a simple radical form.
   Ans. $\sqrt{(150)}$.
2. Let $\frac{1}{5}\sqrt{5a}$ be reduced to a simple radical form.
   Ans. $\sqrt{(\frac{a}{5})}$.
3. Let $\dfrac{2a}{3}\sqrt[3]{\dfrac{9}{4a^2}}$ be reduced to a simple radical form.

   Ans. $\sqrt[3]{\dfrac{2a}{3}}$

CASE II.

To reduce quantities of different indices, to others that shall have a given index.

RULE.—Divide the indices of the proposed quantities by

the given index, and the quotients will be the new indices for those quantities.

Then, over the said quantities, with their new indices, place the given index, and they will be the equivalent quantities required.

EXAMPLES.

1. Reduce $3^{\frac{1}{2}}$ and $2^{\frac{1}{3}}$ to quantities that shall have the index $\frac{1}{6}$.

$$\text{Here } \frac{1}{2} \div \frac{1}{6} = \frac{1}{2} \times \frac{6}{1} = \frac{6}{2} = 3, \text{ the 1st index;}$$

$$\text{And } \frac{1}{3} \div \frac{1}{6} = \frac{1}{3} \times \frac{6}{1} = \frac{6}{3} = 2, \text{ the 2d index.}$$

Whence $(3^3)^{\frac{1}{6}}$ and $(2^2)^{\frac{1}{6}}$, or $27^{\frac{1}{6}}$, and $4^{\frac{1}{6}}$, are the quantities required.

2. Reduce $5^{\frac{1}{2}}$ and $6^{\frac{1}{3}}$ to quantities that shall have the common index $\frac{1}{6}$. Ans. $125^{\frac{1}{6}}$ and $36^{\frac{1}{6}}$.

3. Reduce $2^{\frac{1}{2}}$ and $4^{\frac{1}{4}}$ to quantities that shall have the common index $\frac{1}{8}$. Ans. $16^{\frac{1}{8}}$ and $16^{\frac{1}{8}}$.

4. Reduce a^2 and $a^{\frac{1}{2}}$ to quantities that shall have the common index $\frac{1}{4}$. Ans. $(a^8)^{\frac{1}{4}}$ and $(a^2)^{\frac{1}{4}}$.

5. Reduce $a^{\frac{1}{2}}$ and $b^{\frac{2}{3}}$ to quantities that shall have the common index $\frac{1}{6}$. Ans. $(a^3)^{\frac{1}{6}}$ and $(b^4)^{\frac{1}{6}}$.

Note. Surds may also be brought to a common index, by reducing the indices of the quantities to a common denominator, and then involving each of them to the power denoted by its numerator.

EXAMPLES.

1. Reduce $3^{\frac{1}{2}}$ and $4^{\frac{1}{3}}$ to quantities having a common index.

$$\text{Here } 3^{\frac{1}{2}} = 3^{\frac{3}{6}} = (3^3)^{\frac{1}{6}} = (27)^{\frac{1}{6}}$$

$$\text{And } 4^{\frac{1}{3}} = 4^{\frac{2}{6}} = (4^2)^{\frac{1}{6}} = (16)^{\frac{1}{6}}$$

Whence $(27)^{\frac{1}{6}}$ and $(16)^{\frac{1}{6}}$. Ans.

2. Reduce $4^{\frac{1}{3}}$ and $5^{\frac{1}{4}}$ to quantities that shall have a common index.

Ans. $256^{\frac{1}{12}}$ and $125^{\frac{1}{12}}$.

3. Reduce $a^{\frac{1}{2}}$ and $a^{\frac{1}{3}}$ to quantities that shall have a common index.

Ans. $(a^3)^{\frac{1}{6}}$ and $(a^2)^{\frac{1}{6}}$.

4. Reduce $a^{\frac{1}{3}}$ and $b^{\frac{1}{4}}$ to quantities that shall have a common index.

Ans. $(a^4)^{\frac{1}{12}}$ and $(b^3)^{\frac{1}{12}}$.

5. Reduce $a^{\frac{1}{n}}$ and $b^{\frac{1}{m}}$ to quantities that shall have a common index.

Ans. $(a^m)^{\frac{1}{nm}}$ and $(b^n)^{\frac{1}{mn}}$.

CASE III.

To reduce surds to their most simple forms.

RULE—Resolve the given number, or quantity, into two factors, one of which shall be the greatest power contained in it, and set the root of this power before the remaining part, with the proper radical sign between them.*

EXAMPLES.

1. Let $\sqrt{48}$ be reduced to its most simple form.
Here $\sqrt{48} = \sqrt{(16 \times 3)} = 4\sqrt{3}$. Ans.

2. Let $\sqrt[3]{108}$ be reduced to its most simple form.
Here $\sqrt[3]{108} = \sqrt[3]{(27 \times 4)} = 3\sqrt[3]{4}$. Ans.

Note 1. When any number, or quantity, is prefixed to the surd, that quantity must be multiplied by the root of the factor above mentioned, and the product be then joined to the other part, as before.

EXAMPLES.

1. Let $2\sqrt{32}$ be reduced to its most simple form.
Here $2\sqrt{32} = 2\sqrt{(16 \times 2)} = 8\sqrt{2}$. Ans.

2. Let $5\sqrt[3]{24}$ be reduced to its most simple form.
Here $5\sqrt[3]{24} = 5\sqrt[3]{(8 \times 3)} = 10\sqrt[3]{3}$. Ans.

Note 2. A fractional surd may also be reduced to a more convenient form, by multiplying both the numerator and denominator by such a number, or quantity, as will make the denominator a complete power of the kind required; and then joining its root, with 1 put over it, as a numerator, to the other part of the surd.†

---

* When the given surd contains no factor that is an exact power of the kind required, it is already in its most simple form.

Thus, $\sqrt{15}$ cannot be reduced lower, because neither of its factors, 5 nor 3, is a square.

† The utility of reducing surds to their most simple forms, in order to have the answer in decimals, will be readily perceived from considering

EXAMPLES.

1. Let $\sqrt{\frac{2}{7}}$ be reduced to its most simple form.

Here $\sqrt{\frac{2}{7}} = \sqrt{\frac{14}{49}} = \sqrt{\left(\frac{1}{49} \times 14\right)} = \frac{1}{7}\sqrt{14}$. Ans.

2. Let $3\sqrt[3]{\frac{2}{5}}$ be reduced to its most simple form.

Here $3\sqrt[3]{\frac{2}{5}} = 3\sqrt[3]{\frac{50}{125}} = 3\sqrt[3]{\left(\frac{1}{125} \times 50\right)} = \frac{3}{5}\sqrt[3]{50}$. Ans.

EXAMPLES FOR PRACTICE.

3. Let $\sqrt{125}$ be reduced to its most simple form. Ans. $5\sqrt{5}$.

4. Let $\sqrt{294}$ be reduced to its most simple form. Ans. $7\sqrt{6}$.

5. Let $\sqrt[3]{56}$ be reduced to its most simple form. Ans. $2\sqrt[3]{7}$.

6. Let $\sqrt[3]{192}$ be reduced to its most simple form. Ans. $4\sqrt[3]{3}$.

7. Let $7\sqrt{80}$ be reduced to its most simple form. Ans. $28\sqrt{5}$.

8. Let $9\sqrt[3]{81}$ be reduced to its most simple form. Ans. $27\sqrt[3]{3}$.

9. Let $\frac{3}{121}\sqrt{\frac{5}{6}}$ be reduced to its most simple form. Ans. $\frac{1}{242}\sqrt{30}$.

10. Let $\frac{4}{7}\sqrt[3]{\frac{3}{16}}$ be reduced to its most simple form. Ans. $\frac{1}{7}\sqrt[3]{12}$.

11. Let $\sqrt{98a^2x}$ be reduced to its most simple form. Ans. $7a\sqrt{2x}$.

12. Let $\sqrt{(x^3 - a^2x^2)}$ be reduced to its most simple form. Ans. $x\sqrt{(x - a^2)}$.

---

the first question above given, where it is found that $\sqrt{\frac{2}{7}} = \frac{1}{7}\sqrt{14}$; in which case it is only necessary to extract the square root of the whole number 14, (or to find it in some of the tables that have been calculated for this purpose,) and then divide it by 7; whereas, otherwise, we must have first divided the numerator by the denominator, and then have found the root of the quotient, for the surd part; or else have determined the root both of the numerator and denominator, and then divided one by the other; which are each of them troublesome processes when performed by the common rules; and in the next example for the cube root, the labour would be much greater

CASE IV.

To add surd quantities together.

RULE.—When the surds are of the same kind, reduce them to their simplest forms as in the last case; then, if the surd part be the same in them all, annex it to the sum of the rational parts, and it will give the whole sum required.

But if the quantities have different indices, or the surd part be not the same in each of them, they can only be added together by the signs + and —.

EXAMPLES.

1. It is required to find the sum of $\sqrt{27}$ and $\sqrt{48}$.

Here $\sqrt{27} = \sqrt{(9 \times 3)} = 3\sqrt{3}$
And $\sqrt{48} = \sqrt{(16 \times 3)} = 4\sqrt{3}$

Whence $7\sqrt{3}$ the sum.

2. It is required to find the sum of $\sqrt[3]{500}$ and $\sqrt[3]{108}$.

Here $\sqrt[3]{500} = \sqrt[3]{(125 \times 4)} = 5\sqrt[3]{4}$
And $\sqrt[3]{108} = \sqrt[3]{(27 \times 4)} = 3\sqrt[3]{4}$

Whence $8\sqrt[3]{4}$ the sum.

3. It is required to find the sum of $4\sqrt{147}$ and $3\sqrt{75}$

Here $4\sqrt{147} = 4\sqrt{(49 \times 3)} = 28\sqrt{3}$
And $3\sqrt{75} = 3\sqrt{(25 \times 3)} = 15\sqrt{3}$

Whence $43\sqrt{3}$ the sum.

4. It is required to find the sum of $3\sqrt{\frac{2}{5}}$ and $2\sqrt{\frac{1}{10}}$.

Here $3\sqrt{\frac{2}{5}} = 3\sqrt{\frac{10}{25}} = \frac{3}{5}\sqrt{10}$

And $2\sqrt{\frac{1}{10}} = 2\sqrt{\frac{10}{100}} = \frac{2}{10}\sqrt{10}$

Whence $\frac{4}{5}\sqrt{10}$ the sum.

EXAMPLES FOR PRACTICE.

5. It is required to find the sum of $\sqrt{72}$ and $\sqrt{128}$.
Ans. $14\sqrt{(2)}$.

6. It is required to find the sum of $\sqrt{180}$ and $\sqrt{405}$.
Ans. $15\sqrt{(5)}$.

7. It is required to find the sum of $3\sqrt[3]{40}$ and $\sqrt[3]{135}$.
Ans. $9\sqrt[3]{(5)}$.

8. It is required to find the sum of $4\sqrt[3]{54}$ and $5\sqrt[3]{128}$.

Ans. $32\sqrt[3]{(2)}$.

9. It is required to find the sum of $9\sqrt{243}$ and $10\sqrt{363}$

Ans. $191\sqrt{(3)}$.

10. It is required to find the sum of $3\sqrt{\frac{2}{3}}$ and $7\sqrt{\frac{27}{50}}$.

Ans. $3\frac{1}{10}\sqrt{(6)}$.

11. It is required to find the sum of $12\sqrt[3]{\frac{1}{4}}$ and $3\sqrt[3]{\frac{1}{32}}$.

Ans. $6\frac{3}{4}\sqrt[3]{(2)}$.

12. It is required to find the sum of $\frac{1}{2}\sqrt{a^2b}$ and $\frac{1}{3}\sqrt{4bx^4}$.

Ans. $\left(\frac{a}{2}+\frac{2x^2}{3}\right)\sqrt{b}$

CASE V.

To find the difference of surd quantities.

Rule.—When the surds are of the same kind, prepare the quantities as in the last rule; then the difference of the rational parts annexed to the common surd, will give the whole difference required.

But if the quantities have different indices, or the surd part be not the same in each of them, they can only be subtracted by means of the sign $-$.

1. It is required to find the difference of $\sqrt{448}$ and $\sqrt{112}$.

Here $\sqrt{448}=\sqrt{(64\times 7)}=8\sqrt{7}$

And $\sqrt{112}=\sqrt{(16\times 7)}=4\sqrt{7}$

Whence $4\sqrt{7}$ the difference.

2. It is required to find the difference of $\sqrt[3]{192}$ and $\sqrt[3]{24}$.

Here $\sqrt[3]{192}=\sqrt[3]{(64\times 3)}=4\sqrt[3]{3}$

And $\sqrt[3]{24}=\sqrt[3]{(8\times 3)}=2\sqrt[3]{3}$

Whence $2\sqrt[3]{3}$ the difference.

3. It is required to find the difference of $5\sqrt{20}$ and $3\sqrt{45}$.

Here $5\sqrt{20}=5\sqrt{(4\times 5)}=10\sqrt{5}$

And $3\sqrt{45}=3\sqrt{(9\times 5)}=9\sqrt{5}$

Whence $\sqrt{5}$ the difference.

4 It is required to find the difference of $\frac{3}{4}\sqrt{\frac{2}{3}}$, and $\frac{2}{5}\sqrt{\frac{1}{6}}$.

$$\text{Here } \frac{3}{4}\sqrt{\frac{2}{3}} = \frac{3}{4}\sqrt{\frac{6}{9}} = \frac{3}{12}\sqrt{6} = \frac{1}{4}\sqrt{6}$$

$$\text{And } \frac{2}{5}\sqrt{\frac{1}{6}} = \frac{2}{5}\sqrt{\frac{6}{36}} = \frac{2}{30}\sqrt{6} = \frac{1}{15}\sqrt{6}$$

Whence $\frac{11}{60}\sqrt{6}$ the difference, or answer required.

EXAMPLES FOR PRACTICE.

1. It is required to find the difference of $2\sqrt{50}$ and $\sqrt{18}$.
Ans. $7\sqrt{(2)}$.

2. It is required to find the difference of $\sqrt[3]{320}$ and $\sqrt[3]{40}$.
Ans. $2\sqrt[3]{(5)}$.

3. It is required to find the difference of $\sqrt{\frac{3}{5}}$ and $\sqrt{\frac{5}{27}}$.
Ans. $\frac{4}{45}\sqrt{(15)}$.

4. It is required to find the difference of $2\sqrt{\frac{1}{2}}$ and $\sqrt{8}$.
Ans. $\sqrt{(2)}$.

5. It is required to find the difference of $3\sqrt[3]{\frac{1}{3}}$ and $\sqrt[3]{72}$.
Ans. $\sqrt[3]{(9)}$.

6. It is required to find the difference of $\sqrt[3]{\frac{2}{3}}$ and $\sqrt[3]{\frac{9}{32}}$.
Ans. $\frac{1}{12}\sqrt[3]{(18)}$.

7. It is required to find the difference of $\sqrt{80a^4x}$ and $\sqrt{20a^2x^3}$.
Ans. $(4a^2 - 2ax)\sqrt{(5x)}$.

8. It is required to find the difference of $8\sqrt[3]{a^3b}$ and $2\sqrt[3]{a^6b}$.
Ans. $(8a - 2a^2)\sqrt[3]{(b)}$.

Note. The two last answers may be written thus,

$(2ax - 4a^2)\sqrt{(5x)}$, and
$(2a^2 - 8a)\sqrt[3]{(b)}$;
or $(4a^2 \quad 2ax)\sqrt{5x}$
$(8a \quad 2a^2)\sqrt[3]{b}$.

CASE VI.

To multiply surd quantities together.

RULE.—When the surds are of the same kind, find the product of the rational parts, and the product of the surds, and the two joined together, with their common radical sign between them, will give the whole product required; which may be reduced to its most simple form by Case III.

But if the surds are of different kinds, they must be reduced to a common index, and then multiplied together as usual.

It is also to be observed, as before mentioned, that the product of different powers, or roots of the same quantity, is found by adding their indices.

EXAMPLES.

1. It is required to find the product of $3\sqrt{8}$ and $2\sqrt{6}$.

Here $3\sqrt{8}$

Multiplied $2\sqrt{6}$

Gives $6\sqrt{48} = 6\sqrt{16 \times 3} = 24\sqrt{3}$. Ans.

2. It is required to find the product of $\frac{1}{2}\sqrt[3]{\frac{2}{3}}$ and $\frac{3}{4}\sqrt[3]{\frac{5}{6}}$.

Here $\frac{1}{2}\sqrt[3]{\frac{2}{3}}$

Multiplied $\frac{3}{4}\sqrt[3]{\frac{5}{6}}$

Gives $\frac{3}{8}\sqrt[3]{\frac{10}{18}} = \frac{3}{8}\sqrt[3]{\frac{5}{9}} = \frac{3}{8}\sqrt[3]{\frac{15}{27}} = \frac{1}{8}\sqrt[3]{15}$.

3. It is required to find the products of $2^{\frac{1}{2}}$ and $3^{\frac{1}{3}}$

Here $2^{\frac{1}{2}} = 2^{\frac{3}{6}} = (2^3)^{\frac{1}{6}} = 8^{\frac{1}{6}}$

And $3^{\frac{1}{3}} = 3^{\frac{2}{6}} = (3^2)^{\frac{1}{6}} = 9^{\frac{1}{6}}$

Whence $(72)^{\frac{1}{6}}$. Ans.

4. It is required to find the product of $5\sqrt{a}$ and $3\sqrt[3]{a}$.

Here $5\sqrt{a} = 5a^{\frac{1}{2}} = 5a^{\frac{3}{6}}$

And $3\sqrt[3]{a} = 3a^{\frac{1}{3}} = 3a^{\frac{2}{6}}$

Whence $15a^{\frac{5}{6}} = 15(a^5)^{\frac{1}{6}}$ or $15\sqrt[6]{a^5}$. Ans.

EXAMPLES FOR PRACTICE.

5. It is required to find the product of $5\sqrt{8}$ and $3\sqrt{5}$.

Ans. $30\sqrt{(10)}$.

6. It is required to find the product of $\sqrt[3]{18}$ and $5\sqrt[3]{4}$.

Ans. $10\sqrt[3]{(9)}$

7. Required the product of $\frac{1}{4}\sqrt{6}$ and $\frac{2}{15}\sqrt{9}$.

Ans. $\frac{1}{10}\sqrt{(6)}$.

8. Required the product of $\frac{1}{2}\sqrt{18}$ and $5\sqrt{20}$.

Ans. $15\sqrt{(10)}$.

9. Required the product of $2\sqrt{3}$ and $13\frac{1}{2}\sqrt{5}$.
Ans. $27\sqrt{(15)}$.

10. Required the product of $72\frac{1}{4}a^{\frac{2}{3}}$ and $120\frac{1}{2}a^{\frac{1}{4}}$.
Ans. $8706\frac{1}{8}a^{\frac{11}{12}}$.

11 Required the product of $4+2\sqrt{2}$ and $2-\sqrt{2}$.
Ans 4.

12. Required the product of $(a+b)^{\frac{1}{n}}$ and $(a+b)^{\frac{1}{m}}$.
Ans. $(a+b)^{\frac{m+n}{mn}}$.

CASE VII.

To divide one surd quantity by another.

RULE.—When the surds are of the same kind, find the quotient of the rational parts, and the quotient of the surds, and the two joined together, with their common radical sign between them, will give the whole quotient required.

But if the surds are of different kinds, they must be reduced to a common index, and then be divided as before.

It is also to be observed, that the quotient of different powers or roots of the same quantity, is found by subtracting their indices.

EXAMPLES.

1. It is required to divide $8\sqrt{108}$ by $2\sqrt{6}$.

Here $\dfrac{8\sqrt{108}}{2\sqrt{6}}=4\sqrt{18}=4\sqrt{(9\times2)}=12\sqrt{2}$. Ans.

2. It is required to divide $8\sqrt[3]{512}$ by $4\sqrt[3]{2}$.

Here $\dfrac{8\sqrt[3]{512}}{4\sqrt[3]{2}}=2\sqrt[3]{256}=2\sqrt[3]{(64\times4)}=8\sqrt[3]{4}$. Ans.

3. It is required to divide $\dfrac{1}{2}\sqrt{5}$ by $\dfrac{1}{3}\sqrt{10}$. Ans.

Here $\dfrac{\frac{1}{2}\sqrt{5}}{\frac{1}{3}\sqrt{2}}=\dfrac{3}{2}\sqrt{\dfrac{5}{2}}=\dfrac{3}{2}\sqrt{\dfrac{10}{4}}=\dfrac{3}{4}\sqrt{10}$. Ans.

4. It is required to divide $\sqrt{7}$ by $\sqrt[3]{7}$.

Here $\dfrac{\sqrt{7}}{\sqrt[3]{7}}=\dfrac{7^{\frac{1}{2}}}{7^{\frac{1}{3}}}=\dfrac{7^{\frac{3}{6}}}{7^{\frac{2}{6}}}=7^{\frac{3}{6}-\frac{2}{6}}=7^{\frac{1}{6}}$. Ans.

5. It is required to divide $6\sqrt{54}$ by $3\sqrt{2}$. Ans. $6\sqrt{3}$.

6. It is required to divide $4\sqrt[3]{72}$ by $2\sqrt[3]{18}$. Ans. $2\sqrt[3]{4}$.

7. It is required to divide $5\dfrac{3}{4}\sqrt{\dfrac{1}{135}}$ by $\dfrac{2}{3}\sqrt{\dfrac{1}{5}}$.
Ans. $\frac{23}{24}\sqrt{3}$.

8. It is required to divide $4\frac{5}{7}\sqrt{\frac{2}{3}}$ by $2\frac{2}{5}\sqrt{\frac{3}{4}}$.

Ans. $\frac{55}{42}\sqrt{2}$.

9. It is required to divide $4\frac{1}{2}\sqrt[2]{a}$ by $2\frac{2}{3}\sqrt[3]{ab}$.

Ans. $\frac{27}{16}\left(\frac{a}{b^2}\right)^{\frac{1}{6}}$.

10. It is required to divide $32\frac{2}{5}\sqrt{a}$ by $13\frac{3}{4}\sqrt[3]{a}$.

Ans. $\frac{648}{275}a^{\frac{1}{6}}$

11. It is required to divide $9\frac{3}{8}a^{\frac{1}{n}}$ by $4\frac{9}{11}a^{\frac{1}{m}}$.

Ans. $\frac{825}{424}a^{\frac{m-n}{mn}}$.

12. It is required to divide $\sqrt{20}+\sqrt{12}$ by $\sqrt{5}+\sqrt{3}$.

Ans. $\sqrt{4}$ or 2.

Note. Since the division of surds is performed by subtracting their indices, it is evident that the denominator of any fraction may be taken into the numerator, or the numerator into the denominator, by changing the sign of its index.

Also, since $\frac{a^m}{a^m}=1$, or $=a^{m-m}=a^0$, it follows, that the expression a^0 is a symbol equivalent to unity, and consequently, that it may always be replaced by 1 whenever it occurs.*

---

* To what is above said, we may also farther observe,

1. That 0 added to or subtracted from any quantity, makes it neither greater nor less; that is,

$$a+0=a, \text{ and } a-0=a.$$

2. Also, if nought be multiplied or divided by any quantity, both the product and the quotient will be nought; because any number of times 0, or any part of 0, is 0; that is,

$$0\times a, \text{ or } a\times 0=0, \text{ and } \frac{0}{a}=0.$$

3. From this it likewise follows, that nought divided by nought, is a finite quantity, of some kind or other.

For since $0\times a=0$, or $0=0\times a$, it is evident that $\frac{0}{0}=a$.

4. Farther, if any finite quantity be divided by 0, the quotient will be infinite.

For let $\frac{b}{a}=q$, then if b remain the same, it is plain, the less a is, the greater will be the quotient q; whence, if a be indefinitely small, q will

EXAMPLES.

1. Thus, $\frac{1}{a}=\frac{a^{-1}}{1}$ or a^{-1}; and $\frac{1}{a^n}=\frac{a^{-n}}{1}$, or a^{-n}.

2. Also, $\frac{b}{a^2}=\frac{ba^{-2}}{1}$, or ba^{-2}, and $\frac{a^{-n}}{b^{-m}}=\frac{1}{a^nb^{-m}}$, or $\frac{b^m}{a^n}$.

3. Let $\frac{1}{a^2}$ be expressed with a negative index. Ans. a^{-2}

4 Let $a^{-\frac{1}{2}}$ be expressed with a positive index.

Ans. $\frac{1}{\frac{1}{2}a}$.

5. Let $\frac{1}{a+x}$ be expressed with a negative index.

Ans. $(a+x)^{-1}$.

6. Let $a(a^2-x^2)^{-\frac{1}{3}}$ be expressed with a positive index.

Ans. $\frac{a}{(a^2-x^2)^{\frac{1}{3}}}$.

CASE VIII.

To involve, or raise surd quantities to any power.

RULE.—When the surd is a simple quantity, multiply its index by 2 for the square, by 3 for the cube, &c., and it will give the power of the surd part, which, being annexed to the proper power of the rational part, will give the whole power required. And if it be a compound quantity, multiply it by

---

be indefinitely great: and consequently, when a is 0, the quotient q will be infinite; that is,

$$\frac{b}{0}, \text{ or } \frac{1}{0}=\infty.$$

Which properties are of frequent occurrence in some of the higher parts of the science, and should be carefully remembered.

Since, therefore, $\frac{1}{a+b}$ is the same as $(a+b)^{-1}$. Let us suppose, in the general formula, $n=-1$; and we shall have for the coefficients $n=-1$; $\frac{n-1}{2}=-1$; $\frac{n-2}{3}=-1$; $\frac{n-3}{4}=-1$, &c. and for the powers of a we have $a^n=a^{-1}=\frac{1}{a}$; $a^{n-1}=a^{-2}=\frac{1}{a^2}$; $a^{n-2}=\frac{1}{a^3}$; $a^{n-3}=\frac{1}{a^4}$, &c.; so that $(a+b)^{-1}=\frac{1}{a+b}=\frac{1}{a}-\frac{b}{a^2}+\frac{b^2}{a^3}-\frac{b^3}{a^4}+\frac{b^4}{a^5}-\frac{b^5}{a}$, &c., which is the same series that is found by division. For more on this subject see the Binomial Theorem, (further on) or Euler's Algebra.

itself the proper number of times according to the usual rule.*

EXAMPLES.

1. It is required to find the square of $\frac{2}{3}a^{\frac{1}{3}}$.

Here $\left(\frac{2}{3}a^{\frac{1}{3}}\right)^2 = \frac{4}{9}a^{\frac{1}{3}\times 2} = \frac{4}{9}a^{\frac{2}{3}} = \frac{4}{9}\sqrt[3]{a^2}$. Ans.

2. It is required to find the cube of $\frac{2}{3}\sqrt{3}$.

Here $\frac{8}{27}\times 3^{\frac{3}{2}} = \frac{8}{27}\sqrt{27} = \frac{8}{27}\sqrt{(9\times 3)} = \frac{8}{9}\sqrt{3}$. Ans.

3. It is required to find the square of $3\sqrt[3]{3}$. Ans. $9\sqrt[3]{9}$.

4. It is required to find the cube of $17\sqrt{21}$.
Ans. $103173\sqrt{(21)}$.

5. It is required to find the 4th power of $\frac{1}{6}\sqrt{6}$. Ans. $\frac{1}{36}$.

6. It is required to find the square of $3+2\sqrt{5}$.
Ans. $29+12\sqrt{5}$.

7. It is required to find the cube of $\sqrt{x}+3\sqrt{y}$.
Ans. $x\sqrt{x}+27y\sqrt{x}+9x\sqrt{y}+27y\sqrt{y}$.

8. It is required to find the 4th power of $\sqrt{3}-\sqrt{2}$.
Ans. $49-20\sqrt{6}$.

CASE IX.

To find the roots of surd quantities.

RULE.—When the surd is a simple quantity, multiply its index by $\frac{1}{2}$ for the square root, by $\frac{1}{3}$ for the cube root, &c., and it will give the root of the surd part; which being annexed to the root of the rational part, will give the whole root required. And if it be a compound quantity, find its root by the usual rule.†

* When any quantity that is affected by the sign of the square root is to be raised to the second power, or squared, it is done by suppressing the sign. Thus,
$(\sqrt{a})^2$, or $\sqrt{a}\times\sqrt{a}=a$; and $(\sqrt{a+b})^2$, or $\sqrt{(a+b)}\times\sqrt{(a+b)}=a+b$.

† The nth root of the mth power of any number a, or the mth power of the nth root of a, is $a^{\frac{m}{n}}$.

Also, the nth root of the mth root of any number a, or the mth root of the nth root of a, is $a^{\frac{1}{mn}}$.

From which last expression, it appears, that the square root of the square root of a is the 4th root of a; and that the cube root of the square root of a, or the square root of the cube root of a, is the 6th root of a; and so on for the fourth, fifth, or any other numerical root of this kind.

EXAMPLES.

1. It is required to find the square root of $9\sqrt[3]{3}$.

Here $(9\sqrt[3]{3})^{\frac{1}{2}} = 9^{\frac{1}{2}} \times 3^{\frac{1}{3}\times\frac{1}{2}} = 9^{\frac{1}{2}} \times 3^{\frac{1}{6}} = 3\sqrt[6]{3}$. Ans.

2. It is required to find the cube root of $\frac{1}{8}\sqrt{2}$.

Here $\left(\frac{1}{8}\sqrt{2}\right)^{\frac{1}{3}} = \left(\frac{1}{8}\right)^{\frac{1}{3}} \times (2^{\frac{1}{2}\times\frac{1}{3}}) = \frac{1}{2}(2^{\frac{1}{6}}) = \frac{1}{2}\sqrt[6]{2}$. Ans.

3. It is required to find the square root of 10^3.

Ans. $10\sqrt{(10)}$

4. It is required to find the cube root of $\frac{8}{27}a^4$.

Ans. $\frac{2}{3}a\sqrt[3]{a}$.

5. It is required to find the 4th root of $\frac{16}{81}a^{\frac{2}{3}}$. Ans. $\frac{2}{3}a^{\frac{1}{6}}$.

6. It is required to find the cube root of $\frac{a}{3}\sqrt{\frac{a}{3}}$.

Ans. $\sqrt{\frac{a}{3}}$, or $\frac{1}{3}\sqrt{(3a)}$.

7. It is required to find the square root of $x^2 - 4x\sqrt{a} + 4a$.

Ans. $x - 2\sqrt{a}$.

8. It is required to find the square root of $a + 2\sqrt{ab} + b$.

Ans. $\sqrt{a} + \sqrt{b}$.

CASE X.

To transform a binomial, or a residual surd, into a general surd.

Rule.—Involve the given binomial, or residual, to a power corresponding with that denoted by the surd; then set the radical sign of the same root over it, and it will be the general surd required.

EXAMPLES.

1. It is required to reduce $2 + \sqrt{3}$ to a general surd.

Here $(2 + \sqrt{3})^2 = 4 + 3 + 4\sqrt{3} = 7 + 4\sqrt{3}$; therefore $2 + \sqrt{3} = \sqrt{(7 + 4\sqrt{3})}$, the answer.

2. It is required to reduce $\sqrt{2} + \sqrt{3}$ to a general surd.

Here $(\sqrt{2} + \sqrt{3})^2 = 2 + 3 + 2\sqrt{6} = 5 + 2\sqrt{6}$; therefore $\sqrt{2} + \sqrt{3} = \sqrt{(5 + 2\sqrt{6})}$, the answer.

3. It is required to reduce $\sqrt[3]{2} + \sqrt[3]{4}$ to a general surd.

Here $(\sqrt[3]{2} + \sqrt[3]{4})^3 = 6 + 6\sqrt[3]{2} + 6\sqrt[3]{4}$; therefore $\sqrt[3]{2} + \sqrt[3]{4} = \sqrt[3]{6(1 + \sqrt[3]{2} + \sqrt[3]{4})}$, the answer.

4. It is required to reduce $3 - \sqrt{5}$ to a general surd.

Ans. $\sqrt{(14 - 6\sqrt{5})}$.

5. It is required to reduce $\sqrt{2} - 2\sqrt{6}$ to a general surd.
Ans. $\sqrt{(26 - 8\sqrt{3})}$.

6. It is required to reduce $4 - \sqrt{7}$ to a general surd.
Ans. $\sqrt{(23 - 8\sqrt{7})}$.

7. It is required to reduce $2\sqrt[3]{3} - 3\sqrt[3]{9}$ to a general surd.
Ans. $\sqrt[3]{(162\sqrt[3]{9} - 108\sqrt[3]{3} - 219)}$.

CASE XI.

To extract the square root of a binomial, or residual surd.

RULE.*—Substitute the numbers, or parts, of which the given surd is composed, in the place of the letters, in one of

---

* Prop. 1. *The square root of a quantity cannot be partly rational and partly a quadratic surd.* If possible, let $\sqrt{n} = a + \sqrt{m}$; then, by squaring both sides, $n = a^2 + 2a\sqrt{m} + m$, and by transposition $2a\sqrt{m} = n - a^2 - m$; therefore $\sqrt{m} = \frac{n - a^2 - m}{2a}$, a rational quantity, which is contrary to the supposition. A quantity of the form $\sqrt{a}$, is called a *quadratic surd.*

Prop. 2. *In any equation* $x + \sqrt{y} = a + \sqrt{b}$, *consisting of rational quantities and quadratic surds, the rational parts on each side are equal, and also the irrational parts.*

If x be not equal to a, let $x = a + m$; then $a + m + \sqrt{y} = a + \sqrt{b}$, or $m + \sqrt{y} = \sqrt{b}$; that $\sqrt{b}$ is partly rational and partly a quadratic surd, which is impossible, (Prop. 1.);

$\therefore x = a$, and $\sqrt{y} = \sqrt{b}$.

In like manner, if $x - \sqrt{y} = a - \sqrt{b}$; then $x = a$, and $-\sqrt{y} = -\sqrt{b}$.

Prop. 3. *If two quadratic surds,* $\sqrt{x}$ *and* $\sqrt{y}$, *cannot be reduced to others which have the same irrational part, their product is irrational.*

If possible, let $\sqrt{xy} = rx$. where r is a whole number or a fraction: Then $xy = r^2x^2$, and $y = r^2x$; $\therefore \sqrt{y} = r\sqrt{x}$; that is, $\sqrt{y}$ and $\sqrt{x}$ may be so reduced as to have the same rational part, which is contrary to the supposition.

Prop. 4. *One quadratic surd,* $\sqrt{x}$, *cannot be made up of two others,* $\sqrt{m}$ *and* $\sqrt{n}$, *which have not the same irrational part.*

If possible, let $\sqrt{x} = \sqrt{m} + \sqrt{n}$; then by squaring both sides, $x = m + 2\sqrt{mn} + n$, and $x - m - n = 2\sqrt{mn}$, a rational quantity equal to an irrational, which is absurd.

Prop. 5. *The square root of a binomial, one of whose terms is a quadratic surd, and the other rational, may sometimes be expressed by a binomial, one or both of whose terms are quadratic surds..*

Let $a + \sqrt{b}$ be the given binomial, and suppose $\sqrt{(a + \sqrt{b})} = x + y$; where x and y are one or both quadratic surds; then, (see Ryan's Elementary Treatise on Algebra, Art. 367,) $\sqrt{(a - \sqrt{b})} = x - y$; $\therefore$ by multiplication, $\sqrt{(a^2 - b)} = x^2 - y^2$.

Also, by squaring both sides of the first equation, $a + \sqrt{b} = x^2 + 2xy + y^2$, and (Prop. 2.) $\because a = x^2 + y^2$.

Hence by addition, $a + \sqrt{(a^2 - b)} = 2x^2$,

and by subtraction, $a - \sqrt{(a^2 - b)} = 2y^2$;

$\therefore$ The root $x + y = \sqrt{\left\{\frac{1}{2}a + \frac{1}{2}\sqrt{(a^2 - b)}\right\}} + \sqrt{\left\{\frac{1}{2}a - \frac{1}{2}\sqrt{(a^2 - b)}\right\}}$

From this conclusion it appears, that the square root of $a + \sqrt{b}$ can

the two following formulæ, according as it is a binomial or a residual, and it will give the root required.

$$\sqrt{(a+\sqrt{b})}=\sqrt{[\tfrac{1}{2}a+\tfrac{1}{2}\sqrt{(a^2-b)}]}+\sqrt{[\tfrac{1}{2}a-\tfrac{1}{2}\sqrt{(a^2-b)}]}$$
$$\sqrt{(a-\sqrt{b})}=\sqrt{[\tfrac{1}{2}a+\tfrac{1}{2}\sqrt{(a^2-b)}]}-\sqrt{[\tfrac{1}{2}a-\tfrac{1}{2}\sqrt{(a^2-b)}]}$$

And if the second part of the binomial, or residual, in this case, be an imaginary surd, the same theorems will still hold, by only changing $-b$ into $+b$, as below.

$$\sqrt{(a+\sqrt{-b})}=\sqrt{[\tfrac{1}{2}a+\tfrac{1}{2}\sqrt{(a^2+b)}]}+\sqrt{[\tfrac{1}{2}a-\tfrac{1}{2}\sqrt{(a^2+b)}]}$$
$$\sqrt{(a-\sqrt{-b})}=\sqrt{[\tfrac{1}{2}+\tfrac{1}{2}\sqrt{(a^2+b)}]}-\sqrt{[\tfrac{1}{2}a-\tfrac{1}{2}\sqrt{(a^2+b)}]}$$

Where it is to be observed, that the only cases that are useful in this extraction, are, when a is rational, and a^2-b in the first of these formulæ, or a^2+b in the latter, is a complete square.

EXAMPLES.

1. It is required to find the square root of $11+\sqrt{72}$, or $\sqrt{(11+6\sqrt{2})}$.

Here,

$$\sqrt{\tfrac{1}{2}a+\tfrac{1}{2}\sqrt{(a^2-b)}}=\sqrt{\tfrac{11}{2}+\tfrac{1}{2}\sqrt{(121-72)}}=\sqrt{\tfrac{11}{2}+\tfrac{7}{2}}=3;$$

and

$$\sqrt{\tfrac{1}{2}a-\tfrac{1}{2}\sqrt{a^2-b}}=\sqrt{\tfrac{11}{2}-\tfrac{1}{2}\sqrt{121-72}}=\sqrt{\tfrac{11}{2}-\tfrac{7}{2}}=\sqrt{2}.$$

Whence $\sqrt{(11+6\sqrt{2})}=3+\sqrt{2}$, the answer required.

2. It is required to find the square root of $3-2\sqrt{2}$

Here,

$$\sqrt{\tfrac{1}{2}a+\tfrac{1}{2}\sqrt{(a^2-b)}}=\sqrt{\tfrac{3}{2}+\tfrac{1}{2}\sqrt{(9-8)}}=\sqrt{\tfrac{3}{2}-\tfrac{1}{2}}=\sqrt{2};\text{ and}$$
$$-\sqrt{\tfrac{1}{2}a-\tfrac{1}{2}\sqrt{(a^2-b)}}=-\sqrt{\tfrac{3}{2}-\tfrac{1}{2}\sqrt{(9-8)}}=-\sqrt{(\tfrac{3}{2}-\tfrac{1}{2})}=-1;$$

Whence $\sqrt{(3-2\sqrt{2})}=\sqrt{2}-1$, the answer required.

3. It is required to find the square root of $6\pm2\sqrt{5}$.

Ans. $\sqrt{5}\pm1$.

4. It is required to find the square root of $23\pm8\sqrt{7}$.

Ans. $4\pm\sqrt{7}$.

5. It is required to find the square root of $36\pm10\sqrt{11}$.

Ans. $5\pm\sqrt{(11)}$.

6 It is required to find the square root of $33\pm12\sqrt{6}$.

Ans. $2\sqrt{6}\pm3$.

7. It is required to find the square root of $1+4\sqrt{-3}$, or $1+\sqrt{-48}$.

Ans. $2+\sqrt{-3}$.

---

be expressed by a binomial of the form $x+y$, one or both of which are quadratic surds, when a^2-b is a perfect square.

By a similar process it might be shown that the square root of $a-\sqrt{b}$, or

$$\sqrt{(a-\sqrt{b})}=\sqrt{\left\{\tfrac{1}{2}a+\tfrac{1}{2}\sqrt{(a^2-b)}\right\}}-\sqrt{\left\{\tfrac{1}{2}a-\tfrac{1}{2}\sqrt{(a^2-b)}\right\}}$$

subject to the same limitation.—Ed.

8. It is required to find the square root of $3 \pm 4 \surd -1$, or $3 \pm \surd -16$. Ans. $2 \pm \surd +1$.

9. It is required to find the square root of $-1 + \surd -8$. Ans. $1 + \surd -2$.

10. It is required to find the square root of $a^2 + 2x \surd (a^2 - x^2)$. Ans. $x + \surd (a^2 - x^2)$.

11. It is required to find the square root of $6 + 2 \surd 2 - \surd (12) - \surd (24)$. Ans. $1 + \surd 2 - \surd 3$.

FOR TRINOMIAL, QUADRINOMIAL SURDS, &c.

RULE.—Divide half the product of any two radicals by a third, gives the square of one radical part of the root; this repeated with different quantities, will give the squares of all the parts of the root, to be connected by + and —. But if any quantity occur oftener than once, it must be taken but once.

For if $x + y + z$ be any trinomial surd, its square will be $x^2 + y^2 + z^2 + 2xy + 2xz + 2yz$; then if half the product of any two rectangles as $2xy \times 2xz$ (or $2x^2yz$) be divided by some third $2yx$, the quotient $\frac{2x^2yz}{2yz} = x^2$, must needs be the square of one of the parts; and the like for the rest.

EXAMPLE 1.

To extract the square root of $10 + \surd (24) + \surd (40) + \surd (60)$.

Here $\frac{\surd (24) \times \surd (40)}{2 \surd (60)} = 2$, and $\frac{\surd (24) \times \surd (60)}{2 \surd (40)} = \surd 9 = 3$, and $\frac{\surd (40) \times \surd (60)}{2 \surd (24)} = \surd (25) = 5$. And the root is $\surd 2 + \surd 3 + \surd 5$.

EXAMPLE 2.

It is required to find the square root of $14 + \surd (32) - \surd (48) + \surd (80) - \surd (24) + \surd (40) - \surd (60)$.

Here $\frac{\surd (32 \times 48)}{2 \surd (80)} = \frac{\surd (24)}{\surd 5}$, this produces nothing.

Again, $\frac{\surd (32 \times 48)}{2 \surd (24)} = \surd (16) \quad 4.$ And $\frac{\surd (40 \times 60)}{2 \surd (24)} = \surd (25) = 5$; and $\frac{\surd (32 \times 40)}{2 \surd (60)} = \surd 4 = 2$; and $\frac{\surd (48 \times 24)}{2 \surd (32)} = \surd 9 = 3$; and $\frac{\surd (32 \times 80)}{2 \surd (40)} = \surd (16) = 4$, &c., therefore the parts of the root are $\sqrt{4}$, $\sqrt{5}$, $\sqrt{3}$, $\sqrt{2}$, $\sqrt{4}$, &c., and the root is $2+\sqrt{2}-\sqrt{3}+\sqrt{5}$; for, being squared, it produces the surd quantity given.

CASE XII.

To extract any root (c) of a binomial surd.

RULE 1.*—Let the quantity be $A \pm B$, whereof A is the greater part and c the exponent of the root required. Seek

---

* *Let the sum or difference of two quantities,* x *and* y, *be raised to a power whose exponent is* c, *and let the* 1*st*, 3*d*, 5*th*, 7*th*, *&c., terms of that power, collected into one sum, be called* A, *and the rest of the terms, in the even places, called* B; *the difference of the squares of* A *and* B *shall be equal to the difference of the squares of* x *and* y *raised to the same power* c.

For the terms in the power c of $x+y$, writing for their coefficients respectively, 1, c, d, e, &c., are $x^c+cx^{c-1}y+dx^{c-2}y^2+ex^{c-3}y^3+$, &c. $=A+B$; and the same power of $x-y$ (changing the signs in the even places) is $x^c-cx^{c-1}y+dx^{c-2}y^2-ex^{c-3}y^3+$, &c. $=A-B$.

And therefore, $(x+y)^c(x-y)^c$ or $(x^2-y^2)^c=(A+B)(A-B)=A^2-B^2$.

Let one, or both of the quantities, x, y, be a quadratic *surd*, that is, let $x+y$, the c root of the proposed binomial $A+B$ belong to one of these forms, $p+l\sqrt{q}$, $k\sqrt{p}+q$, or $k\sqrt{p}+l\sqrt{q}$. And it follows that,

1. If $x+y=p+l\sqrt{q}$, c being any whole number, A, the sum of the odd terms, will be a rational number; and B, the sum of the terms in the even places, each of which involves an odd power of y, will be a rational number multiplied into the quadratic surd $\sqrt{q}$.

2. Let c, the exponent of the root sought, be an odd number, as we may always suppose it, because if it is even, it may be halved by the extraction of the square root, till it becomes odd: and let $x+y=k\sqrt{p}+q$. Then A will involve the surd $\sqrt{p}$, and B will be rational.

3. But if both members of the root are irrational, $x+y=(k\sqrt{p}+l\sqrt{q})$ A and B are both irrational, the one involving $\sqrt{p}$, and the other the surd $\sqrt{q}$. And in all these cases, it is easily seen that when x is greater than y, A will be greater than B. From this composition of the binomial $A+B$, we are led to its resolution, as in the above rule, by these steps.

I.

When A is *rational*, and A^2-B^2 is a perfect power.

1. By the *theorem* just demonstrated, $A^2-B^2=(x^2-y^2)^c$ *accurately*; and therefore extracting the c root of A^2-B^2, it will be x^2-y^2; call this root n.

2. Extract in the nearest integer, the c root of $A+B$, it will be (*nearly*) $x+y$; which put $=r$.

3. Divide $x^2-y^2\,(=n)$ by $x+y\,(=r)$ the quotient is (*nearly*) $x-y$; and the sum of the divisor and quotient is (*more nearly*) $2x$; that is, if an integer value of x is to be found, it will be the nearest to $\dfrac{r+\frac{n}{r}}{2}$.

the least number n whose power n^c is divisible by $A^2 - B^2$, the quotient being Q compute $\sqrt[c]{[(A+B) \times \sqrt{Q}]}$ in the nearest integer number, which suppose to be r. Divide

---

4. $x^2-(x^2-y^2)=y^2$; or, $\left(\frac{r+\frac{n}{r}}{2}\right)^2-n=y^2$: whence

$y=\sqrt{\left\{\left(\frac{r+\frac{n}{r}}{2}\right)^2-n\right\}}$; and therefore, putting $t=\frac{r+\frac{n}{r}}{2}$, the root sought

$x+y=t+\sqrt{(t^2-n)}$; the same expression as in the rule, when $Q=1$, $s=1$; that is, when A^2-B^2 is a perfect c power, and the greater number, A, is rational.

II.

When A is *irrational*, and $Q=1$. By the same process, $x=\frac{r+\frac{n}{r}}{2}$ $(=T)$ and $y=\sqrt{(T^2-n)}$. But seeing A is supposed irrational, and c an odd number, x will be irrational likewise: and they will both involve the same irreducible surd $\sqrt{p}$, or s, which is found by dividing A by its greatest rational divisor. Write, therefore, for x or T, its value $t\times s$, and $x+y=ts+\sqrt{(t^2s^2-n)}$.

III.

If the c root of A^2-B^2 cannot be taken, multiply A^2-B^2 by a number Q, such that the product may be the (*least*) perfect c power n^c $(=A^2Q-B^2Q)$. And now (instead of $A+B$) extract the c root of $(A+B)\times\sqrt{Q}$, which found as above, will be $ts+\sqrt{(t^2s^2-n)}$; and consequently the c root of $A+B$ will be $ts+\sqrt{(t^2s^2-n)}$, divided by the c root of $\sqrt{Q}$; that is, $\frac{ts+\sqrt{(t^2s^2-n)}}{\sqrt[c]{\sqrt{Q}}}$.

In the operation, it is required to find a number Q, such, that $(A^2-B^2)\times Q$ may be a perfect c power; this will be the case if Q be taken equal to $(A^2-B^2)^{c-1}$; but to find a less number which will answer this condition, let A^2-B^2 be divisible by $a, a, \ldots\ldots (m)$; $b, b, \ldots\ldots (n)$; $d, d, \ldots\ldots (r)$; &c. in succession, that is, let $A^2-B^2=a^m b^n d^r$, &c.; also, let $Q=a^x b^y d^z$, &c.; $(A^2-B^2)\times Q=a^{m+x}\times b^{n+y}\times d^{r+z}$, &c.
which is a perfect cth power, if x, y, z, &c., be so assumed that $m+x$, $n+y$, $r+z$, are respectively equal to c, or some multiple of c. Thus, to find a number which multiplied by 180 will produce a perfect cube, divide 180 as often as possible by 2, 3, 5, &c., and it appears that $2.2.3.3.5=180$; if, therefore, it be multiplied by $2.3.5.5$, it becomes $2.^3\ 3.^3 5^3$, or $(2.3.5)^3$, a perfect cube.

If A and B be divided by their greatest common measure, either integer or quadratic surd, in all cases where the cth root can be obtained by this method, Q will either be unity, or some power of 2 less than $2c$.

If the residual $A-B$ be given, it is evident from its genesis by involution, that the same rule gives its root $x-y$. See Universal Arithmetic, p. 139, Dr. Waring's Med. Alg. p. 287, or Maclaurin's Alg. p. 124.

$A\sqrt{Q}$ by its greatest divisor, and let the quotient be s, and let $\frac{r+\frac{n}{r}}{2s} = t$, the nearest integer. Then the root $= \frac{ts \pm \sqrt{(t^2s^2 - n)}}{\sqrt[2c]{Q}}$, if the c root of $A \pm B$ can be extracted.

It is proper to observe that this rule, which was first given by NEWTON in the *Universal Arithmetic*, fails, when $t = \frac{1}{2}$ exactly; in which case, instead of taking t the nearest integer value of $\frac{r+\frac{n}{r}}{2s}$, it must be taken equal to $\frac{1}{2}$. See Ryan's Key to the second New York edition of Bonnycastle's Algebra.

EXAMPLE.

What is the cube root of $\sqrt{968} + 25$.

We have $A^2 - B^2 = 343 = 7 \times 7 \times 7$. $Q \times 7^3 = n^3$, whence $n = 7$, and $Q = 1$. Then $\sqrt[3]{[(A+B) \times \sqrt{Q}]} = \sqrt[3]{56} + = r = 4$. $A\sqrt{Q} = \sqrt{968} = 22\sqrt{2}$, and the radical part $\sqrt{2} = s$, and $\frac{r+\frac{n}{r}}{2s} = \frac{4+\frac{7}{4}}{2\sqrt{2}} = t = 2$, in the nearest integer. And $ts = 2\sqrt{2}$, $\sqrt{(t^2s^2 - n)} = \sqrt{(8-7)} = 1$ $\sqrt[6]{(Q = 1)}$. And the root is $\frac{2\sqrt{2}+1}{1} = 2\sqrt{2}+1$, whose cube, upon trial, I find to be $\sqrt{968} + 25$.

RULE 2.*—Let the surd, that is to have its root extracted,

---

* Thus, let $\sqrt[n]{(a+\sqrt{b})} = x + \sqrt{y}$; and we shall have by involution, $a + \sqrt{b} = (x+\sqrt{y})^n$.

An equation, which, by expanding the righthand member, and comparing the rational and irrational parts, gives

$$a = x^n + \frac{n(n-1)}{2}x^{n-2}y + \frac{n(n-1)(n-2)(n-3)}{2.3.4}x^{n-4}y^2 +, \text{\&c.}$$

$$\sqrt{b} = nx^{n-1}\sqrt{y} + \frac{n(n-1)(n-2)}{2.3}x^{n-3}y\sqrt{y} + \ldots\ldots +, \text{\&c.}$$

Or, which is the same thing under a different form,

$$a = \tfrac{1}{2}\left\{(x+\sqrt{y})^n + (x-\sqrt{y})^n\right\},$$

$$\sqrt{b} = \tfrac{1}{2}\left\{(x+\sqrt{y})^n - (x-\sqrt{y})^n\right\}.$$

Whence, by squaring each of these equations, and subtracting the latter from the former, we shall have

$$a^2 - b = \tfrac{1}{4}\left\{(x+\sqrt{y})^{2n} + 2(x^2-y)^n + (x-\sqrt{y})^{2n}\right\},$$

$$-\tfrac{1}{4}\left\{(x+\sqrt{y})^2 - 2(x^2-y)^n + (x-\sqrt{y})^{2n}\right\}.$$

be of the form $\sqrt[3]{(a+\sqrt{b})}$, or $\sqrt[3]{(a-\sqrt{b})}$. Then if a^2-b be a perfect integral cube, and some whole number can be found, that, when substituted for n, will make

$$n^3 - 3\,[\sqrt[3]{(a^2-b)}]\,n = 2a,$$

the roots of the two expressions, in this case, will be

$$\sqrt[3]{(a+\sqrt{b})} = \tfrac{1}{2}n + \tfrac{1}{2}\sqrt{[n^2 - 4\sqrt[3]{(a^2-b)}]}$$
$$\sqrt[3]{(a-\sqrt{b})} = \tfrac{1}{2}n - \tfrac{1}{2}\sqrt{[n^2 - 4\sqrt[3]{(a^2-b)}]}$$

And if the second part of the binomial, or residual, be an imaginary surd, and a^2+b be a perfect integral cube, the extraction may be effected by finding the integral value of n in the following equation as before :—

$$n^3 - 3\,[\sqrt[3]{(a^2+b)}]\,n = 2a.$$

In which last case, the roots of the two expressions will be,

$$\sqrt[3]{(a+\sqrt{-b})} = \tfrac{1}{2}n + \tfrac{1}{2}\sqrt{[n^2 - 4\sqrt[3]{(a^2+b)}]}$$
$$\sqrt[3]{(a-\sqrt{-b})} = \tfrac{1}{2}n - \tfrac{1}{2}\sqrt{[n^2 - 4\sqrt[3]{(a^2+b)}]}$$

each of which formulæ may be obtained by barely changing the sign of b in the former.

EXAMPLE.

It is required to find the cube root of $10 \pm 6\sqrt{3}$, or $10 \pm \sqrt{(108)}$.

Here $a = 10$, and $b = 108$; whence $\sqrt[3]{(a^2-b)} = \sqrt[3]{(100-108)} = -2$, and $n^3 - 3\,[\sqrt[3]{(a^2-b)}]\,n = 20$,

$$\text{or } n^3 + 6n = 20,$$

where it readily appears from inspection, that $n = 2$. Whence $\sqrt[3]{(10+\sqrt{108})} = \frac{2}{2} + \frac{1}{2}\sqrt{(4 - 4 \times -2)} = 1 + \frac{1}{2}\sqrt{(12)} = 1 + \sqrt{3}$, and $\sqrt[3]{(10-\sqrt{108})} = \frac{2}{2} - \frac{1}{2}\sqrt{(4 - 4 \times -2)} = 1 - \frac{1}{2}\sqrt{12} = 1 - \sqrt{3}$.

EXAMPLES FOR PRACTICE.

1. Required the cube root of $68 - \sqrt{4374}$.

Ans. $\dfrac{4-\sqrt{6}}{\sqrt[3]{2}}$.

Or, by rejecting the terms that destroy each other, and then multiplying by $\frac{1}{4}$,

$$a^2 - b = (x^2-y)^n, \text{ or } x^2 - y = (a^2-b)^{\frac{1}{n}}.$$

Where, supposing a^2-b to be a complete power of the nth degree, let $(a^2-b)^{\frac{1}{n}}$ be put $= c$.

Then, since $x^2 - y = c$, and consequently $y = x^2 - c$, if this value be substituted for y, in the equation $x^n + \frac{n(n-1)}{2}x^{n-2}y + \frac{n(n-1)(n-2)(n-3)}{2.3.4}x^{n-4}y^2 +$, &c., $= a$, we shall obtain an equation, in which the value of x, as before mentioned, is irrational, when the extraction required is possible. See Wood, or Ryan's Algebra.—Ed.

2. Required the cube root of $11 + 5\sqrt{7}$

Ans. $\dfrac{\sqrt{7}+1}{\sqrt[3]{2}}$

3 Required the cube root of $2\sqrt{7} + 3\sqrt{3}$.

Ans. $\dfrac{\sqrt{6}+\sqrt{3}}{2}$.

4. Required the fifth root of $29\sqrt{6} + 41\sqrt{3}$.

Ans. $\dfrac{\sqrt{6}+\sqrt{3}}{\sqrt[5]{9}}$.

5. Required the cube root of $45 \pm 29\sqrt{2}$.

Ans. $3 + \sqrt{2}$, and $3 - \sqrt{2}$.

6. Required the cube root of $9 \pm 4\sqrt{5}$, or $9 \pm \sqrt{80}$.

Ans. $\frac{3}{2} + \frac{1}{2}\sqrt{5}$, and $\frac{3}{2} - \frac{1}{2}\sqrt{5}$.

7. Required the cube root of $20 \pm 68\sqrt{-7}$.

Ans. $5 + \sqrt{-7}$, and $5 - \sqrt{-7}$.

8. It is required to find the cube root of $35 \pm 69\sqrt{-6}$.

Ans. $5 + \sqrt{-6}$, and $5 - \sqrt{-6}$.

9. It is required to find the cube root of $81 \pm \sqrt{-2700}$.*

Ans. $-3 + 2\sqrt{-3}$, and $-3 - 2\sqrt{3}$.

CASE XIII.

To find such a multiplier, or multipliers, as will make any binomial surd rational.

RULE.†—1. When one or both of the terms are any even roots, multiply the given binomial or residual, by the same

* Whenever it can be done, the operation, in cases of this kind, ought to be abridged, by dividing the given binomial by the greatest cube that it contains, and then finding the root of the quotient; which, being multiplied by the root of the cube by which the binomial was divided, will give the root required. Thus, in the example above given, $81+\sqrt{-2700} = 27 \times (3+\sqrt{-\frac{100}{27}})$, where the root of $3+\sqrt{-\frac{100}{27}}$ being now more easily found to be $-1+2\sqrt{-\frac{1}{3}}$, or $-1+\frac{2}{3}\sqrt{-3}$, we shall have, by multiplying by 3, (which is the cube root of 27), $-3+2\sqrt{-3}$, as above.

Also this is useful, in Cardan's rule for cubic equations; thus, $\sqrt[3]{(81+\sqrt{(-2700)})}+\sqrt[3]{(81-\sqrt{(-2700)})} = -3\times 2 = -6$, or $= -\frac{3}{2}\times 2 = -3$, or $\frac{9}{2}\times 2 = 9$, the imaginary parts vanishing, by the contrariety of their signs. *See* De Moivre's Appendix to Sanderson's Algebra, Universal Arithmetic, or Maclaurin's Algebra.

† If a multiplier be required, that shall render any binomial surd, whether it consist of *even* or *odd* roots, rational, it may be found by substituting the given numbers, or letters, of which it is composed, in the places of their equals, in the following general formula:—

Binomial $\sqrt[n]{a} \pm \sqrt[n]{b}$.

Multiplier $\sqrt[n]{a^{n-1}} \mp \sqrt[n]{a^{n-2}b} + \sqrt[n]{a^{n-3}b^2} \mp \sqrt[n]{a^{n-4}b^3} +$, &c.; where the upper sign of the multiplier must be taken with the upper expression, with the sign of one of its terms changed; and repeat the operation in the same way, as long as there are surds, when the last result will be rational.

2. When the terms of the binomial surd are odd roots, the rule becomes more complicated; but for the sum or difference of two cube roots, which is one of the most useful cases, the multiplier will be a trinomial surd, consisting of the squares of the two given terms and their product, with its sign changed.

---

sign of the binomial, and the lower with the lower; and the series continued to n terms.

This multiplier may be derived from observing the quotient which arises from the actual division of the numerator by the denominator of the following fractions: thus,

I. $\frac{x^n - y^n}{x+y} = x^{n-1} + x^{n-2}y + x^{n-3}y^2 +$, &c., ... $+ y^{n-1}$ to n terms, whether n be *even* or *odd*.

II. $\frac{x^n - y^n}{x+y} = x^{n-1} - x^{n-2}y + x^{n-3}y^2 -$, &c., ... $- y^{n-1}$ to n terms, where n is an *even* number.

III. $\frac{x^n + y^n}{x+y} = y^{n-1} - x^{n-2}y + x^{n-3}y^2 -$, &c., ... $+ y^{n-1}$ to n terms, when n is an *odd* number.

Now, let $x^n = a$, $y^n = b$; then $x = \sqrt[n]{a}$, $y = \sqrt[n]{b}$, and these fractions severally become $\frac{a-b}{\sqrt[n]{a} - \sqrt[n]{b}}$, $\frac{a-b}{\sqrt[n]{a} + \sqrt[n]{b}}$, and $\frac{a+b}{\sqrt[n]{a} + \sqrt[n]{b}}$:

And, since $x^{n-1} = \sqrt[n]{a^{n-1}}$, $x^{n-2} = \sqrt[n]{a^{n-2}}$, &c., also $y^2 = \sqrt[n]{b^2}$, $y^3 = \sqrt[n]{b^3}$, &c., therefore,

$$\frac{a-b}{\sqrt[n]{a} - \sqrt[n]{b}} = \sqrt[n]{a^{n-1}} + \sqrt[n]{a^{n-2}b} + \sqrt[n]{a^{n-3}b^2} +, \text{ \&c. } \ldots + \sqrt[n]{b^{n-1}}$$

to n terms; where n may be any whole number whatever. And,

$$\frac{a+b}{\sqrt[n]{a} + \sqrt[n]{b}} = \sqrt[n]{a^{n-1}} - \sqrt[n]{a^{n-2}b} + \sqrt[n]{a^{n-3}b^2} -, \text{ \&c., } \ldots \pm \sqrt[n]{b^{n-1}}$$

to n terms; where the terms b and $\sqrt[n]{b^{n-1}}$ have the sign $+$, when n is an *odd* number; and the sign $-$, when n is an even number.

Now, since the *divisor* multiplied by the *quotient* gives the *dividend*, it appears from the foregoing operations, that, *if a binomial surd of the form* $\sqrt[n]{a} - \sqrt[n]{b}$ *be multiplied by* $\sqrt[n]{a^{n-1}b}$ $\sqrt[n]{a^{n-2}b} +$, &c., ... $+ \sqrt[n]{b^{n-1}}$, (n being any whole number whatever), *the product will be* $a - b$, *a rational quantity; and if a binomial surd of the form* $\sqrt[n]{a} + \sqrt[n]{b}$ *be multiplied by* $\sqrt[n]{a^{n-1}} - \sqrt[n]{a^{n-2}b} + \sqrt[n]{a^{n-3}b^2} -$, &c., ... $\pm \sqrt[n]{b^{n-1}}$, *the product will be* $a+b$, or $a-b$; according as the index n is an *odd* or an *even* number. See my Elementary Treatise on Algebra, Theoretical and Practical.—ED.

EXAMPLES.

1. To find a multiplier that shall render $5 + \sqrt{3}$ rational.

Given surd $5 + \sqrt{3}$
Multiplier $5 - \sqrt{3}$

Product $25 - 3 = 22$, as required.

2. To find a multiplier that shall make $\sqrt{5} + \sqrt{3}$ rational

Given surd $\sqrt{5} + \sqrt{3}$
Multiplier $\sqrt{5} - \sqrt{3}$

Product $5 - 3 = 2$, as required.

3. To find multipliers that shall make $\sqrt[4]{5} + \sqrt[4]{3}$ rational

Given surd $\sqrt[4]{5} + \sqrt[4]{3}$
1st multiplier $\sqrt[4]{5} - \sqrt[4]{3}$

1st product $\sqrt{5} - \sqrt{3}$
2d multiplier $\sqrt{5} + \sqrt{3}$

2d product $5 - 3 = 2$, as required.

4. To find a multiplier that shall make $\sqrt[3]{7} + \sqrt[3]{3}$ rational.

Given surd $\sqrt[3]{7} + \sqrt[3]{3}$
Multiplier $\sqrt[3]{7^2} - \sqrt[3]{(7 \times 3)} + \sqrt[3]{3^2}$

$$7 + \sqrt[3]{(3 \times 7^2)}$$
$$- \sqrt[3]{(3 \times 7^2)} - \sqrt[3]{(7 \times 3^2)}$$
$$+ \sqrt[3]{(7 \times 3^2)} + 3$$

Product $7 + 3 = 10$, as was required.

5. To find a multiplier that shall make $\sqrt{5} - \sqrt{x}$ rational.
Ans. $\sqrt{5} + \sqrt{x}$.

6. To find a multiplier that shall make $\sqrt{a} + \sqrt{b}$ rational.
Ans. $\sqrt{a} - \sqrt{b}$.

7. To find a multiplier that shall make $a + \sqrt{b}$ rational.
Ans. $a - \sqrt{b}$.

8. It is required to find a multiplier that shall make $1 - \sqrt[3]{2a}$ rational. Ans. $1 + \sqrt[3]{2a} + \sqrt[3]{4a^2}$.

9. It is required to find a multiplier that shall make $\sqrt[3]{3} - \frac{1}{2}\sqrt[3]{2}$ rational. Ans. $\sqrt[3]{9} + \frac{1}{2}\sqrt[3]{6} + \frac{1}{4}\sqrt[3]{4}$.

10. It is required to find a multiplier that shall make $\sqrt[4]{(a^3)} + \sqrt[4]{(b^3)}$, or $a^{\frac{3}{4}} + b^{\frac{3}{4}}$ rational.
Ans $\sqrt[4]{a^9} - \sqrt[4]{(a^6b^3)} + \sqrt[4]{(a^3b^6)} - \sqrt[4]{b^9}$.

CASE XIV.

To reduce a fraction, whose denominator is either a simple or a compound surd, to another that shall have a rational denominator.

RULE.—1.—When any simple fraction is of the form $\frac{b}{\sqrt{a}}$, multiply each of its terms by $\sqrt{a}$, and the resulting fraction will be $\frac{b\sqrt{a}}{a}$.

Or, when it is of the form $\frac{b}{\sqrt[3]{a}}$, multiply them by $\sqrt[3]{a^2}$, and the result will be $\frac{b\sqrt[3]{a^2}}{a}$.

And for the general form $\frac{b}{\sqrt[n]{a}}$. multiply by $\sqrt[n]{a^{n-1}}$, and the result will be $\frac{b\sqrt[n]{a^{n-1}}}{a}$.

2. If it be a compound surd, find sucn a multiplier, by the last rule, as will make the denominator rational; and multiply both the numerator and denominator by it, and the result will be the fraction required.

EXAMPLES

1. Reduce the fractions $\frac{2}{\sqrt{3}}$ and $\frac{3}{\frac{1}{2}\sqrt[4]{5}}$, to others that shall have rational denominators.

Here $\frac{2}{\sqrt{3}} = \frac{2}{\sqrt{3}} \times \frac{\sqrt{3}}{\sqrt{3}} = \frac{2\sqrt{3}}{3}$; and $\frac{3}{\frac{1}{2}\sqrt[4]{5}} = \frac{3}{\frac{1}{2}\sqrt[4]{5}} \times \frac{\sqrt[4]{5^3}}{\sqrt[4]{5^3}} =$ $\frac{3\sqrt[4]{5^3}}{\frac{1}{2} \times 5} = \frac{6\sqrt[4]{5^3}}{5} = \frac{6}{5}\sqrt[4]{125}$ the answer required.

2. Reduce $\frac{3}{\sqrt{5} - \sqrt{2}}$ to a fraction whose denominator shall be rational.

Here $\frac{3}{\sqrt{5} - \sqrt{2}} \times \frac{\sqrt{5} + \sqrt{2}}{\sqrt{5} + \sqrt{2}} = \frac{3\sqrt{5} + 3\sqrt{2}}{5 - 2} = \frac{3\sqrt{5} + 3\sqrt{2}}{3}$ $= \frac{\sqrt{5} + \sqrt{2}}{1} = \sqrt{5} + \sqrt{2}$ the answer required.

3. Reduce $\frac{\sqrt{2}}{3 - \sqrt{2}}$ to a fraction, whose denominator shall be rational.

Here $\frac{\sqrt{2}}{3-\sqrt{2}} = \frac{\sqrt{2}\times(3+\sqrt{2})}{(3-\sqrt{2})\times(3+\sqrt{2})} = \frac{3\sqrt{2}+2}{9-2} = \frac{2+3\sqrt{2}}{7}$

$= \frac{2}{7} + \frac{3}{7}\sqrt{2}$ the answer required.

4. Reduce $\frac{\sqrt{6}}{\sqrt{7}+\sqrt{3}}$ to a fraction that shall have a rational denominator. Ans. $\frac{\sqrt{(42)}-\sqrt{(18)}}{4}$.

5. Reduce $\frac{x}{3+\sqrt{x}}$ to a fraction that shall have a rational denominator. Ans. $\frac{3x - x\sqrt{x}}{9-x}$.

6. Reduce $\frac{a-\sqrt{b}}{a+\sqrt{b}}$ to a fraction, the denominator of which shall be rational. Ans. $\frac{a^2+b-2a\sqrt{b}}{a^2-b}$.

7. Reduce $\frac{10}{\sqrt[3]{7}-\sqrt[3]{5}}$ to a fraction that shall have a rational denominator. Ans. $5\times[\sqrt[3]{(49)}+\sqrt[3]{(35)}+\sqrt[3]{(25)}]$

8. Reduce $\frac{\sqrt[3]{3}}{\sqrt[3]{9}+\sqrt[3]{10}}$ to a fraction that shall have a rational denominator. Ans. $\frac{3\sqrt[3]{9}-3\sqrt[3]{(10)}+\sqrt[3]{(300)}}{19}$.

9. Reduce $\frac{4}{\sqrt[4]{4}+\sqrt[4]{5}}$ to a fraction that shall have a rational denominator.

Ans. $4\left\{-\sqrt{10}-2\sqrt{2}+(2+\sqrt{5})\times\sqrt[4]{5}\right\}$.

OF ARITHMETICAL PROPORTION AND PROGRESSION.

Arithmetical Proportion, is the relation which two quantities of the same kind, have to two others, when the difference of the first pair is equal to that of the second.

Hence, three quantities are said to be in arithmetical proportion, when the difference of the first and second is equal to the difference of the second and third.

Thus, 2, 4, 6, and a, $a+b$, $a+2b$, are quantities in arithmetical proportion.

And four quantities are said to be in arithmetical proportion, when the difference of the first and second is equal to the difference of the third and fourth.

Thus, 3, 7, 12, 16, and $a, a+b, c, c+b$, are quantities in arithmetical proportion.

ARITHMETICAL PROGRESSION, is when a series of quantities increase or decrease by the same common difference.

Thus, 1, 3, 5, 7, 9, &c., and $a, a+d, a+2d, a+3d$, &c., are increasing series in arithmetical progression, the common differences of which are 2 and d.

And 15, 12, 9, 6, &c., and $a, a-d, a-2d, a-3d$, &c., are decreasing series in arithmetical progression, the common differences of which are 3 and d.

The most useful properties of arithmetical proportion and progression are contained in the following theorems :—

1. If four quantities are in arithmetical proportion, the sum of the two extremes will be equal to the sum of the two means.

Thus, if the proportionals be 2, 5, 7, 10, or a, b, c, d, then will $2+10=5+7$, and $a+d=b+c$.

2. And if three quantities be in arithmetical proportion, the sum of the two extremes will be double the mean.

Thus, if the proportionals be 3, 6, 9, or a, b, c, then will $3+9=2\times 6=12$, and $a+c=2b$.

3. Hence an arithmetical mean between any two quantities is equal to half the sum of those quantities.

Thus, an arithmetical mean between 2 and 4 is $=\frac{2+4}{2}=3$; and between 5 and 6 it is $=\frac{5+6}{2}=5\frac{1}{2}$.

And an arithmetical mean between a and b is $\frac{a+b}{2}$.*

4. In any continued arithmetical progression, the sum of the two extremes is equal to the sum of any two terms that are equally distant from them, or to double the middle term, when the number of terms is odd.

---

* If two, or more arithmetical means between any two quantities be required, they may be expressed as below.

Thus, $\frac{2a+b}{3}$ and $\frac{a+2b}{3}$ = two arithmetical means between a and b, a being the less extreme and b the greater.

And $\frac{na+b}{n+1}, \frac{(n-1)a+2b}{n+1}, \frac{(n-2)a+3b}{n+1}$, &c. to $\frac{a+nb}{n+1}$ = any number (n) of arithmetical means between a and b; where $\frac{b-a}{n+1}$ is the common difference; which, being added to a, gives the first of these means; and then again to this last, gives the second; and so on.

Thus, if the series be 2, 4, 6, 8, 10, then will $2 + 10 = 4 + 8 = 2 \times 6 = 12$.

And, if the series be a, $a + d$, $a + 2d$, $a + 3d$, $a + 4d$, then will $a + (a + 4d) = (a + d) + (a + 3d) = 2 \times (a + 2d)$.

5. The last term of any increasing arithmetical series is equal to the first term *plus* the product of the common difference by the number of terms less one; and if the series be decreasing, it will be equal to the first term *minus* that product.

Thus, the nth term of the series a, $a + d$, $a + 2d$, $a + 3d$, $a + 4d$, &c., is $a + (n - 1)d$.

And the nth term of the series a, $a - d$, $a - 2d$, $a - 3d$, $a - 4d$, &c., is $a - (n - 1)d$.

6. The sum of any series of quantities in arithmetical progression is equal to the sum of the two extremes multiplied by half the number of terms.

Thus, the sum of 2, 4, 6, 8, 10, 12, is $= (2 + 12) \times \frac{6}{2} = 14 \times 3 = 42$.

And if the series be $a + (a + d) + (a + 2d) + (a + 3d) + (a + 4d)$, &c. . . . $+ l$, and its sum be denoted by S, we shall have $S = (a + l) \times \frac{n}{2}$, where l is the last term, and n the number of terms.

Or, the sum of any increasing arithmetical series may be found, without considering the last term, by adding the product of the common difference by the number of terms less one to twice the first term, and then multiplying the result by half the number of terms.

And, if the series be decreasing, the sum will be found by subtracting the above product from twice the first term, and then multiplying the result by half the number of terms, as before.

Thus, if the series be $a + (a + d) + (a + 2d) + (a + 3d) + (a + 4d)$, &c., continued to n terms, we shall have

$$S = \left\{ 2a + (n - 1)d \right\} \times \frac{n}{2}.$$

And if the series be $a + (a - d) + (a - 2d) + (a - 3d) + (a - 4d)$, &c., to n terms, we shall have

$$S = \left\{ 2a - (n - 1)d \right\} \times \frac{n}{2} (*).$$

(*) The sum of any number of terms (n) of the series of natural numbers 1, 2, 3, 4, 5, 6, 7, &c., is $= \frac{n(n + 1)}{2}$.

EXAMPLES.

1. The first term of an increasing arithmetical series is 3, the common difference 2, and the number of terms 20; required the sum of the series.

First, $3+2(20-1)=3+2\times 19=3+38=41$, the last term.

And $(3+41)\times\frac{20}{2}=44\times\frac{20}{2}=44\times 10=440$, the sum required.

Or, $\left\{2\times 3+(20-1)\times 2\right\}\times\frac{20}{2}=(6+19\times 2)\times 10=(6+38)\times 10=44\times 10=440$, as before.

2. The first term of a decreasing arithmetical series is 100, the common difference 3, and the number of terms 34; required the sum of the series.

First, $100-3(34-1)=100-3\times 33=100-99=1$, the last term.

And $(100+1)\times\frac{34}{2}=101\times\frac{34}{2}=101\times 17=1717$, the sum required.

Or, $\left\{2\times 100-(34-1)\times 3\right\}\times\frac{34}{2}=(200-33\times 3)\times 17=(200-99)\times 17=101\times 17=1717$, as before.

3. Required the sum of the natural numbers 1, 2, 3, 4, 5, 6, &c., continued to 1000 terms. Ans. 500500.

4. Required the sum of the odd numbers 1, 3, 5, 7, 9, &c., continued to 101 terms. Ans. 10201.

5. How many strokes do the clocks of Venice, which go on to 24 o'clock, strike in a day? Ans. 300.

6. Required the 365th term of the series of even numbers 2, 4, 6, 8, 10, 12, &c Ans. 730.

7. The first term of a decreasing arithmetical series is 10,

Thus, $1+2+3+4+5$, &c., continued to 100 terms, is $=\frac{100\times 101}{2}=50\times 101=5050$.

Also the sum of any number of terms (n) of the series of odd numbers 1, 3, 5, 7, 9, 11, &c., is $=n^2$.

Thus, $1+3+5+7+9$, &c., continued to 50 terms, is $=50^2=2500$.

And if any three of the quantities, a, d, n, S, be given, the fourth may be found from the equation

$$S=\left\{2a\pm(n-1)\,d\right\}\times\frac{n}{2}, \text{ or } (a+l)\times\frac{n}{2}.$$

Where the upper sign $+$ is to be used when the series is increasing, and the lower sign $-$ when it is decreasing; also the last term $l=a\pm(n-1)d$, as above.

the common difference $\frac{1}{3}$, and the number of terms 21; required the sum of the series. Ans. 140.

8. One hundred stones being placed on the ground, in a straight line, at the distance of a yard from each other, how far will a person travel, who shall bring them one by one, to a basket, placed at the distance of a yard from the first stone? Ans. 5 miles and 1300 yards.

OF GEOMETRICAL PROPORTION AND PROGRESSION.

* Geometrical Proportion, is the relation which two quantities of the same kind have to two others, when the

---

* If there be taken any four proportionals, a, b, c, d, which it has been usual to express by means of points: thus,

$$a : b :: c : d,$$

this relation will be denoted by the equation $\frac{a}{b} = \frac{c}{d}$; where the equal ratios are represented by fractions, the numerators of which are the antecedents, and the denominators the consequents. Hence, if each of the two members of this equation be multiplied by bd, there will arise $ad = bc$. From which it appears, as in the common rule, that the product of the two extremes of any four proportionals is equal to that of the means. And if the third c, in this case, be the same as the second, or $c = b$, the proportion is said to be continued, and we have $ad = b^2$, or $b = \sqrt{ad}$; where it is evident, that the product of the extremes of three proportionals is equal to the square of the mean; or, that the mean is equal to the square root of the product of the two extremes.

Also, if each member of the equation $ad = bc$ be successively divided by bd, dc, ae, &c., the results will give

$$\left.\begin{array}{l} \frac{a}{b} = \frac{c}{d} \\ \frac{a}{c} = \frac{b}{d} \\ \frac{b}{a} = \frac{d}{c} \end{array}\right\} \text{Or the proportions} \left\{\begin{array}{l} a : b :: c : d \\ a : c :: b : d \\ b : a :: d : c \end{array}\right.$$

&c. &c.

So that, by following this method, we can easily obtain all the transformations of the terms of the proportion, that can be made to agree with the equations $ad = bc$.

In like manner, from the same equality $\frac{a}{b} = \frac{c}{d}$, there will result, by multiplication, the following equivalent forms: $\frac{ma}{mb} = \frac{nc}{nd}$, $\frac{ma}{nb} = \frac{mc}{nd}$; Which, being converted into proportions, become $ma : mb :: ne : nd$, and $ma : nb :: mc : nd$. And, by taking any like powers, or roots, of the different sides of the same equation, we have $\frac{a^m}{b^n} = \frac{c^m}{d^n}$. Or, putting the

antecedents, or leading terms of each pair, are the same parts of their consequents, or the consequents of the antecedents.

---

terms in the form of a proportion, $a^m : b^n :: c^m : d^n$. In which cases m and n may be any whole or fractional numbers whatever.

Again, if there be taken the several equations,

$$\left.\begin{array}{l}\frac{a}{b}=\frac{c}{d}\\ \frac{e}{f}=\frac{g}{h}\\ \frac{i}{k}=\frac{l}{m}\end{array}\right\}\text{ which correspond with the proportions }\left\{\begin{array}{l}a:b::c:d\\ e:f::g:h\\ i:k::l:m\end{array}\right.$$

&c. &c.

we shall have, by multiplying their like terms, $\frac{a\times e\times i, \&c.}{b\times f\times k, \&c.}=\frac{c\times g\times l, \&c.}{d\times h\times m, \&c.}$

Or, by putting the expression in the form of a proportion, aei, &c., $: bfk$, &c., $:: cgl$, &c., $: dhm$, &c. Also, taking $\frac{a}{b}=\frac{c}{d}$, as before, we shall have, by multiplication, $\frac{ma}{nb}=\frac{mc}{nd}$; and by augmenting or diminishing each side of the equation by 1; $\frac{ma}{nb}\pm 1=\frac{mc}{nd}\pm 1$; or $\frac{ma\pm nb}{nb}=\frac{mc\pm nd}{nd}$; which, being expressed in the form of a proportion, gives $ma\pm nb : nb :: mc\pm nd : nd$; or $ma\pm nb : mc\pm nd :: nb : nd$.

And if the abovementioned equation $\frac{a}{b}=\frac{c}{d}$, be put, by a similar multiplication of its terms, under the form $\frac{pa}{qb}=\frac{pc}{qd}$, and then augmented or diminished by 1, as in the last case, there will arise $pa\pm qb : pc\pm qd :: qb : qd$. Whence, dividing each of the antecedents of these two analogies by their consequents, the result will give $\frac{ma\pm nb}{mc\pm nd}=\frac{nb}{nd}=\frac{b}{d}$; and $\frac{pa\pm qb}{pc\pm qd}=\frac{qb}{qd}=\frac{b}{d}$. And, consequently, as the two righthand members of these expressions are each $=\frac{b}{d}$, we shall have $\frac{ma\pm nb}{mc\pm nd}=\frac{pa\pm qb}{pc\pm qd}$.

Or, by converting the corresponding terms of this equation into a proportion $ma\pm nb : mc\pm nd :: pa\pm qb : pc\pm qd$. Also, because the common equation $\frac{a}{b}=\frac{c}{d}$ gives $\frac{a}{c}=\frac{b}{d}$, if the latter be put under the equivalent forms $\frac{ma}{nc}=\frac{mb}{nd}$, and $\frac{pa}{qc}=\frac{pb}{qd}$, we shall obtain, by a similar process, $ma\pm nc : pa\pm qc :: mb\pm nd : pb\pm qd$; which two analogies may be considered as general formulæ for changing the term sof the proportion $a : b :: c : d$, without altering its nature. Thus, by supposing m, n, p, q, to be each $=1$, and taking the antecedents with the superior signs, and

And if two quantities only are to be compared together, the part or parts, which the antecedent is of its consequent, or the consequent of the antecedent, is called the ratio; observing, in both cases, always to follow the same method.

Hence, three quantities are said to be in geometrical proportion, when the first is to the same part, or multiple, of the second, as the second is of the third.

Thus, 3, 6, 12, and a, ar, ar^2, are quantities in geometrical proportion.

And four quantities are said to be in geometrical proportion, when the first is the same part, or multiple, of the second, as the third is of the fourth.

Thus, 2, 8, 3, 12, and a, ar, b, br, are geometrical proportionals.

Direct proportion, is when the same relation subsists between the first of four terms and the second, as between the third and fourth.

Thus, 3, 6, 5, 10, and a, ar, b, br, are in direct proportion.

Inverse, or reciprocal proportion, is when the first and second of four quantities are directly proportional to the reciprocals of the third and fourth.

---

the consequents with the inferior, we have $a+b : a-b :: c+d : c-d$, and $a+c : a-c :: b+d : b-d$; which forms, together with several of those already given, are the usual transformations of the common analogy pointed out above.

In like manner, by taking m, n, and p each $=1$, and $q=0$, there will arise $a \pm b : a :: c \pm d : c$, and $a \pm c : a :: b \pm d : b$; each of which proportions may be verified by making the product of the extremes equal to that of the means, and observing that $ad=bc$.

Lastly, taking any number of equations of the form before used, for expressing proportions, to $\frac{a}{b}=\frac{c}{d}=\frac{e}{f}=\frac{g}{h}=$, &c.; which, according to the common method, are called a series of equal ratios, and are usually denoted by $a : b :: c : d :: e : f :: g : h ::$, &c., we shall necessarily have, from the fractions being all equal to each other, $\frac{a}{b}=q$, $\frac{c}{d}=q$, $\frac{e}{f}=q$, $\frac{g}{h}=q$, &c. And, by multiplying q by each of the denominators, $a=bq$, $c=dq$, $e=fq$, $g=hq$, &c.

Whence, equating the sum of all the terms on the lefthand side of these equations, with those on the right, we have $a+c+e+g+$, &c. $=(b+d+f+h+$, &c.,$)q$. And consequently, by division, and the properties of proportionals before shown,

$$\frac{a+c+e+g+, \text{ \&c.}}{b+d+f+h+, \text{ \&c.}}=\frac{a}{b}=\frac{a+c}{b+d}=\frac{a+c+e}{b+d+f}=, \text{ \&c.},$$

which results show that, in a series of equal ratios, the sum of any number of the antecedents is to that of their consequents, as one, or more of the antecedents, is to one, or the same number of consequents.

Thus, 2, 6, 9, 3, and a, ar, br, b, are inversely proportional; because 2, 6, $\frac{1}{9}$, $\frac{1}{3}$, and a ar, $\frac{1}{br}$, $\frac{1}{b}$, are directly proportional.

Geometrical Progression, is when a series of quantities have the same constant ratio; or which increase, or decrease, by a common multiplier, or divisor.

Thus, 2, 4, 8, 16, 32, 64, &c., and a, ar, ar^2, ar^3, ar^4, &c., are series in geometrical progression.

The most useful properties of geometrical proportion and progression are contained in the following theorems :—

1. If three quantities be in geometrical proportion, the product of the two extremes will be equal to the square of the mean.

Thus, if the proportionals be 2, 4, 8, or a, b, c, then will $2 \times 8 = 4^2$, and $a \times c = b^2$.

2. Hence, a geometrical mean proportional, between any two quantities, is equal to the square root of their product.

Thus, a geometric mean between 4 and 9 is $= \sqrt{36} = 6$.

And a geometric mean between a and b is $= \sqrt{ab}$.*

3. If four quantities be in geometrical proportion, the product of the two extremes will be equal to that of the means.

Thus, if the proportionals be 2, 4, 6, 12, or a, b, c, d; then will $2 \times 12 = 4 \times 6$, and $a \times d = b \times c$.

4. Hence, the product of the means of four proportional quantities, divided by either of the extremes, will give the other extreme; and the product of the extremes, divided by either of the means, will give the other mean.

Thus, if the proportionals be 3, 9, 5, 15, or a, b, c, d; then will $\frac{9 \times 5}{3} = 15$, and $\frac{3 \times 15}{5} = 9$: also, $\frac{b \times c}{a} = d$, and $\frac{a \times d}{c} = b$.

---

* If two or more geometrical means between any two quantities be required, they may be expressed as below:

$\sqrt[3]{a^2b}$ and $\sqrt[3]{ab^2}$ = two geometrical means between a and b.

$\sqrt[4]{a^3b}$, $\sqrt[4]{a^2b^2}$, and $\sqrt[4]{ab^3}$ = three geometrical means between a and b.

And generally,

$(a^nb)^{\frac{1}{n+1}}$, $(a^{n-1}b^2)^{\frac{1}{n+1}}$, $(a^{n-2}b^3)^{\frac{1}{n+1}}$ = any number (n) of geometrical means between a and b. Where $\left(\frac{b}{a}\right)^{\frac{1}{n+1}}$ is the ratio; so that if a be multiplied by this, it will give the first of these means; and this last being again multiplied by the same, will give the second; and so on.

5. Also, if any two products be equal to each other, either of the terms of one of them, will be to either of the terms of the other, as the remaining term of the last is to the remaining term of the first.

Thus, if $ad = bc$, or $2 \times 15 = 6 \times 5$, then will any of the following forms of these quantities be proportional:—

Directly, $a : b :: c : d$, or $2 : 6 :: 5 : 15$.

Invertedly, $b : a :: d : c$, or $6 : 2 :: 15 : 5$.

Alternately, $a : c :: b : d$, or $2 : 5 :: 6 : 15$.

Conjunctly, $a : a + b :: c : c + d$, or $2 : 8 :: 5 : 20$.

Distinctly, $a : b \sim a :: c : d \sim c$, or $2 : 4 :: 5 : 10$.

Mixedly, $b + a : b \sim a :: d + c : d \sim b$, or $8 : 4 :: 20 : 10$.

In all of which cases, the product of the two extremes is equal to that of the two means.

6. In any continued geometrical series, the product of the two extremes is equal to the product of any two means that are equally distant from them; or to the square of the mean, when the number of terms is odd.

Thus, if the series be 2, 4, 8, 16, 32; then will

$$2 \times 32 = 4 \times 16 = 8^2.$$

7. In any geometrical series, the last term is equal to the product arising from multiplying the first term by such a power of the ratio as is denoted by the number of terms less one.

Thus, in the series 2, 6, 18, 54, 162, we shall have $2 \times 3^4 = 2 \times 81 = 162$.

And in the series, a, ar, ar^2, ar^3, ar^4, &c., continued to n terms, the last term will be

$$l = ar^{n-1}.$$

8. The sum of any series of quantities in geometrical progression, either increasing or decreasing, is found by multiplying the last term by the ratio, and then dividing the difference of this product and the first term by the difference between the ratio and unity.

Thus, in the series 2, 4, 8, 16, 32, 64, 128, 256, 512, we shall have $\frac{512 \times 2 - 2}{2 - 1} = 1024 - 2 = 1022$, the sum of the terms.

Or the same rule, without considering the last term, may be expressed thus:—

Find such a power of the ratio as is denoted by the number of terms of the series; then divide the difference between this power and unity, by the difference between the ratio and unity, and the result, multiplied by the first term, will be the sum of the series.

Thus, in the series $a + ar \surd ar^2 + ar^3 + ar^4$, &c., to ar^{n-1}, we shall have

$$S = a\left(\frac{r^n \sim 1}{r \sim 1}\right).$$

Where it is to be observed, that if the ratio, or common multiplier, r, in this last series, be a proper fraction, and consequently the series a decreasing one, we shall have, in that case,

$$a + ar + ar^2 + ar^3 + ar^4, \text{ \&c., } \textit{ad infinitum} \ \frac{a}{1-r}.$$

9. Three quantities are said to be in harmonical proportion, when the first is to the third, as the difference between the first and second is to the difference between the second and third.

Thus, a, b, c, are harmonically proportional, when $a : c :: a - b : b - c$, or $a : c :: b - a : c - b$.

And c is a third harmonical proportion to a and b, when $c = \frac{ab}{2a - b}$.

10. Four quantities are in harmonical proportion, when the first is to the fourth, as the difference between the first and second is to the difference between the third and fourth.

Thus, a, b, c, d, are in harmonical proportion, when $a : d :: a - b : c - d$, or $a : d :: b - a : d - c$. And d is a fourth harmonical proportional to a, b, c, when $d = \frac{ac}{2a - b}$, in each of which cases it is obvious, that twice the first term must be greater than the second, or otherwise the proportionality will not subsist.

11. Any number of quantities, a, b, c, d, e, &c., are in harmonical progression, if $a : c :: a - b : b - c$; $b : d :: b - c : c - d$; $c : e :: c - d : d - e$; &c.

12. The reciprocals of quantities in harmonical progression, are in arithmetical progression.

Thus, if a, b, c, d, e, &c., are in harmonical progression, $\frac{1}{a}$, $\frac{1}{b}$, $\frac{1}{c}$, $\frac{1}{d}$, $\frac{1}{e}$, &c., will be in arithmetical progression.

13. An harmonical mean between any two quantities, is equal to twice their product divided by their sum.

Thus $\frac{2ab}{a+b}$ = an harmonical mean between a and b.*

* In addition to what is here said, it may be observed, that the ratio of two squares is frequently called *duplicate ratio;* of two square roots, *sub-duplicate ratio;* of two cubes, *triplicate ratio;* and of two cube roots, *sub-triplicate ratio;* &c.

EXAMPLES.

1. The first term of a geometrical series is 1, the ratio 2, and the number of terms 10; what is the sum of the series?

Here $1 \times 2^9 = 1 \times 512 = 512$, the last term.

And $\frac{512 \times 2 - 1}{2 - 1} = \frac{1024 - 1}{1} = 1023$, the sum required.

2. The first term of a geometrical series is $\frac{1}{2}$, the ratio $\frac{1}{3}$, and the number of terms 5; required the sum of the series.

Here $\frac{1}{2} \times \left(\frac{1}{3}\right)^4 = \frac{1}{2} \times \frac{1}{81} = \frac{1}{162}$, the last term.

And $\frac{\frac{1}{2} - \frac{1}{162} \times \frac{1}{3}}{1 - \frac{1}{3}} = \frac{\frac{1}{2} - \frac{1}{486}}{\frac{2}{3}} = \frac{121}{243} \times \frac{3}{2} = \frac{121}{162}$, the sum.

3. Required the sum of 1, 2, 4, 8, 16, 32, &c., continued to 20 terms. Ans. 1048575.

4. Required the sum of 1, $\frac{1}{2}$, $\frac{1}{4}$, $\frac{1}{8}$, $\frac{1}{16}$, $\frac{1}{32}$, &c., continued to 8 terms. Ans. $1\frac{127}{128}$.

5. Required the sum of 1, $\frac{1}{3}$, $\frac{1}{9}$, $\frac{1}{27}$, $\frac{1}{81}$, &c., continued to 10 terms. Ans. $1\frac{9841}{19683}$.

6. A person being asked to dispose of a fine horse, said he would sell him on condition of having a farthing for the first nail in his shoes, a halfpenny for the second, a penny for the third, twopence for the fourth, and so on, doubling the price of every nail, to 32, the number of nails in his four shoes; what would the horse be sold for at that rate?

Ans. 4473924*l*. 5*s*. $3\frac{3}{4}$*d*.

OF EQUATIONS.

THE DOCTRINE OF EQUATIONS, is that branch of algebra which treats of the methods of determining the values of unknown quantities by means of their relations to others which are known.

This is done by making certain algebraic expressions equal to each other (which formula, in that case, is called an equation), and then working by the rules of the art, till the quantity sought, is found equal to some given quantity, and consequently becomes known.

The terms of an equation are the quantities of which it is

composed; and the parts that stand on the right and left of the sign $=$, are called the two members, or sides, of the equation.

Thus, if $x = a + b$, the terms are x, a, and b; and the meaning of the expression is, that some quantity x, standing on the lefthand side of the equation, is equal to the sum of the quantities a and b on the righthand side.

A simple equation is that which contains only the first power of the unknown quantity: as,

$$x + a = 3b, \text{ or } ax = bc, \text{ or } 2x + 3a^2 = 5b^2;$$

Where x denotes the unknown quantity, and the other letters, or numbers, the known quantities.

A compound equation is that which contains two or more different powers of the unknown quantity: as,

$$x^2 + ax = b, \text{ or } x^3 - 4x^2 + 3x = 25.$$

Equations are also divided into different orders, or receive particular names, according to the highest power of the unknown quantity contained in any one of their terms: as quadratic equations, cubic equations, biquadratic equations, &c.

Thus, a quadratic equation is that in which the unknown quantity is of two dimensions, or which rises to the second power: as,

$$x^2 = 20; \; x^2 + ax = b, \text{ or } 3x^2 + 10x = 100.$$

A cubic equation is that in which the unknown quantity is of three dimensions, or which rises to the third power: as,

$$x^3 = 27; \; 2x^3 - 3x = 35; \text{ or } x^3 - ax^2 + bx = c.$$

A biquadratic equation is that in which the unknown quantity is of four dimensions, or which rises to the fourth power: as $x^4 = 25$; $5x^4 - 4x = 6$; or $x^4 - ax^3 + bx^2 - cx = d$.

And so on for equations of the 5th, 6th, and other higher orders, which are all denominated according to the highest power of the unknown quantity contained in any one of their terms.

The root of an equation is such a number or quantity, as, being substituted for the unknown quantity, will make both sides of the equation vanish, or become equal to each other.

A simple equation can have only one root; but every compound equation has as many roots as it contains dimensions, or as is denoted by the index of the highest power of the unknown quantity, in that equation.

Thus, in the quadratic equation $x^2 + 2x = 15$, the root, or value of x, is either $+3$ or -5; and, in the cubic equation $x^3 - 9x^2 + 26x = 24$, the roots are 2, 3, and 4, as will be found by substituting each of these numbers for x.

In an equation of an odd number of dimensions, one of its

roots will always be real: whereas, in an equation of an even number of dimensions, all its roots may be imaginary; as roots of this kind always enter into an equation by pairs.

Such are the equations $x^2 - 6x + 14 = 0$, and $x^4 - 2x^3 - 9x^2 + 10x + 50 = 0$.*

OF THE RESOLUTION OF SIMPLE EQUATIONS,

Containing only one unknown Quantity:

THE resolution of simple, as well as of other equations, is the disengaging the unknown quantity, in all such expressions, from the other quantities with which it is connected, and making it stand alone, on one side of the equation, so as to be equal to such as are known on the other side; for the performing of which, several axioms and processes are required, the most useful and necessary of which are the following:—†

CASE I.

Any quantity may be transposed from one side of an equation to the other, by changing its sign; and the two members, or sides, will still be equal.

Thus, if $x + 3 = 7$; then will $x = 7 - 3$, or $x = 4$.

---

* To the properties of equations abovementioned, we may here farther add:—

1. That the sum of all the roots of any equation is equal to the coefficient of the second term of that equation, with the sign changed.

2. The sum of the products of every two of the roots, is equal to the coefficient of the third term, without any change in its sign.

3. The sum of the products of every three terms of the roots, is equal to the coefficient of the fourth term, with its sign changed.

4. And so on, to the last, or absolute term, which is equal to the product of all the roots, with the sign changed or not, according as the equation is of an odd or an even number of dimensions. See, for a more particular account of the general theory of equations, Vol. II. of Bonnycastle's Treatise on Algebra, 8vo., 1820; or Ryan's Elementary Treatise on Algebra, 12mo., 1824.—ED.

† The operations required for the purpose here mentioned, are chiefly such as are derived from the following simple and evident principles:—

1. If the same quantity be added to, or subtracted from, each of two equal quantities, the results will still be equal; which is the same, in effect, as taking any quantity from one side of an equation, and placing it on the other side, with a contrary sign.

2. If all the terms of any two equal quantities, be multiplied or divided, by the same quantity, the products, or quotients thence arising, will be equal.

3. If two quantities, either simple or compound, be equal to each other, any like powers, or roots, of them will also be equal.

All of which axioms will be found sufficiently illustrated by the processes arising out of the several examples annexed to the six different cases given in the text.

And, if $x - 4 + 6 = 8$; then will $x = 8 + 4 - 6 = 6$.

Also, if $x - a + b = c - d$; then will $x = a - b + c - d$.

And, if $4x - 8 = 3x + 20$; then $4x - 3x = 20 + 8$, and consequently $x = 28$.

From this rule it also follows, that if a quantity be found on each side of an equation, with the same sign, it may be left out of both of them; and that the signs of all of the terms of any equation may be changed from $+$ to $-$, or from $-$ to $+$, without altering its value.

Thus, if $x + 5 = 7 + 5$; then, by cancelling, $x = 7$.

And if $a - x = b - c$; then, by changing the signs, $x - a = c - b$, or $x = a + c - b$.

EXAMPLES FOR PRACTICE.

1. Given $2x + 3 = x + 17$ to find x. Ans. $x = 14$.
2. Given $5x - 9 = 4x + 7$ to find x. Ans. $x = 16$.
3. Given $x + 9 - 2 = 4$ to find x. Ans. $x = -3$.
4. Given $9x - 8 = 8x - 5$ to find x. Ans. $x = 3$.
5. Given $7x + 8 - 3 = 6x + 4$ to find x. Ans. $x = -1$.

CASE II.

If the unknown quantity, in any equation, be multiplied by any number, or quantity, the multiplier may be taken away, by dividing all the rest of the terms by it; and if it be divided by any number, the divisor may be taken away, by multiplying all the other terms by it.

Thus, if $ax = 3ab - c$; then will $x = 3b - \frac{c}{a}$.

And, if $2x + 4 = 16$; then will $x + 2 = 8$,
or $x = 8 - 2 = 6$.

Also, if $\frac{x}{2} = 5 + 3$; then will $x = 10 + 6 = 16$.

And, if $\frac{2x}{3} - 2 = 4$; then $2x - 6 = 12$, or by division, $x - 3 = 6$, or $x = 9$.

EXAMPLES FOR PRACTICE.

1. Given $16x + 2 = 34$ to find x. Ans. $x = 2$.
2. Given $4x - 8 = -3x + 13$ to find x. Ans. $x = 3$.
3. Given $10x - 19 = 7x + 17$ to find x. Ans. $x = 12$.
4. Given $8x - 3 + 9 = -7x + 9 + 27$ to find x. Ans. $x = 2$
5. Given $3ax - 3ab = 12d$. Ans. $x = b + \frac{4d}{ab}$.

CASE III.

Any equation may be cleared of fractions, by multiplying each of its terms, successively, by the denominators of those fractions, or by multiplying both sides by the product of all the denominators, or by any quantity that is a multiple of them.

Thus, if $\frac{x}{3}+\frac{x}{4}=5$, then, multiplying by 3, we have $x+\frac{3x}{4}=15$; and this, multiplied by 4, gives $4x+3x=60$; whence, by addition, $7x=60$, or $x=\frac{60}{7}=8\frac{4}{7}$.

And, if $\frac{x}{4}+\frac{x}{6}=10$; then, multiplying by 12, (which is a multiple of 4 and 6,) $3x+2x=120$, or $5x=120$, or $x=\frac{120}{5}=24$.

It also appears, from this rule, that if the same number, or quantity, be found in each of the terms of an equation, either as a multiplier or divisor, it may be expunged from all of them, without altering the result.

Thus, if $ax=ab+ac$; then by cancelling, $x=b+e$.

And if $\frac{x}{a}+\frac{b}{a}=\frac{c}{a}$; then, $x+b=c$, or $x=c-b$.

EXAMPLES FOR PRACTICE.

1. Given $\frac{3x}{2}=\frac{x}{4}+24$ to find x. Ans. $x=19\frac{1}{5}$.

2. Given $\frac{x}{3}+\frac{x}{5}+\frac{x}{2}=62$ to find x. Ans. $x=60$.

3. Given $\frac{x-3}{2}+\frac{x}{3}=20-\frac{x+19}{2}$ to find x. Ans. $x=9$.

4. Given $\frac{x+1}{2}+\frac{x+2}{3}=16-\frac{x+3}{4}$ to find x. Ans. $x=13$.

5. Given $\frac{x+a}{b}+\frac{x}{c}=\frac{2x}{a}+\frac{a+b}{d}$ to find x.

Ans. $x=\frac{a^2cb+acb^2-a^2cd}{acd+abd-2cbd}$

CASE IV

If the unknown quantity, in any equation, be in the form of a surd, transpose the terms so that this may stand alone, on one side of the equation, and the remaining terms on the other (by Case I.); then involve each of the sides to such a power as corresponds with the index of the surd, and the equation will be rendered free from any irrational expression.

Thus, if $\sqrt{x} - 2 = 3$; then will $\sqrt{x} = 3 + 2 = 5$, or, by squaring, $x = 5^2 = 25$.

And if $\sqrt{(3x + 4)} = 5$; then will $3x + 4 = 25$, or $3x = 25 - 4 = 21$, or $x = \frac{21}{3} = 7$.

Also, if $\sqrt[3]{(2x + 3)} + 4 = 8$; then will $\sqrt[3]{(2x + 3)} = 8 - 4 = 4$, or $2x + 3 = 4^3 = 64$, and consequently $2x = 64 - 3 = 61$, or $x = \frac{61}{2} = 30\frac{1}{2}$.

EXAMPLES FOR PRACTICE.

1. Given $2\sqrt{x} + 3 = 9$ to find x. Ans. $x = 9$.
2. Given $\sqrt{(x + 1)} - 2 = 3$ to find x. Ans. $x = 24$.
3. Given $\sqrt[3]{(3x + 4)} + 3 = 6$ to find x. Ans. $x = 7\frac{2}{3}$
4. Given $\sqrt{(4 + x)} = 4 - \sqrt{x}$ to find x. Ans. $x = 2\frac{1}{4}$.
5. Given $\sqrt{(4a^2 + x^2)} = \sqrt[4]{(4b^4 + x^4)}$ to find x.

Ans. $x = \sqrt{\left(\frac{b^4 - 4a^4}{2a^2}\right)}$

CASE V.

If that side of the equation which contains the unknown quantity, be a complete power, the equation may be reduced to a lower dimension, by extracting the root of the said power on both sides of the equation.

Thus, if $x^2 = 81$; then $x = \sqrt{81} = 9$; and if $x^3 = 27$, then $x = \sqrt[3]{27} = 3$.

Also, if $3x^2 - 9 = 24$; then $3x^2 = 24 + 9 = 33$, or $x^3 = \frac{33}{3} = 11$, and consequently $x = \sqrt{11}$.

And, if $x^2 + 6x + 9 = 27$; then, since the lefthand side of the equation is a complete square, we shall have, by extracting the roots, $x + 3 = \sqrt{27} = \sqrt{(9 \times 3)} = 3\sqrt{3}$, or $x = 3\sqrt{3} - 3$.

EXAMPLES FOR PRACTICE.

1. Given $9x^2 - 6 = 30$ to find x. Ans. $x = 2$.
2. Given $x^3 + 9 = 36$ to find x. Ans. $x = 3$
3. Given $x^2 + x + \frac{1}{4} = \frac{81}{4}$ to find x. Ans. $x = 4$
4. Given $x^2 + ax + \frac{a^2}{4} = b^2$ to find x. Ans. $x = b - \frac{a}{2}$.
5. Given $x^2 + 14x + 49 = 121$ to find x. Ans. $x = 4$.

CASE VI.

Any analogy, or proportion, may be converted into an equation, by making the product of the two extreme terms equal to that of the two means.

Thus, if $3x : 16 :: 5 : 6$; then $3x \times 6 = 16 \times 5$, or $18x = 80$, or $x = \frac{80}{18} = \frac{40}{9} = 4\frac{4}{9}$.

And if $\frac{2x}{3} : a :: b : c$; then will $\frac{2cx}{3} = ab$, or $2cx = 3ab$; or, by division, $x = \frac{3ab}{2c}$.

Also, if $12 - x : \frac{x}{2} :: 4 : 1$; then $12 - x = \frac{4x}{2} = 2x$, or $2x + x = 12$, and consequently $x = \frac{12}{3} = 4$.

EXAMPLES FOR PRACTICE.

1. Given $\frac{3}{4}x : a :: 5bc : cd$ to find x. Ans. $x = \frac{20ab}{3d}$.
2. Given $10 - x : \frac{2}{3}x :: 3 : 1$ to find x. Ans. $x = 3\frac{1}{3}$.
3. Given $8 + 8x : 4x :: 8 : 2$ to find x. Ans. $x = 1$.
4. Given $x : 6 - x :: 2 : 4$ to find x. Ans. $x = 2$.
5. Given $4x : a :: 9\sqrt{x} : 9$ to find x. Ans. $x = \frac{a^2}{16}$.

MISCELLANEOUS EXAMPLES.

1. Given $5x - 15 = 2x + 6$ to find the value of x.

Here $5x - 2x = 6 + 15$, or $3x = 6 + 15 = 21$; and therefore $x = \frac{21}{3} = 7$.

2. Given $40 - 6x - 16 = 120 - 14x$, to find the value of x.

Here $14x - 6x = 120 - 40 + 16$; or $8x = 136 - 40 = 96$; and therefore $x = \frac{96}{8} = 12$.

3. Given $3x^2 - 10x = 8x + x^2$, to find the value of x.

Here $3x - 10 = 8 + x$, by dividing by x; or $3x - x = 8 + 10 = 18$, by transposition.

And consequently $2x = 18$, or $x = \frac{18}{2} = 9$.

4. Given $6ax^3 - 12abx^2 = 2ax^3 + 6ax^2$, to find the value of x.

Here $2x - 4b = x + 2$, by dividing by $3ax^2$; or $2x - x = 2 + 4b$; and therefore $x = 4b + 2$.

5. Given $x^2 + 2x + 1 = 16$, to find the value of x.

Here $x + 1 = 4$, by extracting the square root of each side.

And therefore, by transposition, $x = 4 - 1 = 3$.

6. Given $5ax - 3b = 2dx + c$, to find the value of x.

Here $5ax - 2dx = c + 3b$; or $(5a - 2d)x = c + 3b$; and therefore, by division, $x = \frac{c + 3b}{5a - 2d}$.

7. Given $\frac{x}{2} - \frac{x}{3} + \frac{x}{4} = 10$, to find the value of x.

Here $x - \frac{2x}{3} + \frac{2x}{4} = 20$; and $3x - 2x + \frac{6x}{4} = 60$; or $12x - 8x + 6x = 240$; whence $10x = 240$, or $x = 24$.

8. Given $\frac{x-3}{2} + \frac{x}{3} = 20 - \frac{x-19}{2}$, to find the valueof x.

Here $x - 3 + \frac{2x}{3} = 40 - x + 19$; or $3x - 9 + 2x = 120 - 3x + 57$; whence $3x + 2x + 3x = 120 + 57 + 9$; that is, $8x = 186$, or $x = 23\frac{1}{4}$

9. Given $\sqrt{\frac{2x}{3}} + 5 = 7$, to find the value of x.

Here $\sqrt{\frac{2x}{3}} = 7 - 5 = 2$; whence, by squaring, $\frac{2x}{3} = 2^2 = 4$, and $2x = 12$, or $x = 6$.

10. Given $x + \sqrt{(a^2 + x^2)} = \frac{2a^2}{\sqrt{(a^2 + x^2)}}$, to find the value of x.

Here $x\sqrt{(a^2 + x^2)} + a^2 + x^2 = 2a^2$; or $x\sqrt{(a^2 + x^2)} = a^2 - x^2$, and $x^2 . (a^2 + x^2) = a^4 - 2a^2x^2 + x^4$; whence $a^2x^2 + x^4 = a^4 - 2a^2x^2 + x^4$, and $a^2x^2 = a^4 - 2a^2x^2$; therefore $3a^2x^2 = a^4$, or $x^2 = \frac{a^4}{3a^2} = \frac{a^2}{3}$; and consequently $x = \sqrt{\frac{a^2}{3}} = a\sqrt{\frac{1}{3}} = a\sqrt{\frac{3}{9}} = \frac{a}{3}\sqrt{3}$, the answer required.

EXAMPLES FOR PRACTICE.

1. Given $3x - 2 + 24 = 31$, to find the value of x.
Ans. $x = 3$.

2. Given $4 - 9y = 14 - 11y$, to find the value of y.
Ans. $y = 5$.

3 Given $x + 18 = 3x - 5$, to find the value of x.
Ans. $x = 11\frac{1}{2}$.

4. Given $x + \frac{x}{2} + \frac{x}{3} = 11$, to determine the value of x.
Ans. $x = 6$.

5. Given $2x - \frac{x}{2} + 1 = 5x - 2$, to find the value of x.

Ans. $x = \frac{6}{7}$.

6. Given $\frac{x}{2} + \frac{x}{3} - \frac{x}{4} = \frac{7}{10}$, to determine the value of x.

Ans. $x = 1\frac{1}{5}$.

7. Given $\frac{x+3}{2} + \frac{x}{3} = 4 - \frac{x-5}{4}$, to find the value of x.

Ans. $x = 3\frac{6}{13}$.

8. Given $2 + \sqrt{3x} = \sqrt{(4 + 5x)}$, to find the value of x.

Ans. $x = 12$.

9. Given $x + a = \frac{x^2}{a + x}$, to find the value of x.

Ans. $x = -\frac{a}{2}$.

10. Given $\sqrt{x} + \sqrt{(a + x)} = \frac{2a}{\sqrt{(a + x)}}$, to find the value of x.

Ans. $x = \frac{a}{3}$.

11. Given $\frac{ax - b}{4} + \frac{a}{3} = \frac{bx}{2} - \frac{bx - a}{3}$, to find the value of x.

Ans. $x = \frac{3b}{3a - 2b}$.

12. Given $\sqrt{(a^2 + x^2)} = \sqrt[4]{(b^4 + x^4)}$, to find the value of x.

Ans. $x = \sqrt{\frac{b^4 - a^4}{2a^2}}$.

13. Given $\sqrt{(a + x)} + \sqrt{(a - x)} = \sqrt{ax}$, to find the value of x.

Ans. $x = \frac{4a^2}{a^2 + 4}$.

14. Given $\frac{a}{1 + x} + \frac{a}{1 - x} = b$ to determine the value of x.

Ans. $x = \sqrt{\frac{b - 2a}{b}}$,

15. Given $a + x = \sqrt{[a^2 + x\sqrt{(b^2 + x^2)}]}$, to find the value of x.

Ans. $x = \frac{b^2}{4a} - a$.

16. Given $\frac{1}{2}\sqrt{(x^2 + 3a^2)} - \frac{1}{2}\sqrt{(x^2 - 3a^2)} = x\sqrt{a}$, to find the value of x.

Ans. $x = \sqrt[4]{\frac{9a^3}{4 - 4a}}$.

17. Given $\sqrt{(a+x)} + \sqrt{(a-x)} = b$, to find the value of x

Ans. $x = \frac{b}{2}\sqrt{(4a - b^2)}$.

18. Given $\sqrt[3]{(a+x)} + \sqrt[3]{(a-x)} = b$, to find the value of x

Ans. $x = \sqrt{a^2 - \left(\frac{b^3 - 2a}{3b}\right)^3}$.

19. Given $\sqrt{a} + \sqrt{x} = \sqrt{ax}$, to find the value of x.

Ans. $x = \frac{a}{(\sqrt{a} - 1)^2}$.

20. Given $\sqrt{\left(\frac{x+1}{x-1}\right)} + \sqrt{\left(\frac{x-1}{x+1}\right)} = a$, to determine the value of x.

Ans. $x = \frac{a}{\sqrt{(a^2 - 4)}}$.

21. Given $\sqrt{(a^2 + ax)} = a - \sqrt{(a^2 - ax)}$, to find the value of x.

Ans. $x = \frac{a}{2}\sqrt{3}$.

22. Given $\sqrt{(a^2 - x^2)} + x\sqrt{(a^2 - 1)} = a^2\sqrt{(1 - x^2)}$, to find the value of x.

Ans. $x = \sqrt{\left(\frac{a^2 - 1}{a^2 + 3}\right)}$.

23 Given $\sqrt{(x + a)} = c - \sqrt{(x + b)}$, to find the value of x.

Ans. $x = \left(\frac{c^2 + b - a}{2c}\right)^2 - b$.

24. Given $\sqrt{\frac{b}{a+x}} + \sqrt{\frac{c}{a-x}} = \sqrt[4]{\frac{4bc}{a^2 - x^2}}$, to find the value of x.

Ans. $x = a\left(\frac{b+c}{b-c}\right)$.

Of the resolution of simple equations, containing two unknown quantities.

When there are two unknown quantities, and two independent simple equations involving them, they may be reduced to one, by any of the three following rules :—

Rule 1.—Observe which of the unknown quantities is the least involved, and find its value in each of the equations, by the methods already explained; then let the two values, thus found, be put equal to each other, and there will arise a new equation with only one unknown quantity in it, the value of which may be found as before.*

* This rule depends upon the well-known axiom, that things which are equal to the same thing, are equal to each other; and the two following methods are founded on principles which are equally simple and obvious.

EXAMPLES.

1. Given $\begin{cases} 2x + 3y = 23 \\ 5x - 2y = 10 \end{cases}$ to find the values of x and y.

Here, from the first equation, $x = \frac{23 - 3y}{2}$.

And from the second, $x = \frac{10 + 2y}{5}$;

Whence we have $\frac{23 - 3y}{2} = \frac{10 + 2y}{5}$,

Or $115 - 15y = 20 + 4y$, or $19y = 115 - 20 = 95$.

That is, $y = \frac{95}{19} = 5$, and $x = \frac{23 - 15}{2} = 4$.

2. Given $\begin{cases} x + y = a \\ x - y = b \end{cases}$ to find the values of x and y

Here, from the first equation, $x = a - y$,
And from the second $x = b + y$,
Whence $a - y = b + y$, or $2y = a - b$,

And therefore $y = \frac{a - b}{2}$, and $x = a - y$.

Or by substitution, $x = a - \frac{a - b}{2} = \frac{a + b}{2}$.

3. Given $\begin{cases} \frac{1}{2}x + \frac{1}{3}y = 7 \\ \frac{1}{3}x + \frac{1}{2}y = 8 \end{cases}$ to find the values of x and y

Here, from the first equation, $x = 14 - \frac{2y}{3}$

And from the second, $x = 24 - \frac{3y}{2}$,

Therefore, by equality, $14 - \frac{2y}{3} = 24 - \frac{3y}{2}$,

And consequently, $42 - 2y = 72 - \frac{9y}{2}$,

Or, by multiplication, $84 - 4y = 144 - 9y$;
And, therefore, also $5y = 144 - 84 = 60$,

Or, by division, $y = \frac{60}{5} = 12$, and $x = 14 - \frac{24}{3} = 6$.

EXAMPLES FOR PRACTICE.

1. Given $4x + y = 34$, and $4y + x = 16$, to find the values of x and y. Ans. $x = 8$, $y = 2$.

2. Given $2x + 3y = 16$, and $3x - 2y = 11$, to find the values of x and y. Ans. $x = 5$, $y = 2$.

3. Given $\frac{2x}{5}+\frac{3y}{4}=\frac{9}{20}$, and $\frac{3x}{4}+\frac{2y}{5}=\frac{61}{120}$, to find the values of x and y. Ans. $x=\frac{1}{2}$, $y=\frac{1}{3}$.

4. Given $\left\{\begin{array}{l}\frac{1}{2}x+2y=a\\ \frac{1}{2}x-2y=b\end{array}\right\}$ to find x and y.

Ans. $x=a+b$, and $y=\frac{1}{4}a-\frac{1}{4}b$

5. Given $\left\{\begin{array}{l}\frac{x}{2}+\frac{y}{3}=8\\ \frac{x}{3}-\frac{y}{2}=1\end{array}\right\}$ to find x and y.

Ans. $x=12$, and $y=6$.

6. Given $\left\{\begin{array}{l}\frac{x}{2}+\frac{y}{3}=9\\ x:y::4:3\end{array}\right\}$ to find x and y.

Ans. $x=12$, and $y=9$.

7. Given $x+y=80$, and $\frac{2x}{3}=\frac{3y}{4}$, to find x and y.

Ans. $x=42\frac{6}{17}$, and $y=37\frac{11}{17}$.

8 Given $y-6=\frac{x}{2}$, and $x=y+6$, to find x and y.

Ans. $x=24$, and $y=18$.

RULE 2.—Find the values of either of the unknown quantities in that equation in which it is the least involved; then substitute this value in the place of its equal in the other equation, and there will arise a new equation with only one unknown quantity in it; the value of which may be found as before.

EXAMPLES.

1. Given $\left\{\begin{array}{l}x+2y=17\\ 3x-y=2\end{array}\right\}$ to find the values of x and y.

From the first equation, $x=17-2y$; which value, being substituted for x, in the second,

gives $3(17-2y)-y=2$,

Or $51-6y-y=2$, or $7y=51-2=49$.

Whence $y=\frac{49}{7}=7$, and $x=17-2y=3$.

2. Given $\left\{\begin{array}{l}x+y=13\\ x-y=3\end{array}\right\}$ to find the values of x and y.

From the first equation, $x=13-y$; which value, being substituted for x, in the second,

Gives $13-y-y=3$, or $2y=13-3=10$.

Whence $y=\frac{10}{2}=5$, and $x=13-y=8$.

3. Given $\left\{\begin{array}{l}x:y::a:b\\ x^2+y^2=c\end{array}\right\}$ to find the values of x and y.

Here the analogy in the first, turned into an equation,

gives $bx = ay$, or $x = \frac{ay}{b}$,

And this value, substituted for x in the second,

gives $\left(\frac{ay}{b}\right)^2 + y^2 = c$, or $\frac{a^2y^2}{b^2} + y^2 = c$,

Whence we have $a^2y^2 + b^2y^2 = b^2c$, or $y^2 = \frac{b^2c}{a^2 + b^2}$.

And, consequently, $y = b\sqrt{\frac{c}{a^2 + b^2}}$, and $x = a\sqrt{\frac{c}{a^2 + b^2}}$.

EXAMPLES FOR PRACTICE.

1. Given $\frac{x}{7} + 7y = 99$, and $\frac{y}{7} + 7x = 51$, to find the values of x and y. Ans. $x = 7$, and $y = 14$.

2. Given $\frac{x}{2} - 12 = \frac{y}{4} + 8$, and $\frac{x+y}{5} + \frac{x}{3} - 8 = \frac{2y - x}{4} + 27$, to find the values of x and y. Ans. $x = 60$, $y = 40$.

3. Given $x + y = s$, and $x^2 - y^2 = d$, to find the values of x and y. Ans. $x = \frac{s^2 + d}{2s}$, $y = \frac{s^2 - d}{2s}$.

4. Given $5x - 3y = 150$, and $10x + 15y = 825$, to find x and y. Ans. $x = 45$, and $y = 25$.

5. Given $x + y = 16$, and $x : y :: 3 : 1$, to find x and y. Ans. $x = 12$, and $y = 4$.

6. Given $x + \frac{y}{2} = 12$, and $y + \frac{x}{2} = 9$, to find x and y. Ans. $x = 10$, and $y = 4$.

7. Given $x : y :: 3 : 2$, and $x^2 - y^2 = 20$, to find x and y. Ans. $x = 6$, and $y = 4$.

8. Given $\frac{x}{2} - 12 = \frac{y}{4} + 13$ and $\frac{x+y}{5} + \frac{x}{3} + 16 = \frac{2x - y}{4} + 27$, to find x and y. Ans. $x = 60$, and $y = 20$.

RULE 3.—Let one or both of the given equations be multiplied, or divided, by such numbers, or quantities, as will make the term that contains one of the unknown quantities the same in each of them; then, by adding, or subtracting, the two equations thus obtained, as the case may require, there will arise a new equation, with only one unknown quantity in it, which may be resolved as before.*

* The values of the unknown quantities in the two literal equations $ax + by = c$, add $a'x + b'y = c'$, may be found in general terms, by multiplying the first by a', and the second by a, and then working according

EXAMPLES.

1. Given $\left\{\begin{array}{l} 3x+5y=40 \\ x+2y=14 \end{array}\right\}$ to find the values of x and y.

First, multiply the second equation by 3, and it will give $3x+6y=42$.

Then, subtract the first equation from this and it will give $6y-5y=42-40$, or $y=2$.

Whence, also, $x=14-2y=14-4=10$.

2. Given $\left\{\begin{array}{l} 5x-3y=9 \\ 2x+5y=16 \end{array}\right\}$ to find the values of x and y.

Multiply the first equation by 2, and the second by 5; then $10x-6y=18$, and $10x+25y=80$.

And if the former of these be subtracted from the latter, there will arise $31y=62$, or $y=\frac{62}{31}=2$.

Whence, by the first equation, $x=\frac{9+3y}{5}=\frac{15}{5}=3$.

EXAMPLES FOR PRACTICE.

1. Given $\frac{x+8}{4}+6y=21$, and $\frac{y+6}{3}=23-5x$, to find x and y. Ans. $x=4$, and $y=3$.

2. Given $3x+7y=79$, and $2y=9+\frac{x}{2}$, to find x and y.

Ans. $x=10$, and $y=7$.

3. Given $30x+40y=270$, and $50x+30y=340$, to find x and y. Ans. $x=5$, and $y=3$.

4. Given $3x-3y=2x+2y$, and $x+y:xy::3:5$, to find x and y. Ans. $x=10$, and $y=2$.

5. Given $x^2y+xy^2=30$, and $x^3+y^3=35$, to find x and y.

Ans. $x=3$, and $y=2$.

---

to the last rule, when the results, so determined, will be $y=\frac{ac'-ca'}{ab'-ba'}$, and $x=\frac{cb'-bc'}{ab'-ba'}$; which solution may be applied to any particular case of this kind, by substituting the numeral values of a, b, a', b', in the place of the letters, and observing, when either of them is negative, to change the signs accordingly.

Where the numerator is the difference of the products of the opposite coefficients in the order in which y is not found, and the denominator is the difference of the products of the opposite coefficients taken from the orders that involve the two unknown quantities. Coefficients are of the same order which either affect no unknown quantity, as c and c'; or the same unknown quantity in the different equations, as a and a'. Coefficients are opposite when they affect the different unknown quantities in the different equations, as a and b', a' and b.—Ed.

6. Given $\frac{3x-5y}{2}=\frac{2x+y}{5}-3$, and $8-\frac{x-2y}{4}=\frac{x}{2}+\frac{y}{3}$ to find x and y. Ans. $x=12$, and $y=6$.

7. Given $x+y:a::x-y:b$, and $x^2-y^2=c$, to find the values of x and y.

Ans. $x=\frac{a+b}{2}\sqrt{\frac{c}{ab}}$, $y=\frac{a-b}{2}\sqrt{\frac{c}{ab}}$.

8. Given $ax+by=c$, and $dx+ey=f$, to find the values of x and y. Ans. $x=\frac{ce-bf}{ae-bd}$, $y=\frac{af-dc}{ae-bd}$.

9. Given $x+y=a$, and $x^2-y^2=b$, to find the values of x and y. Ans. $x=\frac{a^2+b}{2a}$, $y=\frac{a^2-b}{2a}$.

10. Given $x^2+xy=a$, and $y^2+xy=b$, to find the values of x and y. Ans. $x=\frac{a}{\sqrt{(a+b)}}$, $y=\frac{b}{\sqrt{(a+b)}}$.

Of the resolution of simple equations, containing three or more unknown quantities.

When there are three unknown quantities, and three independent simple equations containing them, they may be reduced to one by the following method.*

Rule.—Find the values of one of the unknown quantities in each of the three given equations, as if all the rest were known; then put the first of these values equal to the second, and either the first or second equal to the third, and there will arise two new equations with only two unknown quantities in them, the values of which may be found as in the former case; and thence the value of the third

Or, multiply each of the equations by such numbers, or quantities, as will make one of their terms the same in them all; then having subtracted any two of these resulting equations from the third, or added them together, as the case may

* The necessity for observing that the given equations in this and other similar cases are so proposed as to be independent of each other, will be obvious from the following example:—

$x-2y+z=5$; $2x+y-z=7$; $x+3y-2z=2$;

where, if it were required to determine the value of x, y, and z, it will be found, by eliminating x from each of them, and then equating the results, that

$5y-3z=-3$, and $5y-3z=-3$;

which equations, being identical, or both the same, furnish no determinate answer. And, in effect, if the three equations be properly examined, it will be found, that the third is merely the difference of the first and second, and consequently involves no condition but what is contained in the other two.

require, there will remain only two equations, which may be resolved by the former rules.

And in nearly the same way may four, five, &c., unknown quantities be exterminated from the same number of independent simple equations; but, in cases of this kind, there are frequently shorter and more commodious methods of operation, which can only be learnt from practice.*

EXAMPLES.

1. Given $\left\{\begin{array}{l} x + y + z = 29 \\ x + 2y + 3z = 62 \\ \frac{1}{2}x + \frac{1}{3}y + \frac{1}{4}z = 10 \end{array}\right\}$ to find x, y, and z.

Here, from the first equation, $x = 29 - y - z$.

From the second, $x = 62 - 2y - 3z$.

And from the third, $x = 20 - \frac{2}{3}y - \frac{1}{2}z$,

Whence $29 - y - z = 62 - 2y - 3z$,

And, also, $29 - y - z = 20 - \frac{2}{3}y - \frac{1}{2}z$,

From the first of which, $y = 33 - 2z$,

And from the second, $y = 27 - \frac{3}{2}z$,

Therefore $33 - 2z = 27 - \frac{3}{2}z$, or $z = 12$,

Whence, also $y = 33 - 2z = 9$,

And $x = 29 - y - z = 8$.

2. Given $\left\{\begin{array}{l} 2x + 4y - 3z = 22 \\ 4x - 2y + 5z = 18 \\ 6x + 7y - z = 63 \end{array}\right\}$ to find x, y, and z.

---

* The values of the unknown quantities in the three literal equations, $ax + by + cz = d$; $a'x + b'y + c'z = d'$; $a''x + b''y + c''z = d''$; may be exhibited in general terms, like those before mentioned, as follows:—

$$x = \frac{db'c'' - dc'b'' + cd'b'' - bd'c'' + bc'd'' - cb'd''}{ab'c'' - ac'b'' + ca'b'' - ba'c'' + bc'a'' - cb'a''},$$

$$y = \frac{ad'c'' - ac'd'' + ca'd'' - da'c'' + dc'a'' - cd'a''}{ab'c'' - ac'b'' + ca'b'' - ba'c'' + bc'a'' - cb'a''},$$

$$z = \frac{ab'd'' - ad'b'' + da'b'' - ba'd'' + bd'a'' - db'a''}{ab'c'' - ac'b'' + ca'b'' - ba'c'' + bc'a'' - cb'a''},$$

which formulæ, by substitution, may be employed for the resolution of any numeral case of this kind, as in the instance of two equations before given. The numerator of any of these equations, such as z, consists of all the different products, which can be made of three opposite coefficients taken from the orders in which z is not found; and the denominator consists of all the products that can be made of the three opposite coefficients taken from the orders which involve the three unknown quantities.

Here multiplying the first equation by 6, the second by 3, and the third by 2, we shall have

$$12x + 24y - 18z = 132,$$
$$12x - 6y + 15z = 54,$$
$$12x + 14y - 2z = 126.$$

And, subtracting the second of these equations successively from the first and third, there will arise

$$30y - 33z = 78,$$
$$20y - 17z = 72.$$

Or, by dividing the first of these two equations by 3, and then multiplying the result by 2,

$$20y - 22z = 52,$$
$$20y - 17z = 72,$$

Whence, by subtracting the former of these from the latter, we have $5z = 20$, or $z = 4$.

And, consequently, by substitution and reduction,

$$y = 7, \text{ and } x = 3.$$

3. Given $x + y + z = 53$, $x + 2y + 3z = 105$, and $x + 3y + 4z = 134$, to find the values of x, y, and z.

Ans. $x = 24$, $y = 6$, and $z = 23$.

4. Given $x + \frac{1}{2}y + \frac{1}{3}z = 32$, $\frac{1}{3}x + \frac{1}{4}y = \frac{1}{5}z = 15$, and $\frac{1}{4}x + \frac{1}{5}y + \frac{1}{6}z = 12$, to find the values of x, y, and z.

Ans. $x = 12$, $y = 20$, $z = 30$.

5. Given $7x + 5y + 2z = 79$, $8x + 7y + 9z = 122$, and $x + 4y + 5z = 55$, to find the values of x, y, and z.

Ans. $x = 4$, $y = 9$, $z = 3$.

6. Given $x + y = a$, $x + z = b$, and $y + z = c$, to find the values of x, y, and z.

Ans. $y = \frac{a - b + c}{2}$, $x = \frac{a + b - c}{2}$, and $z = \frac{b - a + c}{2}$.

7. Given $\frac{x}{2} + \frac{y}{3} + \frac{z}{4} = 62$, $\frac{x}{3} + \frac{y}{4} + \frac{z}{5} = 47$, and $\frac{x}{4} + \frac{y}{5} + \frac{z}{6} = 38$, to find x, y, and z.

Ans. $x = 24$, $y = 60$, and $z = 120$.

8. Given $z + y = x + 100$, $y - 2x = 2z - 100$, and $z + 100 = 3x + 3y$, to find x, y, and z.

Ans. $x = 9\frac{1}{11}$, $y = 45\frac{5}{11}$, and $z = 63\frac{7}{11}$.

9. Given $x + y + z = 7$, $2x - 3 = y + 3z$, and $5x + 5z = 3y + 19$, to find x, y, and z. Ans. $x = 4$, $y = 2$ and $z = 1$.

10. Given $3x + 5y - 4z = 25$, $5x - 2y + 3z = 46$, and $3y + 5z - x = 62$, to find x, y, and z.

Ans. $x = 7$, $y = 8$, and $z = 9$

11. *Given $x+y+z=13$, $x+y+u=17$, $x+z+u=18$, and $y+z+u=21$, to find x, y, z, and u.

Ans. $x=2$, $y=5$, $z=6$, and $u=10$.

MISCELLANEOUS QUESTIONS,

PRODUCING SIMPLE EQUATIONS.

THE usual method of resolving algebraical questions, is first to denote the quantities, that are to be found, by x, y, or some of the other final letters of the alphabet; then, having properly examined the state of the question, perform with these letters, and the known quantities, by means of the common signs, the same operations and reasonings, that it would be necessary to make if the quantities were known, and it was required to verify them, and the conclusion will give the result sought.

Or, it is generality best, when it can be done, to denote only one of the unknown quantities by x or y, and then to determine the expression for the others from the nature of the question; after which the same method of reasoning may be followed, as above. And, in some cases, the substituting for the sums and differences of quantities, or availing ourselves of any other mode, than a proper consideration of the question may suggest, will greatly facilitate the solution.

1. What number is that whose third part exceeds its fourth part by 16?

Let $x=$ the number required.

Then its $\frac{1}{3}$ part will be $\frac{1}{3}x$, and its $\frac{1}{4}$ part $\frac{1}{4}x$.

And therefore $\frac{1}{3}x-\frac{1}{4}x=16$, by the question,

That is, $x-\frac{3}{4}x=48$, or $4x-3x=192$,

Hence $x=192$, the number required.

2. It is required to find two numbers such, that their sum shall be 40 and their difference 16.

Let x denote the least of the two numbers required,

Then will $x+16=$ the greater number,

And $x+x+16=40$, by the question,

* This can be resolved by proceeding after the same manner as equations involving three unknown quantities; but the resolution of it may be greatly facilitated, by introducing into the calculation, beside the principal unknown quantities, a new unknown quantity arbitrarily assumed, such as, for example, the sum of all the rest; and when a little practised in such calculations, they become easy.

That is, $2x = 40 - 16$, or $x = \frac{24}{2} = 12 =$ least number.

And $x + 16 = 12 + 16 = 28 =$ the greater number required.

3. Divide 1000*l.* between A, B, and C, so that A shall have 72*l.* more than B, and C 100*l.* more than A.

Let $x =$ B's share of the given sum,
Then will $x + 72 =$ A's share,
And $x + 172 =$ C's share,
Hence their sum is $x + x + 72 + x + 172$,
Or $3x + 244 = 1000$, by the question,
That is, $3x = 1000 - 244 = 756$,
Or $x = \frac{756}{3} = 252l. =$ B.'s share,
Hence $x + 72 = 324l. =$ A's share,
And $x + 172 = 424l. =$ C's share,
Also, as above, $252l. =$ B's share.

Sum of all $= 1000l.$ the proof.

4. It is required to divide 1000*l.* between two persons, so that their shares of it shall be in the proportion of 7 to 9

Let $x =$ the first person's share,
Then will $1000 - x =$ second person's share,
And $x : 1000 - x :: 7 : 9$, by the question,
That is, $9x = (1000 - x) \times 7 = 7000 - 7x$,
Or $9x + 7x = 7000$, or $x = \frac{7000}{16} = 437l.\ 10s. =$ 1st share,
and $1000 - x = 1000 - 437l.\ 10s. = 562l.\ 10s. =$ 2d share.

5. The paving of a square court with stones, at 2*s.* a yard, will cost as much as the enclosing it with pallisades, at 5*s.* a yard; required the side of the square.

Let $x =$ length of the side of the square sought,
Then $4x =$ number of yards of enclosure,
And $x^2 =$ number of yards of pavement,
Hence $4x \times 5 = 20x =$ price of enclosing it,
And $x^2 \times 2 = 2x^2 =$ the price of the paving,
Therefore $2x^2 = 20x$, by the question,
Or, $2x = 20$, and $x = 10$, the length of the side required.

6. Out of a cask of wine, which had leaked away a third part, 21 gallons were afterwards drawn, and the cask being then gauged, appeared to be half full; how much did it hold?

Let $x =$ the number of gallons the cask is supposed to have held.

Then it would have leaked away $\frac{1}{3}x$ gallons.

Whence there had been taken out of it, altogether,

$$21 + \frac{1}{3}x \text{ gallons},$$

And therefore $21 + \frac{1}{3}x = \frac{1}{2}x$ by the question,

That is, $63 + x = \frac{3}{2}x$, or $126 + 2x = 3x$,

Consequently, $3x - 2x = 126$, or $x = 126$, the number of gallons required.

7. What fraction is that, to the numerator of which if 1 be added, its value will be $\frac{1}{3}$, but if 1 be added to the denominator, its value will be $\frac{1}{4}$.

Let the fraction required be represented $\frac{x}{y}$,

Then $\frac{x+1}{y} = \frac{1}{3}$, and $\frac{x}{y+1} = \frac{1}{4}$, by the question.

Hence $3x + 3 = y$, and $4x = y + 1$, or $x = \frac{y+1}{4}$,

Therefore $3\left(\frac{y+1}{4}\right) + 3 = y$, or $3y + 3 + 12 = 4y$.

That is, $y = 15$, and $x = \frac{y+1}{4} = \frac{15+1}{4} = \frac{16}{4} = 4$,

Whence the fraction that was to be found is $\frac{4}{15}$.

8. A market-woman bought in a certain number of eggs at 2 a penny, and as many others at 3 a penny, and having sold them out again, altogether, at the rate 5 for $2d.$, found she had lost $4d.$; how many eggs had she?

Let $x =$ the number of eggs of each sort,

Then will $\frac{1}{2}x =$ the price of the first sort,

And $\frac{1}{3}x =$ the price of the second sort,

But $5 : 2 :: 2x$ (the whole number of eggs) $: \frac{4x}{5}$,

Whence $\frac{4x}{5} =$ the price of both sorts, when mixed together at the rate of 5 for $2d$

And consequently $\frac{1}{2}x + \frac{1}{3}x - \frac{4x}{5} = 4$, by the question.

That is, $15x + 10x - 24x = 120$, or $x = 120$, the number of eggs of each sort, as required.

9. If A can perform a piece of work in 10 days, and B in 13; in what time will they finish it, if they are both set about it together?

Let the time sought be denoted by x.

Then $\frac{x}{10} =$ the part done by A in one day,

And $\frac{x}{13} =$ the part done by B in one day.

Consequently $\frac{x}{10} + \frac{x}{13} = 1$ (the whole work),

That is, $13x + 10x = 130$, or $23x = 130$.

Whence $x = \frac{130}{23} = 5\frac{15}{23}$ days, the time required.

10. If one agent A, alone, can produce an effect e, in the time a, and another agent B, alone in the time b; in what time will both of them together produce the same effect?

Let the time sought be denoted by x,

Then $a : e :: x : \frac{ex}{a} =$ part of the effect produced by A.

And $b : e :: x : \frac{ex}{b} =$ part of the effect produced by B.

Hence $\frac{ex}{a} + \frac{ex}{b} = e$, (the whole effect) by the question.

Or $\frac{x}{a} + \frac{x}{b} = 1$ by dividing each side by e.

Therefore, $x + \frac{ax}{b} = a$, or $bx + ax = ab$,

Consequently, $x = \frac{ab}{a+b} =$ time required.

11. How much rye, at 4*s*. 6*d*. a bushel, must be mixed with 50 bushels of wheat, at 6*s*. a bushel, so that the mixture may be worth 5*s*. a bushel?

Let $x =$ the number of bushels required.
Then $9x$ is the price of the rye in sixpences,
And 600 the prices of the wheat in ditto,
Also $(50 + x) \times 10$ the price of the wheat in ditto,
Whence $9x + 600 = 500 + 10x$, by the question.

Or, by transposition, $10x - 9x = 600 - 500$.

Consequently $x = 100$, the number of bushels required.

12. A labourer engaged to serve for 40 days, on condition that for every day he worked he should receive 20*d*., but for every day he was absent he should forfeit 8*d*.: now, at the end of the time he had to receive 1*l*. 11*s*. 8*d*.: how many days did he work, and how many was he idle?

Let the number of days that he worked be denoted by x.

Then will $40 - x$ be the number of days he was idle,

Also $20x$ the sum earned, and $(40 - x) \times 8$,

Or $320 - 8x$ the sum forfeited,

Whence $20x - (320 - 8x) = 380d.$ (= 1*l*. 11*s*. 8*d*.), by the question,

That is $20x - 320 + 8x = 380$,

Or $28x = 380 + 320 = 700$,

Consequently, $x = \frac{700}{28} = 25$, the number of days he worked, and $40 - x = 40 - 25 = 15$, the number of days he was idle.

QUESTIONS FOR PRACTICE.

1. It is required to divide a line, of 15 inches in length, into two such parts, that one may be three fourths of the other.
Ans. $8\frac{4}{7}$ and $6\frac{3}{7}$.

2. My purse and money together are worth 20*s*., and the money is worth 7 times as much as the purse, how much is there in it? Ans. 17*s*. 6*d*.

3. A shepherd being asked how many sheep he had in his flock, said, if I had as many more, half as many more, and 7 sheep and a half, I should have just 500; how many had he?
Ans. 197.

4. A post is one fourth of its length in the mud, one third in the water, and 10 feet above the water, what is its whole length? Ans. 24 feet.

5. After paying away $\frac{1}{4}$ of my money, and then $\frac{1}{5}$ of the remainder, I had 72 guineas left; what had I at first?
Ans. 120 guineas.

6. It is required to divide 300*l*. between A, B, and C, so that A may have twice as much as B, and C as much as A and B together. Ans. A 100*l*., B 50*l*., C 150*l*.

7. A person, at the time he was married, was 3 times as old as his wife: but after they had lived together 15 years, he was only twice as old; what were their ages on the wedding-day? Ans. Bride's age 15, bridegroom, 45.

8. What number is that from which, if 5 be subtracted, two thirds of the remainder will be 40? Ans. 65.

9. At a certain election, 1296 persons voted, and the successful candidate had a majority of 120; how many voted for each? Ans. 708 for one, and 588 for the other.

10. A's age is double of B's, and B's is triple of C's, and the sum of all their ages is 140; what is the age of each?

Ans. A's 84, B's 42, and C's 14

11. Two persons, A and B, lay out equal sums of money in trade; A gains 126*l.*, and B loses 87*l.*, and A's money is now double of B's; what did each lay out? Ans. 300*l.*

12. A person bought a chaise, horse, and harness for 60*l.*; the horse came to twice the price of the harness, and the chaise to twice the price of the horse and harness; what did he give for each? Ans. 13*l.* 6*s.* 8*d.* for the horse, 6*l.* 13*s.* 4*d.* for the harness, and 40*l.* for the chaise.

13. A person was desirous of giving 3*d.* a piece to some beggars, but found he had not money enough in his pocket by 8*d.*. he therefore gave them each 2*d.*, and had then 3*d.* remaining; required the number of beggars? Ans. 11.

14. A servant agreed to live with his master for 8*l.* a year, and a livery, but was turned away at the end of seven months, and received only 2*l.* 13*s.* 4*d.* and his livery; what was its value? Ans. 4*l.* 16*s.*

15. A person left 560*l.* between his son and daughter, in such a manner, that for every half crown the son should have, the daughter was to have a shilling; what were their respective shares? Ans. Son 400*l.*, daughter 160*l.*

16. There is a certain number, consisting of two places of figures, which is equal to four times the sum of its digits; and if 18 be added to it, the digits will be inverted; what is the number? Ans. 24.

17. Two persons, A and B, have both the same income; A saves a fifth of his yearly, but B, by spending 50*l.* per annum more than A, at the end of four years finds himself 100*l.* in debt; what was their income? Ans. 125*l.*

18. When a company at a tavern came to pay their reckoning, they found, that if there had been three persons more, they would have had a shilling apiece less to pay, and if there had been two less, they would have had a shilling apiece more to pay; required the number of persons, and the quota of each. Ans. 12 persons, quota of each 5*s.*

19. A person at a tavern borrowed as much money as he had about him, and out of the whole spent 1*s.*; he then went to a second tavern, where he also borrowed as much as he

had now about him, and out of the whole spent 1*s.*; and going on, in this manner, to a third and fourth tavern, he found, after spending his shilling at the latter, that he had nothing left; how much money had he at first?

Ans. $11\frac{1}{4}d.$

20. It is required to divide the number 75 into two such parts, that three times the greater shall exceed seven times the less by 15. Ans. 54 and 21.

21. In a mixture of British spirits and water, $\frac{1}{2}$ of the whole plus 25 gallons was spirits, and $\frac{1}{3}$ part minus 5 gallons was water; how many gallons were there in each?

Ans. 85 of wine, 35 of water.

22. A bill of 120*l.* was paid in guineas and moidores, and the number of pieces of both sorts that were used were just 100; how many were there of each, reckoning the guineas at 21*s.*, and the moidores at 27*s.*? Ans. 50.

23. Two travellers set out at the same time from London and York, whose distance is 197 miles: one of them goes 14 miles a day, and the other 16: in what time will they meet?

Ans. 6 days $13\frac{3}{5}$ hours.

24. There is a fish whose tail weighs 9*lb.*, his head weighs as much as his tail and half his body, and his body weighs as much as his head and his tail; what is the whole weight of the fish? Ans. 72*lb.*

25. It is required to divide the number 10 into three such parts, that if the first be multiplied by 2, the second by 3, and the third by 4, the three products shall be all equal.

Ans. $4\frac{8}{13}$, $3\frac{1}{13}$, $2\frac{4}{13}$.

26. It is required to divide the number 36 into three such parts, that $\frac{1}{2}$ the first, $\frac{1}{3}$ of the second, and $\frac{1}{4}$ of the third, shall be all equal to each other. Ans. The parts are 8, 12, and 16.

27. A person has two horses, and a saddle, which of itself is worth 50*l.*; now, if the saddle be put on the back of the first horse, it will make his value double that of the second, and if it be put on the back of the second, it will make his value triple that of the first; what is the value of each horse?

Ans. One 30*l.* and the other 40*l.*

28. If A give B 5*s.* of his money, B will have twice as much as the other has left; and if B give A 5*s.* of his money, A will have three times as much as the other has left: how much has each? Ans. A 13*s.* and B. 11*s.*

29. What two numbers are those whose difference, sum, and product, are to each other, as the numbers 2, 3, and 5 respectively? Ans. 10 and 2.

30. A person in play lost a fourth of his money, and then

won back 3*s*., after which he lost a third of what he now had, and then won back 2*s*.; lastly, he lost a seventh of what he then had, and after this found he had but 12*s*. remaining; what had he at first? Ans. 20*s*.

31. A hare is 50 leaps before a greyhound, and takes 4 leaps to the greyhound's 3, but 2 of the greyhound's leaps are as much as 3 of the hare's; how many leaps must the greyhound take to catch the hare? Ans. 300.

32. It is required to divide the number 90 into four such parts, that if the first part be increased by 2, the second diminished by 2, the third multiplied by 2, and the fourth divided by 2, the sum, difference, product, and quotient, shall be all equal? Ans. The parts are 18, 22, 10, and 40.

33. There are three numbers whose differences are equal, (that is, the second exceeds the first as much as the third exceeds the second), and the first is to the third as 5 to 7; also the sum of the three numbers is 324; what are those numbers? Ans. 90, 108, and 126.

34. A man and his wife usually drank out a cask of beer in 12 days, but when the man was from home it lasted the woman 30 days; how many days would the man alone be in drinking it? Ans. 20 days.

35. A general, ranging his army in the form of a solid square, finds he has 284 men to spare, but increasing the side by one man, he wants 25 to fill up the square; how many soldiers had he? Ans. 24000.

36. If A and B together can perform a piece of work in 8 days, A and C together in 9 days, and B and C in 10 days, how many days will it take each person to perform the same work alone. Ans. A $14\frac{34}{49}$ days, B $17\frac{23}{41}$, and C $23\frac{7}{31}$.

QUADRATIC EQUATIONS.

A Quadratic Equation, as before observed, is that in which the unknown quantity is of two dimensions, or which rises to the second power, and is generally divided into *simple* or *pure*, and *compound* or *adfected*.

A simple or pure quadratic equation, is that which contains only the square, or second power, of the unknown quantity, as,

$$ax^2 = b, \text{ or } x^2 = \frac{b}{a}; \text{ where } x = \sqrt{\frac{b}{a}}.$$

A compound or adfected quadratic equation, is that which contains both the first and second power of the unknown quantity, as,

$$ax^2 + bx = c, \text{ or } x^2 + \frac{b}{a}x = \frac{c}{a}.$$

In which case it is to be observed, that every equation of this kind, having any real positive root, will fall under one or other of the three following forms :—

1. *$x^2 + ax = b$ . . . where $x = -\frac{a}{2} \pm \sqrt{\left(\frac{a^2}{4} + b\right)}$.

2. $x - ax = b$ . . . where $x = +\frac{a}{2} \pm \sqrt{\left(\frac{a^2}{4} + b\right)}$.

3. $x^2 - ax = -b$ . . where $x = +\frac{a}{2} \pm \sqrt{\left(\frac{a^2}{4} - b\right)}$.

Or, if the second and last terms be taken either positively or negatively, as they may happen to be, the general equation

$$ax^2 \pm bx = \pm c, \text{ or } x^2 \pm \frac{b}{a}x = \pm \frac{c}{a},$$

which comprehends all the three cases above-mentioned, may be resolved by means of the following rule :—

RULE.—Transpose all the terms that involve the unknown quantity to one side of the equation, and the known terms to

---

* It may be observed, with respect to these forms, that

In the case $x^2 + ax - b = o$, where $x = -\frac{1}{2}a + \sqrt{(\frac{1}{4}a^2 + b)}$, or $-\frac{1}{2}a - \sqrt{(\frac{1}{4}a^2 + b)}$, the first value of x must be positive, because $\sqrt{(\frac{1}{4}a^2 + b)}$ is greater than $\sqrt{\frac{1}{4}a^2}$, or its equal $\frac{1}{2}a$; and its second value will evidently be negative, because each of the terms of which it is composed is negative.

2. In the case $x^2 - ax - b = o$, where $x = \frac{1}{2}a + \sqrt{(\frac{1}{4}a^2 + b)}$ or $\frac{1}{2}a - \sqrt{(\frac{1}{4}a^2 + b)}$, the first value of x is manifestly positive, being the sum of two positive terms: and the second value will be negative, because $\sqrt{(\frac{1}{4}a^2 + b)}$ is greater than $\sqrt{\frac{1}{4}a^2}$, or its equal $\frac{1}{2}a$.

3. In the case $x^2 - ax + b = o$, where $x = \frac{1}{2}a + \sqrt{(\frac{1}{4}a^2 - b)}$, or $\frac{1}{2}a - \sqrt{(\frac{1}{4}a^2 - b)}$, both the values of x will be positive, when $\frac{1}{4}a^2$ is greater than b; for its first value is then evidently positive, being composed of two positive terms; and its second value will also be positive; because $\sqrt{(\frac{1}{4}a^2 - b)}$ is less than $\sqrt{\frac{1}{4}a^2}$, or its equal $\frac{1}{2}a$.

But if $\frac{1}{4}a^2$, in this case, be less than b, the solution of the proposed equation is impossible; because the quantity $\frac{1}{4}a^2 - b$, under the radical, is then negative; and consequently $\sqrt{(\frac{1}{4}a^2 - b)}$ will be imaginary, or of no assignable value.

4. It may be also further observed, that there is a fourth case of the form $x^2 + ax + b = o$, where $x = -\frac{1}{2}a + \sqrt{(\frac{1}{4}a^2 - b)}$, or $x = -\frac{1}{2}a - \sqrt{(\frac{1}{4}a^2 - b)}$, the two values of x will be both negative, or both imaginary, according as $\frac{1}{4}a^2$ is greater or less than b; the imaginary roots, when they occur, being here of the forms $-(a' + c'\sqrt{-1})$ and $-(a' - c'\sqrt{-1})$.

From which it follows, that if all the terms of a quadratic equation, when brought to the lefthand side, be positive, its two roots will be both negative, or both imaginary; and conversely, if each of the roots be negative, or each imaginary, the signs of all the terms will be positive.

So that, of all quadratic equations, which can have any real positive root, that of the third form, $x^2 - ax + b = o$, is the only one, where the solution for certain numeral values of a and b will become impossible

10*

the other; observing to arrange them so that the term which contains the square of the unknown quantity may be positive, and stand first in the equation.

Then, if this square has any coefficient prefixed to it, let all the rest of the terms be divided by it, and the equation will be brought to one of the three forms above-mentioned.

In which case, the value of the unknown quantity x is always equal to half the coefficient, or multiplier of x, in the second term of the equation, taken with a contrary sign, together with $\pm$ the square root of the square of this number and the known quantity that forms the absolute or third term of the equation.*

Note.—All equations, which have the index of the unknown quantity, in one of their terms, just double that of the other, are resolved like quadratics, by first finding the value of the square root of the first term, according to the method used in the above rule, and then taking such a root, or power of the result, as is denoted by the reduced index of the unknown quantity.

Thus, if there be taken any general equation of this kind, as,

$$x^{2m} + ax^m = b,$$

we shall have, by taking the square root of x^{2m}, and observing the latter part of the rule,

---

* This rule, which is more commodious in its practical application than that usually given, is founded upon the same principle; being derived from the well-known property, that in any quadratic,

$x^2 \pm ax = \pm b$, if the square of half the coefficient a of the second term of the equation be added to each of its sides, so as to render it of the form

$$x^2 \pm ax + \tfrac{1}{4}a^2 = \tfrac{1}{4}a^2 \pm b,$$

that side which contains the unknown quantity will then be a complete square; and, consequently, by extracting the root of each side, we shall have

$$x \pm \tfrac{1}{2}a = \pm\sqrt{(\tfrac{1}{4}a^2 \pm b)}, \text{ or } x = \mp \tfrac{1}{2}a \pm \sqrt{(\tfrac{1}{4}a^2 \pm b)},$$

which is the same as the rule, taking a and b in $+$ or $-$ as they may happen to be.

It may here also be observed, that the ambiguous sign $\pm$, which denotes both $+$ and $-$, is prefixed to the radical part of the value of x in every expression of this kind, because the square root of any positive quantity, as a^2, is either $+a$ or $-a$; for $(+a) \times (+a)$, or $(-a) \times (-a)$ are each $= +a^2$: but the square root of a negative quantity, as $-a^2$, is imaginary, or unassignable, there being no quantity, either positive or negative, that, when multiplied by itself, will give a negative product.

To this we may also further add, that from the constant occurrence of the double sign before the radical part of the above expression, it necessarily follows, that every quadratic equation must have two roots; which are either both real, or both imaginary, according to the nature of the question.

$$x^m = -\frac{a}{2} \pm \sqrt{\left(\frac{a^2}{4} + b\right)}, \text{ and } x = \left\{-\frac{a}{2} \pm \sqrt{\left(\frac{a^2}{4} + b\right)}\right\}^{\frac{1}{m}}.$$

And if the equation, which is to be resolved, be of the following form,

$$x^m - ax^{\frac{m}{2}} = b,$$

we shall necessarily have, according to the same principle,

$$x^{\frac{m}{2}} = \frac{a}{2} \pm \sqrt{\left(\frac{a^2}{4} + b\right)}, \text{ and } x = \left\{\frac{a}{2} \pm \sqrt{\left(\frac{a^2}{4} + b\right)}\right\}^{\frac{2}{m}}.$$

EXAMPLES.

1. Given $x^2 + 4x = 140$, to find the value of x.
Here $x^2 + 4x = 140$, by the question,
Whence $x = -2 \pm \sqrt{(4 + 140)}$, by the rule,
Or which is the same thing, $x = -2 \pm \sqrt{144}$.
Wherefore $x = -2 + 12 = 10$, or $-2 - 12 = -14$,
Where one of the values of x is positive and the other negative.

2. Given $x^2 - 12x + 30 = 3$, to find the value of x.
Here $x^2 - 12x = 3 - 30 = -27$, by transposition,
Whence $x = 6 \pm \sqrt{(36 - 27)}$, by the rule,
Or, which is the same thing, $x = 6 \pm \sqrt{9}$,
Therefore $x = 6 + 3 = 9$, or $= 6 - 3 = 3$.
Where it appears that x has two positive values.

3. Given $2x^2 + 8x - 20 = 70$, to find the value of x.
Here $2x^2 + 8x = 70 + 20 = 90$, by transposition.
And $x^2 + 4x = 45$, by dividing by 2,
Whence $x = -2 \pm \sqrt{(4 + 45)}$, by the rule,
Or, which is the same thing, $x = -2 \pm \sqrt{49}$.
Therefore, $x = -2 + 7 = 5$, or $= -2 - 7 = -9$.
Where one of the values of x is positive and the other negative.

4. Given $3x^2 - 3x + 6 = 5\frac{1}{3}$, to find the value of x.

Here $3x^2 - 3x = 5\frac{1}{3} - 6 = -\frac{2}{3}$ by transposition.

And $x^2 - x = -\frac{2}{9}$ by dividing by 3,

Whence $x = \frac{1}{2} \pm \sqrt{\left(\frac{1}{4} - \frac{2}{9}\right)}$, by the rule.

Or, by subtracting $\frac{2}{9}$ from $\frac{1}{4}$, $x = \frac{1}{2} \pm \sqrt{\frac{1}{36}}$,

Therefore $x = \frac{1}{2} + \frac{1}{6} = \frac{2}{3}$, or $= \frac{1}{2} - \frac{1}{6} = \frac{1}{3}$,

In which case x has two positive values

5. Given $\frac{1}{2}x^2 - \frac{1}{3}x + 20\frac{1}{2} = 42\frac{2}{3}$, to find the value of x.

Here $\frac{1}{2}x^2 - \frac{1}{3}x = 42\frac{2}{3} - 20\frac{1}{2}, = 22\frac{1}{6}$, by transposition,

And $x^2 - \frac{2}{3}x = 44\frac{1}{3}$, by dividing by $\frac{1}{2}$, or multiplying by 2,

Whence we have $x = \frac{1}{3} \pm \sqrt{\left(\frac{1}{9} + 44\frac{1}{3}\right)}$, by the rule,

Or, by adding $\frac{1}{9}$ and $44\frac{1}{3}$ together, $x = \frac{1}{3} \pm \sqrt{\frac{400}{9}}$,

Therefore $x = \frac{1}{3} + 6\frac{2}{3} = 7$, or $= \frac{1}{3} - 6\frac{2}{3} = -6\frac{1}{3}$,

Where one value of x is positive and the other negative.

6. Given $ax^2 + bx = c$, to find the value of x.

Here $x^2 + \frac{b}{a}x = \frac{c}{a}$, by dividing each side by a.

Whence, by the rule, $x = -\frac{b}{2a} \pm \sqrt{\left(\frac{b^2}{4a^2} + \frac{c}{a}\right)}$,

Or, multiplying c and a by $4a$, $x = -\frac{b}{2a} \pm \sqrt{\frac{b^2 + 4ac}{4a^2}}$.

Therefore $x = -\frac{b}{2a} \pm \frac{1}{2a}\sqrt{(b^2 + 4ac)}$.

7. Given $ax^2 - bx + c = d$, to find the value of x.

Here $ax^2 - bx = d - c$, by transposition,

And $x^2 - \frac{b}{a}x = \frac{d-c}{a}$, by dividing by a.

Whence $x = \frac{b}{2a} \pm \sqrt{\left(\frac{d-c}{a} + \frac{b^2}{4a^2}\right)}$ by the rule,

Or, multg $d - c$ and a by $4a$, $x = \frac{b}{2a} \pm \frac{1}{2a}\sqrt{\left\{4a(d-c) + b^2\right\}}$.

8. Given $x^4 + ax^2 = b$, to find the value of x.

Here $x^4 + ax^2 = b$, by the question,

Or, $x^2 = -\frac{a}{2} \pm \sqrt{\left(\frac{a^2}{4} + b\right)} = -\frac{a}{2} \pm \frac{1}{2}\sqrt{(a^2 + 4b)}$, by the rule,

Whence $x = \pm\sqrt{\left\{-\frac{a}{2} \pm \frac{1}{2}\sqrt{(4b + a^2)}\right\}}$ by extraction of roots.

9. Given $\frac{1}{2}x^6 - \frac{1}{4}x^3 = -\frac{1}{32}$, to find the value of x.

Here $\frac{1}{2}x^6 - \frac{1}{4}x^3 = -\frac{1}{32}$, by the question,

And $x^6 - \frac{1}{2}x^3 = -\frac{1}{16}$, by multiplying by 2.

Whence $x^3 = \frac{1}{4} \pm \sqrt{\left(\frac{1}{16} - \frac{1}{16}\right)}, = \frac{1}{4}$, by the rule,

And consequently $x = \sqrt[3]{\frac{1}{4}} = \sqrt[3]{\frac{2}{8}} = \frac{1}{2}\sqrt[3]{2}$.

10. Given $2x^{\frac{2}{3}} + 3x^{\frac{1}{3}} = 2$, to find the value of x.

Here $2x^{\frac{2}{3}} + 3x^{\frac{1}{3}} = 2$, by the question,

And $x^{\frac{2}{3}} + \frac{3}{2}x^{\frac{1}{3}} = 1$, by dividing by 2,

Whence $x^{\frac{1}{3}} = -\frac{3}{4} \pm \sqrt{\left(\frac{9}{16} + 1\right)} = -\frac{3}{4} \pm \frac{5}{4} = \frac{1}{2}$, or -2.

Therefore, $x = \left(\frac{1}{2}\right)^3 = \frac{1}{8}$, or $(-2)^3 = -8$.

11. Given $x^4 - 12x^3 + 44x^2 - 48x = 9009\ (a)$, to find the value of x.

This equation may be expressed as follows :—

$$*(x^2 - 6x)^2 + 8(x^2 - 6x) = a,$$

---

* The biquadratic equation

$$x^4 - 12x^3 + 44x^2 - 48x = a$$

can be easily exhibited under the form

$$(x^2 - 6x)^2 + 8(x^2 - 6x) = a$$

by the following method:—

$$
\begin{array}{rl}
 & x^4 - 12x^3 + 44x^2 - 48x\,(x^2 - 6x \\
 & x^4 \\
\hline
2x^2 - 6x) & -12x^3 + 44x^2 - 48x \\
 & -12x^3 + 36x^2 \\
\hline
x^2 - 6x) & \qquad 8x^2 - 48x\,(8 \\
 & \qquad 8x^2 - 48x \\
\hline
 & \qquad * \quad *
\end{array}
$$

Consequently, $(x^2 - 6x)^2 + 8(x^2 - 6x) = x^4 - 12x^3 + 44x^2 - 48x = a$; for since, in extracting the square root of any quantity, the square of the root thus found plus the remainder, is always equal to the proposed quantity.

In a similar manner, the biquadratic equation $x^4 + 2ax^3 + 5a^2x^2 + 4a^3x = d$, may be exhibited under the form

$$(x^2 + ax)^2 + 4a^2(x^2 + ax) = d;$$

which can be resolved by the rule, page 126, for resolving quadratic equations.

Hence it follows, that if the remainder, after having found the first two terms of the square root, according to the rule page 50, can be resolved into two such factors, so that the factor containing the unknown

Whence $x^2 - 6x = -4 \pm \surd(16 + a)$, by the common rule,
And, by a second operation, $x = 3 \pm \surd\{9 - 4 \pm \surd(16 + a)\}$,

Therefore, by restoring the value of a, we have

$$x = 3 \pm \surd(5 \pm \surd 9025),$$

Or, by extraction of roots, $x = 13$, the Ans.

EXAMPLES FOR PRACTICE.*

1. Given $x^2 - 8x + 10 = 19$, to find the value of x.
Ans. $x = 9$.

2. Given $x^2 - x - 40 = 170$, to find the value of x.
Ans. $x = 15$.

3. Given $3x^2 + 2x - 9 = 76$, to find the value of x.
Ans. $x = 5$.

4. Given $\frac{1}{2}x^2 - \frac{1}{3}x + 7\frac{3}{8} = 8$, to find the value of x.
Ans. $x = 1\frac{1}{2}$.

5. Given $\frac{1}{2}x - \frac{1}{3}\surd x = 22\frac{1}{6}$, to find the value of x.
Ans. $x = 49$.

6. †Given $x + \surd(5x + 10) = 8$, to find the value of x.
Ans. $x = 3$.

---

quantity shall be equal to the terms of the root thus found; the proposed biquadratic may be always reduced to a quadratic form, as above. See Ryan's Algebra, page 396.—ED.

* The unknown quantity in each of the following examples, as well as in those given above, has always two values, as appears from the common rule; but the negative and imaginary roots being, in general, but seldom used in practical questions of this kind, as here suppressed,

† In some quadratic equations involving radical quantities of the form $\surd(ax+b)$, both the values of x, found by the ordinary process, will not answer the proposed equation, except we take the radical quantity with the double sign $\pm$. In resolving the above example, two values of x, that is 18 and 3 are found: but it appears, that 18 does not answer the condition of the equation, except we take the radical quantity $\surd(5x+10)$ with the sign $-$.

Now, since these two values of x are found from the resolution of the equation $x^2 - 21x = -54$; it necessarily follows that each of them, when substituted for x, must satisfy that equation; which may be verified thus; in the first place, by substituting 18 for x, in the equation $x^2 - 21x = -54$. we have $(18)^2 - 11 \times 18 = -54$, or $324 - 378 = -54$, that is, $-54 = -54$, or, by transposition, $0 = 0$.

Again, substituting 3 for x, we have $(3)^2 - 21 \times 3 = -54$, or $9 - 63 = -54$, $\therefore 54 - 54 = 0$, or $0 = 0$.

And as the equation $x^2 - 21x = -54$, may be deduced from the equation $+\surd(5x+10) = 8 - x$, or $-\surd(5x+10) = 8 - x$; it is evident that the radical quantity $\surd(5x+10)$ must be taken with the double sign $\pm$, in the primitive equation, in order that it would be satisfied by the values, 18 and 3, of x, found above; that is, 18 answers to the sign $-$, and 3 to the sign $+$. See Ryan's Elementary Treatise on Algebra, Theoretical and Practical, where this subject is clearly illustrated.—ED.

7. Given $\sqrt{(10+x)}-\sqrt[4]{(10+x)}=2$, to find the value of x. Ans. $x=6$.

8. Given $2x^4-x^2+96=99$, to find the value of x.

Ans. $x=\frac{1}{2}\sqrt{6}$.

9. Given $x^6+20x^3-10=59$, to find the value of x.

Ans. $x=\sqrt[3]{3}$.

10. Given $3x^{2n}-2x^n+3=11$, to find the value of x.

Ans. $x=\sqrt[n]{2}$.

11. Given $5\sqrt[4]{x}-3\sqrt{x}=1\frac{1}{3}$, to find the value of x.

Ans. $3\frac{13}{81}$, or $\frac{1}{81}$.

12. Given $\frac{2}{3}x\sqrt{(3+2x^2)}=\frac{1}{2}+\frac{2}{3}x^2$, to find the value of x.

Ans. $x=\frac{1}{2}\sqrt{(-3+3\sqrt{2})}$.

13. Given $x\sqrt{\left(\frac{6}{x}-x\right)}=\frac{1+x^2}{\sqrt{x}}$, to find the value of x.

Ans. $x=\sqrt{(1+\frac{1}{2}\sqrt{2})}$.

14. Given $\frac{1}{x}\sqrt{(1-x^3)}=x^2$, to find the value of x.

Ans. $x=\left(\frac{1}{2}\sqrt{5}-\frac{1}{2}\right)^{\frac{1}{3}}$

15. Given $x\sqrt{\left(\frac{a}{x}-1\right)}=\sqrt{(x^2-b^2)}$, to find the value of x.

Ans. $x=\frac{1}{4}a+\frac{1}{4}\sqrt{(8b^2+a^2)}$

16. Given $\sqrt{(1+x-x^2)}-2(1+x-x^2)=\frac{1}{9}$, to find the value of x.

Ans. $x=\frac{1}{2}+\frac{1}{6}\sqrt{41}$

17. Given $\sqrt{\left(x-\frac{1}{x}\right)}+\sqrt{\left(1-\frac{1}{x}\right)}=x$, to find the value of x. Ans. $x=\frac{1}{2}+\frac{1}{2}\sqrt{5}$.

18. Given $x^{4n}-2x^{3n}+x^n=6$, to find the value of x.

Ans. $x=\sqrt[n]{(\frac{1}{2}+\frac{1}{2}\sqrt{13})}$.

19. Given $x^4-2x^3+x=a$, to find the value of x.

Ans. $x=\frac{1}{2}\pm\sqrt{\left\{\frac{3}{4}\pm\sqrt{\left(a+\frac{1}{4}\right)}\right\}}$.

When there are more equations and unknown quantities than one, a single equation, involving only one of the unknown quantities, may sometimes be obtained by the rules before laid down for the solution of simple equations; and, in this case, one of the unknown quantities being determined, the others may be found by substituting its value in the remaining equations.

EXAMPLES.

1. Given $\left\{ \begin{array}{r} x^2 + y^2 = 65 \\ xy = 28 \end{array} \right\}$ to find the values of x and y.

Here, from the second equation, we have $y = \frac{28}{x}$; and by substituting this in the first, $x^2 + \frac{784}{x^2} = 65$, or $x^4 - 65x^2 = -784$.

Whence, by the common rule before given, we have $x = \pm \sqrt{\left\{ \frac{65}{2} \pm \sqrt{\left(\frac{4225}{4} - 784 \right)} \right\}}$.

Or, by reducing the parts under the last radical, and extracting the root, $x = \pm \sqrt{\left(\frac{65}{2} \pm \frac{33}{2} \right)} = 7$, or -4, and consequently, $y = \frac{28}{7}$, or $-\frac{28}{4} = 4$, or -7.

Or the solution, in cases of this kind, may often be more readily obtained by some of the artifices frequently made use of upon these occasions; which can only be learned from experience: thus, taking, as before, (1.) $x^2 + y^2 = 65$, (2.) $xy = 28$, we shall have, as in the former method, by multiplying by 2, $2xy = 56$, and, by adding this equation to the first, and subtracting it from the same, $x^2 + 2xy + y^2 = 121$, and $x^2 - 2xy + y^2 = 9$. Whence, by extracting the square roots of each of these last equations, there will arise $x + y = \pm 11$, and $x - y = \pm 3$, and consequently, by adding and subtracting these, we shall have $2x = \pm 14$, or $x = 7$, or -7, and $y = 4$, or -4. It will also sometimes facilitate the operations, by substituting for one of the unknown quantities the product of the other, and a third unknown quantity; which method may be applied with advantage whenever the sum of the dimensions of the unknown quantities is the same in every term of the equation.

2. Given $\left\{ \begin{array}{r} x^2 + xy = 56 \\ xy + 2y^2 = 60 \end{array} \right\}$ to find the values of x and y.

Here, agreeably to the above observation, let $x = vy$, then $v^2y^2 + vy^2 = 56$, and $vy^2 + 2y^2 = 60$, whence, from the first of these equations, $y^2 = \frac{56}{v^2 + v}$, and from the second, $y^2 = \frac{60}{v+2}$. Therefore, by equating the righthand member of these two expressions, we shall have $\frac{60}{v+2} = \frac{56}{v^2+v}$, or $60v^2 + 60v = 56v + 112$. And, by transposing $56v$, and dividing the result by 60, $v^2 + \frac{1}{15}v = \frac{28}{15}$. Hence, by the common rule for quadratics, we have $v = -\frac{1}{30} \pm \sqrt{\left(\frac{1}{900} + \frac{28}{15}\right)} = -\frac{1}{30} + \frac{41}{30} = \frac{4}{3}$. And, consequently, by the former part of the process, $y^2 = \frac{60}{v+2} = \frac{60}{1\frac{1}{3}+2} = 18$, or $y = \sqrt{(18)} = 3\sqrt{2}$, and $x = vy = \frac{4}{3} \times 3\sqrt{2} = 4\sqrt{2}$. The work may also be sometimes shortened, by substituting for the unknown quantities, the sum and difference of two other quantities; which method may be used when the unknown quantities, in each equation, are similarly involved.

3. Given $\left\{ \begin{array}{l} \frac{x^2}{y} + \frac{y^2}{x} = 18 \\ x + y = 12 \end{array} \right\}$ to find the values of x and y.

Here, according to the above observation, let there be assumed $x = z + v$, and $y = z - v$. Then, by adding these two equations together, we shall have $x + y = 2z = 12$, or $z = 6$, also, since $x = 6 + v$, $y = 6 - v$, and by the first equation $x^3 + y^3 = 18xy$, we shall obtain, by substitution, $(6 + v)^3 + (6 - v)^3 = 18(6 + v)(6 - v)$, or, by involving the two parts of the first member, and multiplying those of the second, $432 + 36v^2 = 648 - 18v^2$, whence, by transposition, $54v^2 = 216$; and by division, $v^2 = \frac{216}{54} = 4$; or $v = \pm 2$.

And therefore, by the first assumption, and the first part of the process, we have $x = z + v = 6 \pm 2 = 8$, or 4, and $y = z - v = 6 \pm 2 = 4$, or 8.

QUESTIONS PRODUCING QUADRATIC EQUATIONS.

The methods of expressing the conditions of questions of this kind, and the consequent reduction of them, till they are brought to a quadratic equation, involving only one unknown quantity and its square, are the same as those already given for simple equations.

1. To find two numbers such that their difference shall be 8, and their product 240.

Let x equal the least number.
Then will $x + 8 =$ the greater,
And $x(x + 8) = x^2 + 8x = 240$, by the question,
Whence $x = -4 + \sqrt{(16 + 240)} = -4 + \sqrt{256}$ by the common rule, before given,
Therefore $x = 16 - 4 = 12$, the less number,
and $x + 8 = 12 + 8 = 20$, the greater.

2. It is required to divide the number 60 into two such parts, that their product shall be 864.

Let $x =$ the greater part,
Then will $60 - x =$ the less,
And $x(60 - x) = 60x - x^2 = 864$, by the question,
Or, by changing the signs on both sides of the equation,
$$x^2 - 60x = -864,$$
Whence $x = 30 \pm \sqrt{(900 - 864)} = 30 \pm \sqrt{36} = 30 \pm 6$, by the rule,
And consequently, $x = 30 + 6 = 36$, or $30 - 6 = 24$, the two parts sought.

3. It is required to find two numbers such, that their sum shall be 10 (a), and the sum of their squares 58 (b).

Let $x =$ the greater of the two numbers,
Then will $a - x =$ the less,
And $x^2 + (a - x)^2 = 2x^2 - 2ax + a^2 = b$, by the question,
Or $2x^2 - 2ax = b - a^2$, by transposition,
And $x^2 - ax = \frac{b - a^2}{2}$, by division.

$$\text{Whence } x = \frac{a}{2} \pm \sqrt{\left(\frac{a^2}{4} + \frac{b - a^2}{2}\right)} = \frac{a}{2} \pm \frac{1}{2}\sqrt{(2b - a^2)}.$$

by the rule,

And if 10 be put for a, and 58 for b, we shall have

$$x = \frac{10}{2} + \frac{1}{2}\sqrt{(116 - 100)} = 7, \text{ the greater number,}$$

$$\text{And } 10 - x = \frac{10}{2} - \frac{1}{2}\sqrt{(116 - 100)} = 3, \text{ the less.}$$

4 Having sold a piece of cloth for 34*l*., I gained as much per cent. as it cost me; what was the price of the cloth?

Let $x =$ pounds the cloth cost,
Then will $24 - x =$ the whole gain,
But $100 : x :: x : 24 - x$, by the question,
Or, $x^2 = 100(24 - x) = 2400 - 100x$,
That is $x^2 + 100x = 2400$,
Whence $x = -50 + \sqrt{(2500 + 2400)} = -50 + 70 = 20$, by the rule,
And consequently, 20*l*. = price of the cloth.

5. A person bought a number of sheep for 80*l*., and if he had bought four more for the same money, he would have paid 1*l*. less for each; how many did he buy?

Let x represent the number of sheep,

Then will $\frac{80}{x}$ be the price of each,

And $\frac{80}{x+4} =$ price of each, if $x + 4$ cost 80*l*.

But $\frac{80}{x} = \frac{80}{x+4} + 1$, by the question,

Or, $80 = \frac{80x}{x+4} + x$, by multiplication.

And $80x + 320 = 80x + x^2 + 4x$, by the same,
Or, by leaving out $80x$ on each side, $x^2 + 4x = 320$,
Whence $x = -2 + \sqrt{(4 + 320)} = -2 + 18$, by the rule,
And consequently, $x = 16$, the number of sheep.

6. It is required to find two numbers, such that their sum, product, and difference of their squares, shall be all equal to each other.

Let $x =$ the greater number and $y =$ the less.

Then $\left\{\begin{array}{l} x + y = xy \\ x + y = x^2 - y^2 \end{array}\right\}$ by the question.

Hence $1 = \frac{x^2 - y^2}{x + y} = x - y$, or $x = y + 1$, by 2d equation.

And $(y + 1) + y = y(y + 1)$, by 1st equation,
That is, $2y + 1 = y^2 + y$; $y^2 + y = 1$.

Whence, $y = \frac{1}{2} + \sqrt{\left(\frac{1}{4} + 1\right)} = \frac{1}{2} + \frac{1}{2}\sqrt{5}$, by the rule,

Therefore, $y = \frac{1}{2} + \frac{1}{2}\sqrt{5} = 1.6180\ldots$

And $x = y + 1 = \frac{3}{2} + \frac{1}{2}\sqrt{5} = 2.6180\ldots$

Where ... denotes that the decimal does not end.

7. It is required to find four numbers in arithmetical progression, such that the product of the two extremes shall be 45, and the product of the means 77.

Let $x =$ least extreme, and $y =$ common difference,
Then $x, x + y, x + 2y$, and $x + 3y$, will be the four numbers,

Hence $\left\{ \begin{array}{l} x(x + 3y) = x^2 + 3xy = 45 \\ (x + y)(x + 2y) = x^2 + 3xy + 2y^2 = 77 \end{array} \right\}$, by the question,

And $2y^2 = 77 - 45 = 32$, by subtraction,

Or, $y^2 = \frac{32}{2} = 16$ by division, and $y = \sqrt{16} = 4$.

Therefore, $x^2 + 3xy = x^2 + 12x = 45$, by the 1st equation,
And consequently, $x = -6 + \sqrt{(36 + 45)} = -6 + 9 = 3$, by the rule.

Whence the numbers are 3, 7, 11, and 15.

8. It is required to find three numbers in geometrical progression, such that their sum shall be 14, and the sum of their squares 84.

Let x, y, and z, be the three numbers,
Then, $xz = y^2$, by the nature of proportion,

And $\left\{ \begin{array}{l} x + y + z = 14 \\ x^2 + y^2 + z^2 = 84 \end{array} \right\}$ by the question,

Hence, $x + z = 14 - y$, by the second equation,
And $x^2 + 2zx + z^2 = 196 - 28y + y^2$, by squaring both sides,
Or $x^2 + z^2 + 2y^2 = 196 - 28y + y^2$ by putting $2y^2$ for its equal $2xz$,
That is, $x^2 + y^2 + z^2 = 196 - 28y$, by subtraction,
Or, $196 - 28y = 84$, by equality,

Hence $y = \frac{196 - 84}{28} = 4$, by transposition and division.

Again, $xz = y^2 = 16$, or $x = \frac{16}{z}$, by the 1st equation,

And $x + y + z = \frac{16}{z} + 4 + z = 14$, by the 2d equation,

Or, $16 + 4z + z^2 = 14z$, or $z^2 - 10z = -16$,
Whence $z = 5 \pm \sqrt{(25 - 16)} = 5 \pm 3 = 8$, or 2 by the rule,
Therefore, the three numbers are 2, 4, and 8.

9. It is required to find two numbers, such that their sum shall be 13 (a), and the sum of their fourth powers 4721 (b).

Let $x =$ the difference of the two numbers sought,

Then will $\frac{1}{2}a + \frac{1}{2}x$, or $\frac{a + x}{2} =$ the greater number,

And $\frac{1}{2}a - \frac{1}{2}x$, or $\frac{a-x}{2} =$ the less,

But $\frac{(a+x)^4}{16} + \frac{(a-x)^4}{16} = b$, by the question,

Or $(a+x)^4 + (a-x)^4 = 16b$, by multiplication,

Or $2a^4 + 12a^2x^2 + 2x^4 = 16b$, by involution and addition,

And $x^4 + 6a^2x^2 = 8b - a^4$, by transposition and division,

Whence $x^2 = -3a^2 + \sqrt{(9a^4 + 8b - a^4)} = -3a^2 + \sqrt{8(a^4+b)}$, by the rule,

And $x = \sqrt{[-3a^2 + 2\sqrt{2(a^4+b)}]}$, by extracting the root,

Where, by substituting, 13 for a, and 4721 for b, we shall have $x = 3$,

Therefore $\frac{13+x}{2} = \frac{16}{2} = 8$, the greater number,

And $\frac{13-x}{2} = \frac{10}{2} = 5$, the less number.

The sum of which is 13, and $8^4 + 5^4 = 4721$.

10. Given the sum of two numbers equal s, and their product $= p$, to find the sum of their squares, cubes, biquadrates, &c.

Let x and y denote the two numbers; then

$$(1.)\ x + y = s, \quad (2.)\ xy = p.$$

From the square of the first of these equations take twice the second, and we shall have

$$(3.)\ x^2 + y^2 = s^2 - 2p = \text{sum of the squares.}$$

Multiply this by the 1st equation, and the product will be

$$x^3 + xy^2 + x^2y + y^3 = s^3 - 2sp.$$

From which subtract the product of the first and second equations, and there will remain

$$(4.)\ x^3 + y^3 = s^3 - 3sp = \text{sum of the cubes.}$$

Multiply this likewise by the 1st, and the product will be $x^4 + xy^3 + x^3y + y^4 = s^4 - 3s^2p$; from which subtract the product of the second and third equations, and there will remain

$$(5.)\ x^4 + y^4 = s^4 - 4s^2p + 2p^2 = \text{sum of the biquadrates.}$$

And, again multiplying this by the 1st equation and subtracting from the result the product of the second and fourth, we shall have

$$(6.)\ x^5 + y^5 = s^5 - 5s^3p + 5sp^2 = \text{sum of the fifth powers.}$$

And so on; the expression for the sum of any powers in general being $x^m + y^m = s^m - ms^{m-2}p + \frac{m(m-3)}{2}s^{m-4}p^2 - \frac{m(m-4)(m-5)}{2\cdot3}s^{m-6}p^3 + \frac{m(m-5)(m-6)(m-7)}{2\cdot3\cdot4}s^{m-8}p^4 -$,

&c. Where it is evident that the series will terminate when the index of s becomes $= 0$.

EXAMPLES FOR PRACTICE.

1. It is required to divide the number 40 into two such parts, that the sum of their squares shall be 818.
Ans. 23 and 17.

2. To find a number such, that if you subtract it from 10, and then multiply the remainder by the number itself, the product shall be 21. Ans. 7 or 3.

3. It is required to divide the number 24 into two such parts, that their product shall be equal to 35 times their difference. Ans. 10 and 14.

4. It is required to divide a line, of 20 inches in length, into two such parts that the rectangle of the whole and one of the parts shall be equal to the square of the other.
Ans. $10\sqrt{5} - 10$, and $30 - 10\sqrt{5}$.

5. It is required to divide the number 60 into two such parts, that their product shall be to the sum of their squares in the ratio of 2 to 5. Ans. 20 and 40.

6. It is required to divide the number 146 into two such parts, that the difference of their square roots shall be 6.
Ans. 25 and 121.

7. What two numbers are those whose sum is 20 and their product 36? Ans. 2 and 18.

8. The sum of two numbers is $1\frac{1}{3}$, and the sum of their reciprocal $3\frac{1}{5}$; required the numbers. Ans. $\frac{1}{2}$ and $\frac{5}{6}$.

9. The difference of two numbers is 15, and half their product is equal to the cube of the less number; required the numbers. Ans. 3 and 18.

10. The difference of two numbers is 5, and the difference of their cubes 1685; required the numbers. Ans. 8 and 13.

11. A person bought cloth for 33*l.* 15*s.*, which he sold again at 2*l.* 8*s.* per piece, and gained by the bargain as much as one piece cost him; required the number of pieces.
Ans. 15.

12. What two numbers are those, whose sum, multiplied by the greater, is equal to 77, and whose difference, multiplied by the less, is equal to 12? Ans. 4 and 7.

13. A grazier bought as many sheep as cost him 60*l.*, and after reserving 15 out of the number, sold the remainder for 54*l.*, and gained 2*s.* a head by them: how many sheep did he buy? Ans. 75.

14. It is required to find two numbers, such that their

product shall be equal to the difference of their squares, and the sum of their squares equal to the difference of their cubes.
Ans. $\frac{1}{2}\sqrt{5}$ and $\frac{1}{4}(5+\sqrt{5})$.

15. The difference of two numbers is 8, and the difference of their fourth powers is 14560; required the numbers.*
Ans. 3 and 11.

16. A company at a tavern had 8*l.* 15*s.* to pay for their reckoning; but before the bill was settled, two of them went away; in consequence of which those who remained had 10*s.* apiece more to pay than before; how many were there in company? Ans. 7.

17. A person ordered 7*l.* 4*s.* to be distributed among some poor people; but before the money was divided, there came in, unexpectedly, two claimants more, by which means the former received a shilling apiece less than they would otherwise have done; what was their number at first?
Ans. 16 persons.

18. It is required to find four numbers in geometrical progression such, that their sum shall be 15, and the sum of their squares 85. Ans. 1, 2, 4, and 8.

19. The sum of two numbers is 11, and the sum of their fifth powers is 17831; required the numbers. Ans. 4 and 7.

20. It is required to find four numbers in arithmetical pro-

---

* In solving this question, the reduced equation, found by the usual methods of operation, will be of the form $x^3+ax=b$; which is a cubic equation, and therefore cannot be resolved by the ordinary rules of quadratics; but such equations may sometimes be reduced to the form of a quadratic, and then resolved according to the rules already given.

Whenever, in a cubic equation of the form $x^3+ax=b$, b can be divided into two factors m and n, so that $m^2+a=n$, then the cubic equation can be resolved as a quadratic: thus, in the cubic equation $x^3+6x=20$, $20=2\times10$, and $2^2+6=10$. Now, multiplying both the sides of the equation by x, we have $x^4+6x^2=10\times2x$, adding $(2x)^2$ to both sides, $x^4+10x^2=(2x)^2+10(2x)$; $\therefore$ completing the square,

$$x^4+10x^2+25=(2x)^2+10(2x)+25,$$

and extracting the root, $x^2+5=2x+5$; $\therefore$ by transposition, $x^2=2x$, and $x=2$, or $=0$.

This method, as well as *some other similar artifices*, is of no utility when the divisor has not integral roots, and even then it can be resolved *more readily by Newton's Method of Divisors.*

It is proper to observe, that cubic equations of the form $x^3+ax^2+bx=c$, may be also exhibited under the form of a quadratic, from which, by completing the square, the value of the unknown quantity will be determined. For instance, the cubic equation $x^3+2ax^2+5a^2x+4a^3=0$, may be reduced to the form $(x^2+ax)^2+4a^2(x^2+ax)=0$; thus, multiply the given equation by x, we have $x^4+2ax^3+5a^2x^2+4a^3x=0$; which may be readily exhibited under the above form. See Ryan's Elementary Treatise on Algebra, Practical and Theoretical. (Art. 423.).—Ed.

gression, such, that their common difference shall be 4, and their continued product 176985. Ans. 15, 19, 23, and 27.

21. Two detachments of foot being ordered to a station at the distance of 39 miles from their present quarters, begin their march at the same time; but one party, by travelling $\frac{1}{4}$ of a mile an hour faster than the other, arrive there an hour sooner; required their rates of marching.

Ans. $3\frac{1}{4}$ and 3 miles per hour.

22. It is required to find two numbers, such, that the square of the first plus their product shall be 140, and the square of the second minus their product 78. Ans. 7 and 13.

23. It is required to find two numbers, such that their difference shall be $13\frac{5}{216}$, and the difference of their cube roots $1\frac{1}{6}$. Ans. $15\frac{5}{8}$, and $2\frac{10}{27}$.

24. It is required to find three numbers in arithmetical progression, such that the sum of their squares shall be 93; and if the first be multiplied by 3, the second by 4, and the third by 5, the sum of the products shall be 66. Ans. 2, 5, and 8.

25. The sum of three numbers in harmonical proportion is 191, and the product of the first and third is 4032; required their numbers. Ans. 72, 63, and 56.

26. It is required to find four numbers in arithmetical progression, such that if they are increased by 2, 4, 8, and 15 respectively, the sums shall be in geometrical progression.

Ans. 6, 8, 10, and 12.

27. It is required to find two numbers, such, that if their difference be multiplied into their sum, the product will be 5; but if the difference of their squares be multiplied into the sum of their squares, the product will be 65. Ans. 3 and 2.

28. It is required to divide the number 10 into two such parts, that if the square root of the greater part be taken from the greater part, the remainder shall be equal to the square root of the less part added to the less part.

Ans. $5+\frac{1}{2}\sqrt{19}$ and $5-\frac{1}{2}\sqrt{19}$.

29. It is required to find two numbers, such that if their product be added to their sum, it shall make 61; and if their sum be taken from the sum of their squares, it shall leave 88.

Ans. $7+\sqrt{2}$ and $7-\sqrt{2}$.

30. It is required to find two numbers, such that their difference multiplied by the difference of their squares, shall be 576; and their sum multiplied by the sum of their squares, shall be 2336. Ans. 5 and 11.

31. It is required to find three numbers in continual proportion, whose sum shall be 20, and the sum of their squares 140. Ans. $6\frac{3}{4}+\sqrt{3\frac{5}{16}}$, $6\frac{1}{2}$, and $6\frac{3}{4}-\sqrt{3\frac{5}{16}}$.

32. It is required to find two numbers whose product shall be 320, and the difference of their cubes to the cube of their difference, as 61 is to unity. Ans. 20 and 16.

33. The sum of 700 dollars was divided among four persons, A, B, C, and D, whose shares were in geometrical progression; and the difference between the greatest and least, was to the difference between the two means, as 37 to 12 What were all the several shares?

Ans. 108, 144, 192, and 256 dollars.

OF CUBIC EQUATIONS.

A cubic equation is that in which the unknown quantity rises to three dimensions; and, like quadratics, or those of the higher orders, is either simple or compound.

A simple or pure cubic equation is of the form

$$ax^3 = b, \text{ or } x^3 = \frac{b}{a}; \text{ where } x = \sqrt[3]{\frac{b}{a}},$$

A compound cubic equation is of the form

$$x^3 + ax = b,\ x^3 + ax^2 = b, \text{ or } x^3 + ax^2 + bx = c,$$

in each of which the known quantities a, b, c, may be either $+$ or $-$.

Or either of the two latter of these equations may be reduced to the same form as the first, by taking away its second term; which is done as follows:—

Rule.—Take some new unknown quantity, and subjoin to it a third part of the coefficient of the second term of the equation with its sign changed; then, if this sum, or difference, as it may happen to be, be substituted for the original unknown quantity and its powers in the proposed equation, there will arise an equation wanting its second term.

Note.—The second term of any of the higher orders of equations may also be exterminated in a similar manner, by substituting for the unknown quantity some other unknown quantity, and the 4th, 5th, &c., part of the coefficient of its second term, with the sign changed, according as the equation is of the 4th, 5th, &c. power.*

* Equations may be transformed into a variety of other new equations, the principal of which are as follows:—

1. The equation $x^4 - 4x^3 - 19x^2 + 106x - 120 = 0$, the roots of which are 2, 3, 4, and -5, by changing the signs of the second and fourth terms, becomes $x^4 + 4x^3 - 19x^2 - 106x - 120 = 0$, the roots of which are 5, -2, -3, and -4.

2. The equation $x^3 + x^2 - 10x + 8 = 0$, is transformed, by assuming $x = y - 4$ into $y^3 - 11y^2 + 30y = 0$, or $y^2 - 11y + 30 = 0$; the roots of which are greater than those of the former by 4.

3. The equation $x^3 - 6x^2 + 9x - 1 = 0$, may be transformed into one

EXAMPLES.

1. It is required to exterminate the second term of the equation $x^3 + 3ax^2 = b$, or $x^3 + 3ax^2 - b = 0$.

$$\text{Here } x = z - \frac{3a}{3} = z - a,$$

$$\text{Then} \begin{cases} x^3 = z^3 - 3az^2 + 3a^2z - a^3 \\ 3ax^2 = \quad + 3az^2 - 6a^2z + 3a^3 \\ -b = \quad -b \end{cases}$$

$$\text{Whence } z^3 - 3a^2z + 2a^3 - b = 0,$$
$$\text{Or } z^3 - 3a^2z = b - 2a^3,$$

in which equation the second power (z^2), of the unknown quantity, is wanting.

2. Let the equation $x^3 - 12x^2 + 3x = -16$, be transformed into another that shall want the second term.

$$\text{Here } x = z + 4.$$

$$\text{Then} \begin{cases} (z+4)^3 = z^3 + 12z^2 + 48z + 64 \\ -12(z+4)^2 = -12z^2 - 96z - 192 \\ +3(z+4) = \quad + 3z + 12 \end{cases}$$

$$\text{Whence } z^3 - 45z - 116 = -16$$
$$\text{Or, } z^3 - 45z = 100$$

which is an equation where z^2, or the second term, is wanting, as before.

3. Let the equation $x^3 - 6x^2 = 10$, be transformed into another that shall want the second term.

Ans. $z^3 - 12z = 26$.

4. Let $y^3 - 15y^2 + 81y = 243$, be transformed into an equation that shall want the second term.

Ans. $x^3 + 6x = 88$.

---

which shall want the third term, by assuming $x = y + e$, and in the resulting equation, let $3e^2 - 12e + 9$, or $e^2 - 4e + 3 = 0$, in which the values of e are 1 and 3; then assume $x = y + 3$, or $y + 1$, and the resulting equation will be $y^3 + 3y^2 - 1 = 0$, an equation wanting the third term.

4. The equation $6x^3 - 11x^2 + 6x - 1 = 0$, by assuming $x = \frac{1}{y}$, may be transformed into $y^3 - 6y^2 + 11y - 6 = 0$; the roots of which are to be reciprocals of the former.

5. The equation $3x^3 - 13x^2 + 14x + 16 = 0$, by assuming $x = \frac{y}{3}$, may be transformed into $y^3 - 13y^2 + 42y + 144 = 0$, the roots of which are three times those of the former.—Ed.

5. Let the equation $x^3 + \frac{3}{4}x^2 + \frac{7}{8}x - \frac{9}{16} = 0$, be transformed into another that shall want the second term.

Ans. $y^3 + \frac{11}{16}y = \frac{3}{4}$.

6. Let the equation $x^4 + 8x^3 - 5x^2 + 10x - 4 = 0$, be transformed into another, that shall want the second term.

Ans. $y^4 - 29y^2 + 94y - 92 = 0$.

7. Let the equation $x^4 - 3x^3 + 3x^2 - 5x - 2 = 0$, be transformed into another, that shall want the third term.

Ans. $y^4 + y^3 - 4y - 6 = 0$.

8. Let the equation $3x^3 - 2x + 1 = 0$, be transformed into another, whose roots are the reciprocals of the former.

Ans. $y^3 - 2y^2 + 3 = 0$.

9. Let the equation $x^4 - \frac{1}{2}x^3 + \frac{1}{3}x^2 - \frac{3}{4}x + \frac{1}{18} = 0$, be transformed into another, in which the coefficient of the highest term shall be unity, and the remaining terms integers. Ans. $y^4 - 3y^3 + 12y^2 - 162y + 72 = 0$.

OF THE SOLUTION OF CUBIC EQUATIONS.

Rule.—Take away the second term of the equation when necessary, as directed in the preceding rule. Then, if the numeral coefficients of the given equation, or of that arising from the reduction above-mentioned, be substituted for a and b in either of the following formulæ, the result will give one of the roots, as required.*

* If, instead of the regular method of reducing a cubic equation of the general form

$$x^3 + ax^2 + bx + c = 0,$$

to another, wanting the second term, as pointed out in the preceding article, there be put $x = \frac{1}{3}(y - a)$, we shall have, by substitution and reduction, $y^3 + (9b - 3a^2)y = 9ab - 27c - 2a^3$; where, since the value of y can be determined by either of the formulæ given in this rule, the value of x will also be known, being $x = \frac{1}{3}(y - a)$. And if $b = 0$, or the original equation be of the following form, $x^3 + ax^2 + c = 0$, the reduced equation will be $y^3 - 3a^2y = -2a^3 - 27c$, where the value of y, being found as above, we shall have, as before, $x = \frac{1}{3}(y - a)$, which formulæ, it may be observed, are more convenient, in some cases, than those resulting from the preceding article; as the coefficients, thus obtained, are always integers; whereas, by the former method, they are frequently fractions.

$$x^3 + ax = b.$$

$$x = \begin{cases} \sqrt[3]{\left\{\frac{b}{2} + \sqrt{\left(\frac{b^2}{4} + \frac{a^3}{27}\right)}\right\}} + \sqrt[3]{\left\{\frac{b}{2} - \sqrt{\left(\frac{b^2}{4} + \frac{a^3}{27}\right)}\right\}} \\ \text{or} \\ \sqrt[3]{\left\{\frac{b}{2} + \sqrt{\left(\frac{b^2}{4} + \frac{a^3}{27}\right)}\right\}} - \dfrac{\frac{1}{3}a}{\sqrt[3]{\left\{\frac{b}{2} + \sqrt{\left(\frac{b^2}{4} + \frac{a^3}{27}\right)}\right\}}} \end{cases}$$

Where, it is to be observed, that when the coefficient a, of the second term of the above equation, is negative, $\frac{a^3}{27}$, as also $\frac{a}{3}$, in the formula, will be negative; and if the absolute term b be negative, $\frac{b}{2}$ in the formula will also be negative; but $\frac{b^2}{4}$ will be positive.*

It may likewise be remarked, that when the equation is of the form

$$x^3 - ax = \pm b,$$

---

* This method of solving cubic equations is usually ascribed to Cardan, a celebrated Italian analyst of the 16th century; but the authors of it were Scipio Ferreus and Nicholas Tartalea, who discovered it about the same time, independently of each other, as is proved by Montucla, in his *Histoire des Mathematiques*, Vol. I. p. 568, and more at large in Hutton's *Mathematical Dictionary*, Art. Algebra.

The rule above given, which is similar to that of Cardan, may be demonstrated as follows:—

Let the equation, whose root is required, be $x^3 + ax = b$,
And assume $y + z = x$, and $3yx = -a$.

Then, by substituting these values in the given equation, we shall have $y^3 + 3y^2z + 3yz^2 + z^3 + a \times (y + z) = y^3 + z^3 + 3yz \times (y + z) + a \times (y + z) = y^3 + z^3 - a \times (y + z) + a \times (y + z) = b$, or $y^3 + z^3 = b$.

And if, from the square of this last equation, there be taken 4 times the cube of the equation $yz = -\frac{1}{3}a$, we shall have $y^6 - 2y^3z^3 + z^6 = b^2 + \frac{4}{27}a^3$, or $y^3 - z^3 = \sqrt{(b^2 + \frac{4}{27}a^3)}$.

But the sum of this equation and $y^3 + z^3 = b$, is $2y^3 = b + \sqrt{(b^2 + \frac{4}{27}a^3)}$, and their difference is $2z^3 = b - \sqrt{(b^3 + \frac{4}{27}a^3)}$; whence $y = \sqrt[3]{[\frac{1}{2}b + \sqrt{(\frac{1}{4}b^3 + \frac{1}{27}a^3)}]}$, and $z = \sqrt[3]{[\frac{1}{2}b - \sqrt{(\frac{1}{4}b^2 + \frac{1}{27}a^3)}]}$.

From which it appears, that $y + z$, or its equal x, is $= \sqrt[3]{[\frac{1}{2}b + \sqrt{(\frac{1}{4}b^2 + \frac{1}{27}a^3)}]} + \sqrt[3]{[\frac{1}{2}b - \sqrt{(\frac{1}{4}b^2 + \frac{1}{27}a^3)}]}$, which is the theorem;

Or, since z is $= -\frac{a}{3y}$, it will be $y + z = y - \frac{a}{3y}$ or $x = \sqrt[3]{[\frac{1}{2}b + \sqrt{(\frac{1}{4}b^2 + \frac{1}{27}a^3)}]} - \dfrac{\frac{1}{3}a}{\sqrt[3]{[\frac{1}{2}b + \sqrt{(\frac{1}{4}b^2 + \frac{1}{27}a^3)}]}}$, the same as the rule.

and $\frac{a^3}{27}$ is greater than $\frac{b^2}{4}$, or $4a^3$ greater than $27b^2$ the solution of it cannot be obtained by the above rule; as the question, in this instance, falls under what is usually called the *Irreducible Case* of cubic equations.*

EXAMPLES.

1. Given $2x^3 - 12x^2 + 36x = 44$, to find the value of x.

Here $x^3 - 6x^2 + 18x = 22$, by dividing by 2.

And, in order to exterminate the second term,

$$\text{Put } x = z + \frac{6}{3} = z + 2,$$

$$\text{Then} \left| \begin{array}{rl} (z+2)^3 = & z^3 + 6z^2 + 12z + 8 \\ -6(z+2)^2 = & \quad -6z^2 - 24z - 24 \\ 18(z+2) = & \qquad\qquad 18z + 36 \end{array} \right| = 22,$$

Whence $z^3 + 6z + 20 = 22$, or $z^3 + 6z = 2$,

And consequently, by substituting 6 for a and 2 for b, in the first formula, we shall have

$$z = \sqrt[3]{\left\{\frac{2}{2} + \sqrt{\left(\frac{4}{4} + \frac{216}{27}\right)}\right\}} + \sqrt[3]{\left\{\frac{2}{2} - \sqrt{\left(\frac{4}{4} + \frac{216}{27}\right)}\right\}} =$$

* It may here be farther observed as a remarkable circumstance in the history of this science, that the solution of the *Irreducible Case* above-mentioned, except by means of a table of sines, or by infinite series, has hitherto baffled the united efforts of the most celebrated Mathematicians in Europe; although it is well known that all the three roots of the equation are, in this case, real; whereas in those that are resolvable by the above formula, only one of the roots is real; so that, in fact, the rule is only applicable to such cubics as have two equal, or two impossible roots.

The reason why the assumptions, made in the note to the former part of this article with respect to the solution of the equation $x^3 - ax = b$, are found to fail in the case in question (and it does not appear that any other can be adopted) is, that the two auxiliary equations, $3yz = -a$ and $y^3 + z^3 = b$, which in this case become $3yz = a$, and $y^3 + z^3 = b$, or $y^3z^3 = \frac{a^3}{27}$, and $y^3 + z^3 = b$, cannot take place together; being inconsistent with each other.

For the greatest product that can be formed of the two quantities $y^3 + z^3$ is when they are all equal to each other; or since $y^3 + z^3 = b$, when each of these $= \frac{1}{2}b$; in which case their product is $= \frac{1}{4}b^2$.

But, as above shown, $y^3z^3 = \frac{a^3}{27}$ by the question; therefore, when $\frac{a^3}{27} > \frac{b^2}{4}$, the two conditions are incompatible with each other; and conequently the solution of the problem, upon that supposition, can only be obtained by imaginary quantities.

$\sqrt[3]{[1+\sqrt{(1+8)}]}+\sqrt[3]{[1-\sqrt{(1+8)}]}=\sqrt[3]{(1+\sqrt[3]{9})}+$
$\sqrt[3]{(1-\sqrt{9})}=\sqrt[3]{(1+3)}+\sqrt[3]{(1-3)}=\sqrt[3]{4}-\sqrt[3]{2}$,
Therefore $x=z+2=\sqrt[3]{4}-\sqrt[3]{2}+2=2+1.587401-1.259921=2.32748$, the answer.

2. Given $x^3-6x=12$, to find the value of x.

Here a being equal to -6, and b equal to 12, we shall have, by the formula,

$$x=\sqrt[3]{[6+\sqrt{(36-8)}]}-\frac{-2}{\sqrt[3]{\{6+\sqrt{(36-8)}\}}}=$$
$$\sqrt[3]{(6+\sqrt{28})}+\frac{2}{\sqrt[3]{(6+\sqrt{28})}}=\sqrt[3]{(6+5.2915)}+$$
$$\frac{2}{\sqrt[3]{(6+5.2915)}}=\sqrt[3]{(11.2915)}+\frac{2}{\sqrt[3]{(11.2915)}}=2.2435+$$
$$\frac{2}{2.2435}=2.2435+.8957=3.1392,$$

Therefore $x=3.1392$, the answer.

3. Given $x^3-2x=-4$, to find the value of x.

Here a being $=-2$, and $b=-4$, we shall have, by the formula,

$$x=\sqrt[3]{\left\{-2+\sqrt{(4-\frac{8}{27})}\right\}}+\sqrt[3]{\left\{-2-\sqrt{(4-\frac{8}{27})}\right\}}, \text{ or}$$

by reduction, $\sqrt[3]{(-2+\frac{10}{9}\sqrt{3})}-\sqrt[3]{(2+\frac{10}{9}\sqrt{3})}=$

$\sqrt[3]{(-2+1.9245)}-\sqrt[3]{(2+1.9245)}=\sqrt[3]{(-.0755}-$
$\sqrt[3]{3.9245}=-.41226-1.5773=-1.9999$, or -2;

Therefore $x=-2$, the answer.*

Note.—When one of the roots of a cubic equation has been found, by the common formula as above, or in any other way, the other two roots may be determined as follows :—

Let the known root be denoted by r, and put all the terms of the equation, when brought to the lefthand side, $=0$; then, if the equation, so formed, be divided by $x\pm r$ according as r is positive or negative, there will arise a quadratic equation,

---

* When the root of the given equation is a whole number, this method only determines it by an approximation of 9s, in the decimal part, which sufficiently indicates the entire integer; but in most instances of this kind, its value may be readily found, by a few trials, from the equation itself.

Or if, as in the above example, the roots, or numeral values of $\sqrt[3]{(-2+\frac{10}{9}\sqrt{3})}$, and $-\sqrt[3]{(2+\frac{10}{9}\sqrt{3})}$, be determined according to the rule laid down in Surds, Case 12, the result will be found equal to -2 as it ought.

the roots of which will be the other two roots of the given cubic equation.

4. Given $x^3 - 15x = 4$, to find the three roots or values of x.

Here x is readily found, by a few trials, to be equal to 4, and therefore

$$\begin{array}{l} x - 4)\ x^3 - 15x - 4(x^2 + 4x + 1 \\ \quad\quad\ \ x^3 - 4x^2 \\ \hline \quad\quad\quad\quad 4x^2 - 15x \\ \quad\quad\quad\quad 4x^2 - 16x \\ \hline \quad\quad\quad\quad\quad\quad\quad x - 4 \\ \quad\quad\quad\quad\quad\quad\quad x - 4 \\ \hline \quad\quad\quad\quad\quad\quad\quad\quad * \end{array}$$

Whence, according to the note already given,

$$x^2 + 4x + 1 = 0, \text{ or } x^2 + 4x = -1;$$

the two roots of which quadratic are $-2 + \sqrt{3}$ and $-2 - \sqrt{3}$; and consequently

$$4, \ -2 + \sqrt{3}, \text{ and } -2 - \sqrt{3},$$

are the three roots of the proposed equation.

EXAMPLES FOR PRACTICE.

1. Given $x^3 + 3x^2 - 6x = 8$, to find the root of the equation, or the value of x. Ans. $x = 2$.

2. Given $x^3 + x^2 = 500$, to find the root of the equation, or the value of x. Ans. $x = 7.616789$.

3. Given $x^3 - 3x^2 = 5$, to find the root of the equation, or the value of x. Ans. $x = 3.425101$.

4. Given $x^3 - 6x = 6$, to find the root of the equation, or the value of x. Ans. $\sqrt[3]{4} + \sqrt[3]{2}$.

5. Given $x^3 + 9x = 6$, to find the root of the equation, or the value of x. Ans. $\sqrt[3]{9} - \sqrt[3]{3}$.

6. Given $x^3 + 2x^2 - 23x = 70$, to find the root of thee quation, or the value of x. Ans. $x = 5.134599$.

7. Given $x^3 - 17x^2 + 54x = 350$, to find the root of the equation, or the value of x. Ans. $x = 14.954068$.

8. Given $x^3 - 6x = 4$, to find the three roots of the equation, or the three values of x.

Ans. -2, $1 + \sqrt{3}$, and $1 - \sqrt{3}$.

9. Given $x^3 - 5x^2 + 2x = -12$, to find the three roots of the equation, or the three values of x.

Ans. -3, $1 + \sqrt{5}$, and $1 - \sqrt{5}$.

OF THE SOLUTION OF CUBIC EQUATIONS, BY CONVERGING SERIES.

THIS method, which, in some cases, will be found more convenient in practice than the former, consists in substituting the numeral parts of the given equation, in the place of the literal, in one of the following general formulæ, according to which it may be found to belong, and then collecting as many terms of the series as are sufficient for determining the value of the unknown quantity, to the degree of exactness required.*

1. $x^3 + ax = b$.†

$$x = \frac{2b}{\sqrt[3]{[2(27b^2 + 4a^3)]}} \left\{ 1 + \frac{2.5}{6.9}\left(\frac{27b^2}{27b^2 + 4a^3}\right) + \frac{2.5.8.11}{6.9.12.15} \right.$$

---

* The method laid down in this article, of solving cubic equations by means of series, was first given by NICOLE, in the *Memoirs of the Academy of Sciences*, an. 1738, p. 99; and afterwards, at greater length, by CLAIRAUT in his *Elemens d'Algebre.*

† With respect to the determination of the roots of cubic equations by means of series, let there be given, as above, the equation $x^3+ax=b$, where the root by transposing the terms of each of the two branches of the common formula, is

$x = \sqrt[3]{\left\{ \sqrt{(\frac{1}{4}b^2+\frac{1}{27}a^3)}+\frac{1}{2}b \right\}} - \sqrt[3]{\left\{ \sqrt{(\frac{1}{4}b^2+\frac{1}{27}a^3)}-\frac{1}{2}b \right\}}$; or by putting, for the sake of greater simplicity, $\sqrt{(\frac{1}{4}b^2+\frac{1}{27}a^3)} = s$, and reducing the expression, $x = s^{\frac{1}{3}}\left\{ \sqrt[3]{(1+\frac{b}{2s})} - \sqrt[3]{(1-\frac{b}{2s})} \right\}$.

Hence, extracting the roots of the righthand member of this equation by the binomial theorem, there will arise $\sqrt[3]{\left(1+\frac{b}{2s}\right)} = 1+\frac{1}{3}\left(\frac{b}{2s}\right) - \frac{2}{3.6}\left(\frac{b}{2s}\right)^2 + \frac{2.5}{3.6.9}\left(\frac{b}{s}\right)^3 - \frac{2.5.8}{3.6.9.12}\left(\frac{b}{2s}\right)^4 +$, &c.

$\sqrt[3]{\left(1+\frac{b}{2s}\right)} = 1-\frac{1}{3}\left(\frac{b}{2s}\right) - \frac{2}{3.6}\left(\frac{b}{2s}\right)^2 - \frac{2.5}{3.6.9}\left(\frac{b}{2s}\right)^3 - \frac{2.5.8}{3.6.9.12}\left(\frac{b}{2s}\right)^4 -$, &c.

And consequently, if the latter of these two series be taken from the former, the result, by making the first term of the remainder a multiplier, will give

$$x = \frac{2bs^{\frac{1}{3}}}{6s}\left\{1+\frac{2.5}{6.9}\left(\frac{b}{2s}\right)^2 +, \text{\&c.}\right\},$$

where, since $s = \sqrt{\left(\frac{1}{4}b^2+\frac{1}{27}a^3\right)}$, we shall have $\left(\frac{b}{2s}\right)^2 = \frac{27b^2}{27b^2+4a^3}$, $\left(\frac{b}{2s}\right)^4 = \left(\frac{27b^2}{27b^2+4a^3}\right)^2$, &c. And $\frac{2bs^{\frac{1}{3}}}{6s} = \frac{2b}{6s^{\frac{2}{3}}} = \frac{2b}{\sqrt{[2(27b^2+4a^3)]}}$,

Whence, also, by substitution, we have the above formula

$$\left(\frac{27b^2}{27b^2+4a^3}\right)^2 + \frac{2.5.8.11.14.17}{6.9.12.15.18.21}\left(\frac{27b^2}{27b^2+4a^3}\right)^3 +, \text{\&c.}\Big\}$$ Or

$$x = \frac{2b}{\sqrt[3]{[2(27b^2+4a^3)]}}\Big\{1 + \frac{2.5}{6.9}\left(\frac{27b^2}{27b^2+4a^3}\right)\text{A} + \frac{8.11}{12.15}\left(\frac{27b^2}{27b^2+4a^3}\right)\text{B} + \frac{14.17}{12.21}\left(\frac{27b^2}{27b^2+4a^3}\right)\text{C} + \frac{20.23}{24.27}\left(\frac{27b^2}{27b^2+4a^3}\right)\text{D} +, \text{\&c.}$$

In which case, as well as in all the following ones, A, B, C, &c., denote the terms immediately preceding those in which they are first found.

2. $x^3 - ax = \pm b$, where $\frac{1}{4}b^2$ is supposed to be greater than $\frac{1}{27}a^3$, or $27b^2 > 4a^3$.

$$x = \pm 2\sqrt[3]{\frac{b}{2}}\Big\{1 - \frac{2}{3.6}\left(\frac{27b^2-4a^3}{27b^2}\right) - \frac{2.5.8}{3.6.9.12}\left(\frac{27b^2-4a^3}{27b^2}\right)^2 - \frac{2.5.8.11.14}{3.6.9.12.15.18}\left(\frac{27b^2-4a^3}{27b^2}\right)^3 \text{\&c.}\Big\}.^*$$ Or,

$$x = \pm 2\sqrt[3]{\frac{b}{2}}\Big\{1 - \frac{2}{3.6}\left(\frac{27b^2-4a^3}{27b^2}\right)\text{A} - \frac{5.8}{9.12}\left(\frac{27b^2-4a^3}{27b^2}\right)\text{B}. - \frac{11.14}{15.18}\left(\frac{27b^2-4a^3}{27b^2}\right)\text{C} - \frac{17.20}{21.24}\left(\frac{27b^2-4a^3}{27b^2}\right)\text{D} -, \text{\&c.}$$

In which case the upper sign must be taken when b is

---

* The root, as found by the common formula, when properly reduced, is $x = \pm\sqrt[3]{\frac{b}{2}}\Big\{\sqrt[3]{\left[1+\frac{2}{b}\sqrt{(\frac{1}{4}b^2-\frac{1}{27}a^2)}\right]} + \sqrt[3]{\left[1-\frac{2}{b}\sqrt{(\frac{1}{4}b^2-\frac{1}{27}a^3)}\right]}\Big\}$.

Or, putting, as in the last case, $\frac{2}{b}\sqrt{(\frac{1}{4}b^2-\frac{1}{27}a^3)}$, or its equal $\left(\frac{27b^2-4a^3}{27b^2}\right)^{\frac{1}{2}} = s$, we shall have $x = \pm\sqrt[3]{\frac{b}{2}}\Big\{\sqrt[3]{(1+s)} + \sqrt[3]{(1-s)}\Big\}$.

Whence, extracting the roots of the righthand member of this equation, there will arise $\sqrt[3]{(1+s)} = 1 + \frac{1}{3}s - \frac{2}{3.6}s^2 + \frac{2.5}{3.6.9}s^3 - \frac{2.5.8}{3.6.9.12}s^4 +$, &c. $\sqrt[3]{(1-s)} = 1 - \frac{1}{3}s - \frac{2}{3.6}s^2 - \frac{2.5}{3.6.9}s^3 - \frac{2.5.8}{3.6.9.12}s^4 -$, &c.

And, consequently, by adding the two series together, and taking the 1st term of the result as a multiplier, we shall have $x = \pm 2\sqrt[3]{\frac{2}{2}}\Big\{1 - \frac{2}{3.6}s^2 - \frac{2.5.8}{3.6.9.12}s^4 - \frac{2.5.8.11.14}{3.6.9.12.15.18}s^6 -, \text{\&c.}\Big\}$ Or, by substituting $\left(\frac{27b^2-4a^3}{27b^2}\right)$ for its equal s, we get the above expression.

positive, and the under sign when it is negative; and the same for the first root in the two following cases:—

$$3.\quad x^3 - ax = \pm b,$$

where $\frac{1}{4}b^2$ is supposed to be less than $\frac{1}{27}a^3$, or $27b^2 < 4a^3$.

$$x = \pm 2\sqrt[3]{\frac{b}{2}}\left\{1 + \frac{2}{3.6}\left(\frac{4a^3 - 27b^2}{27b^2}\right) - \frac{2.5.8}{3.6.9.12}\left(\frac{4a^3 - 27b^2}{27b^2}\right)^2 + \frac{2.5.8.11.14}{3.6.9.12.15.18}\left(\frac{4a^3 - 27b^2}{27b^2}\right)^3 -, \text{\&c.}\right\}.^{*}$$ Or,

* This expression is obtained from the last series, by barely changing the signs of the numerator and denominator in each of its terms; which does not alter their value.

Hence, in order to determine the other two roots of the equation, let that above found, or its equivalent expression, $\sqrt{\left\{\frac{1}{2}b + \sqrt{(\frac{1}{4}b^2 - \frac{1}{27}a^3)}\right\}} + \sqrt[3]{\left\{\frac{1}{2}b - \sqrt{(\frac{1}{4}b^2 - \frac{1}{27}a^3)}\right\}} = \pm r$.

Then, according to the formula that has been before given for these roots in the former part of the present article, we shall have $x = \mp\frac{r}{2} \pm \frac{\sqrt{-3}}{2}\left\{\sqrt[3]{[\frac{1}{2}b + \sqrt{(\frac{1}{4}b^2 - \frac{1}{27}a^3)}]} - \sqrt[3]{[\frac{1}{2}b - \sqrt{(\frac{1}{4}b^2 - \frac{1}{27}x^3)}]}\right\}$. Or, putting $\frac{2}{b}\sqrt{(\frac{1}{4}b^2 - \frac{1}{27}a^3)} = s$, and reducing the expression, $x = \mp\frac{2}{r} \pm \frac{(\frac{1}{4}b)^{\frac{1}{3}}\sqrt{-3}}{2}\left\{\sqrt[3]{(1+s)} - \sqrt[3]{(1-s)}\right\}$. Whence, extracting the cube roots of the righthand member of this equation, there will arise

$$\sqrt[3]{(1+s)} = 1 + \tfrac{1}{3}s - \frac{2}{3.6}s^2 + \frac{2.5}{3.6.9}s^3 - \frac{2.5.8}{3.6.9.12}s^4 +, \text{\&c.}$$

$$\sqrt[3]{(1-s)} = 1 - \tfrac{1}{3}s - \frac{2}{3.6}s^2 - \frac{2.5}{3.6.9}s^3 - \frac{2.5.8}{3.6.9.12}s^4 -, \text{\&c.}$$

And, consequently, by taking the latter of these series from the former, and making the first term of the remainder a multiplier, we shall have $x = \mp$

$$\frac{r}{2} \pm \frac{(s\frac{1}{2}b)^{\frac{1}{3}}\sqrt{-3}}{3}\left\{1 + \frac{2.5}{6.9}s^2 + \frac{2.5.8.11}{6.9.12.15}s^4 + \frac{2.5.8.11.14.17}{6.9.12.15.18.21}s^6 +, \text{\&c.}\right\}.$$

But since $s = \frac{2}{b}\sqrt{(\frac{1}{4}b^2 - \frac{1}{27}a^3)} = \left(\frac{27b^2 - 4a^3}{27b^2}\right)^{\frac{1}{2}}$, $s^2 = \frac{27b^2 - 4a^3}{27b^2} = -\frac{4a^3 - 27b^2}{27b^2}$, $s^4 = \left(\frac{4a^3 - 27b^2}{27b^2}\right)^2$, &c., and also $\sqrt[3]{\frac{1}{2}b} \times \frac{s}{3}\sqrt{-3} = \sqrt[3]{\frac{1}{2}b} \times \frac{\sqrt{-3}}{3}\sqrt{\left(\frac{27b^2 - 4a^3}{27b^2}\right)} = \frac{1}{3}\sqrt[3]{\frac{1}{2}b} \times \sqrt{\left(\frac{4a^3 - 27b^2}{9b^2}\right)} = \frac{\sqrt{(4a^3 - 27b^2)}}{9\sqrt[3]{2b^2}}$, if these values be substituted for their equals, in the last series, the result will give the above expressions for the two remaining roots of the equation.

$$x = \pm 2 \sqrt[3]{\frac{b}{2}} \left\{ 1 + \frac{2}{3.6}\left(\frac{4a^3 - 27b^2}{27b^2}\right) A - \frac{5.8}{9.12}\left(\frac{4a^3 - 27b^2}{27b^2}\right) B + \frac{11.14}{15.18}\left(\frac{4a^3 - 27b^2}{27b^2}\right) C - \frac{17.20}{21.24}\left(\frac{4a^3 - 27b^2}{27b^2}\right) D +, \&c.\right.$$

which series answers to the irreducible case, and must be used when $2a^3$ is less than $27b^2$.

And if the root thus found be put $= r$, the other two roots may be expressed as follows:—

$$x = \mp \frac{r}{2} \pm \frac{\sqrt{(4a^3 - 27^2)}}{9\sqrt[3]{2b^2}} \left\{ 1 - \frac{2.5}{6.9}\left(\frac{4a^3 - 27b^2}{27b^2}\right) + \frac{2.5.8.11}{6.9.12.15}\left(\frac{4a^3 - 27b^2}{27b^2}\right)^2 - \frac{2.5.8.11.14.17}{3.6.9.12.15.18.21}\left(\frac{4a^3 - 27b^2}{27b^2}\right)^3 +, \&c. \right\}.$$

Or,

$$x = \mp \frac{r}{2} \pm \frac{\sqrt{(4a^3 - 27b^2)}}{9\sqrt[3]{2b^2}} \left\{ 1 - \frac{2.5}{6.9}\left(\frac{4a^3 - 27b^2}{27b^2}\right) A + \frac{8.11}{12.15}\left(\frac{4a^3 - 27b^2}{27b^2}\right) B - \frac{14.17}{18.21}\left(\frac{4a^3 - 27b^2}{27b^2}\right) C + \frac{20.23}{24.27}\left(\frac{4a^3 - 27b^2}{27b^2}\right) D -, \&c.\right.$$

Where $-\frac{1}{2}r$, or $+\frac{1}{2}r$, must be taken according as b is positive or negative; and the double signs $\pm$ must be considered as $+$ in one case, and $-$ in the other, as usual.

4. $x^3 - ax = \pm b$,

where $\frac{1}{4}b^2$ is still supposed to be less than $\frac{1}{27}a^3$, or $27b^2 < 4a^3$.

$$x = \pm \frac{2b}{\sqrt[3]{[2(4a - 27b^2)]}} \left\{ 1 - \frac{2.5}{6.9}\left(\frac{27b^2}{4a^3 - 27b^2}\right) + \frac{2.5.8.11}{6.9\ 12.15}\left(\frac{27b^2}{4a^3 - 27b^2}\right)^2 - \frac{2.5.8.11.14.17}{6.9.12.15.18.21}\left(\frac{27b^2}{4a^3 - 27b^2}\right)^3 + \&c. \right\}.$$

*Or,

* By transposing the terms of the common formula, as in the first case, we shall have $x = \sqrt[3]{\left\{\sqrt{(\frac{1}{4}b^2 - \frac{1}{27}a^3)} + \frac{1}{2}b\right\}} - \sqrt[3]{\left\{\sqrt{(\frac{1}{4}b^2 - \frac{1}{27}a^3)} - \frac{1}{2}b\right\}}$. Or, by putting, for the sake of simplicity as before, $\sqrt{(\frac{1}{4}b^2 - \frac{1}{27}a^3)} = s$, and reducing the equation $x = \sqrt[3]{s}\left\{\sqrt[3]{(1 + \frac{b}{2s})} - \sqrt[3]{(1 - \frac{b}{2s})}\right\}$.

Whence, extracting the roots of the righthand number, as in the former instances,

$$\sqrt[3]{\left(1 + \frac{b}{2s}\right)} = 1 + \frac{1}{3}\left(\frac{b}{2s}\right) - \frac{2}{3.6}\left(\frac{b}{2s}\right)^2 + \frac{2.5}{3.6.9}\left(\frac{b}{2s}\right)^3 - \frac{2.5.8}{3.6.9.12}\left(\frac{b}{2s}\right)^4 +, \&c.$$

$$x = \pm \frac{2b}{\sqrt[3]{[2(4a^3 - 27b^2)]}} \left\{ 1 - \frac{2.5}{6.9} \left(\frac{27b^2}{4a^3 - 27b^2}\right) \text{A} + \frac{8.11}{12.15} \left(\frac{27b^2}{4a^3 - 27b^2}\right) \text{B} - \frac{14.17}{18.21} \left(\frac{27b^2}{4a^3 - 27b^2}\right) \text{C} + \frac{20.23}{24.27} \left(\frac{27b^2}{4a^3 - 27b^2}\right) \right.$$

$$\sqrt[3]{\left(1 - \frac{b}{2s}\right)} = 1 - \tfrac{1}{3}\left(\frac{b}{2s}\right) - \frac{2}{3.6}\left(\frac{b}{2s}\right)^2 - \frac{2.5}{3.6.9}\left(\frac{b}{2s}\right)^3 - \frac{2.5.8}{3.6.9.12}\left(\frac{b}{3s}\right)^4 -,$$
&c.

And consequently, by taking the latter of these series from the former, and making the first term of the result a multiplier, we shall have

$$x = \frac{2bs^{\frac{1}{3}}}{6s} \left\{ 1 + \frac{2.5}{6.9}\left(\frac{b}{2s}\right)^2 + \frac{2.5.8.11}{6.9.12.15}\left(\frac{b}{2s}\right)^4 + \frac{2.5.8.11.14.17}{6.9.12.15.18.21}\left(\frac{b}{2s}\right)^6 + \right.$$

&c.$\Big\}$. But since $s = \surd(\tfrac{1}{4}b^2 - \tfrac{1}{27}a^3)$, we shall have $\left(\frac{b}{2s}\right)^2 = \frac{27b^2}{27b^2 - 4a^3}$ $= -\frac{27b^2}{4a^3 - 27b^2}\left(\frac{b}{2s}\right)^4 = \left(\frac{27b^2}{4a^3 - 27b^2}\right)^2$, &c., and $\frac{2bs^{\frac{1}{3}}}{6s} = \frac{2b}{6s^{\frac{2}{3}}} =$ $\frac{2b}{6\sqrt[3]{(\frac{1}{4}b^2 - \frac{1}{27}a^3)}} = - \frac{2b}{\sqrt[3]{\{2(4a^3 - 27b^2)\}}}$.

Whence, if these values be substituted for their equals in the last series, there will arise the above expression for the first root of the equation. And if we put the root thus found, or its equivalent expression

$$\sqrt[3]{\left\{\surd(\tfrac{1}{4}b^2 - \tfrac{1}{27}a^3) + \tfrac{1}{2}b\right\}} - \sqrt[3]{\left\{\surd(\tfrac{1}{4}b^2 - \tfrac{1}{27}a^3) - \tfrac{1}{2}b\right\}} = \pm r,$$

we shall have, according to the formula before given for the other two roots,

$$x = \mp \frac{r}{2} \pm \frac{\surd -3}{2} \left\{ \sqrt[3]{[\surd(\tfrac{1}{4}b^2 - \tfrac{1}{27}a^3) + \tfrac{1}{2}b]} + \sqrt[3]{[\surd(\tfrac{1}{4}b^2 - \tfrac{1}{27}a^3) - \tfrac{1}{2}b]} \right\}.$$

Or, taking, as before, $\surd(\tfrac{1}{4}b^2 - \tfrac{1}{27}b^3) = s$, and simplifying the result, $x = \mp \frac{r}{2} \pm \frac{s^{\frac{1}{3}}\surd -3}{2}\left\{\sqrt[3]{\left(1 + \frac{b}{2s}\right)} + \sqrt[3]{\left(1 - \frac{b}{2s}\right)}\right\}$.

Whence, by extracting the roots of the righthand side of this equation, there will arise

$$\sqrt[3]{\left(1 + \frac{b}{2s}\right)} = 1 + \tfrac{1}{3}\left(\frac{b}{2s}\right) - \frac{2}{3.6}\left(\frac{b}{2s}\right)^2 + \frac{2.5}{3.6.9}\left(\frac{b}{2s}\right)^3 - \frac{2.5.8}{3.6.9.12}\left(\frac{b}{2s}\right)^4 +,$$
&c.

$$\sqrt[3]{\left(1 - \frac{b}{2s}\right)} = 1 - \tfrac{1}{3}\left(\frac{b}{2s}\right) - \frac{2}{3.6}\left(\frac{b}{2s}\right)^2 - \frac{2.5}{3.6.9}\left(\frac{b}{2s}\right)^3 - \frac{2.5.8}{3.6.9.12}\left(\frac{b}{2s}\right)^4 -,$$
&c.

And, consequently, if the latter of these series be added to the former, we shall have, by making the first term of the result a multiplier,

$$x = \mp \frac{r}{2} \pm s^{\frac{1}{3}}\surd -3 \left\{ 1 - \frac{2}{3.6}\left(\frac{b}{2s}\right)^2 - \frac{2.5.8}{3.6.9.12}\left(\frac{b}{2s}\right)^4 - \frac{2.5.8.11.14}{3.6.9.12.15.18}\right.$$

$\left(\frac{b}{2s}\right)^6$ —, &c. But since $s = \surd(\tfrac{1}{4}b^2 - \tfrac{1}{27}a^3) = \left(\frac{27b^2 - 4a^3}{4.27}\right)^{\frac{1}{2}}$, we shall

D —, &c., which series also answers to the irreducible case, and must be used when $2a^3$ is greater than $27b^2$. And if the root thus found, be put $= r$, as before, the other two roots may be expressed thus: $x = \mp \frac{r}{2} \pm \sqrt[6]{\frac{4a^3 - 27b^2}{4}} \Big\{ 1 + \frac{2}{3.6}\left(\frac{27b^2}{4a^3 - 27b^2}\right) - \frac{2.5.8}{3.6.9.12}\left(\frac{27b^2}{4a^3 - 27b^2}\right)^2 + \frac{2.5.8.11.14}{3.6.9.12.15.18}\left(\frac{27b^2}{4a^3 - 27b^2}\right)^3 -$, &c. $\Big\}$. Or,

$$x = \mp \frac{r}{2} \pm \sqrt[6]{\frac{4a^3 - 27b^2}{4}} \Big\{ 1 + \frac{2}{3.6}\left(\frac{27b^2}{4a^3 - 27b^2}\right) \text{A} - \frac{5.8}{9.12}\left(\frac{27b^2}{4a^3 - 27b^2}\right) \text{B} + \frac{11.14}{15.18}\left(\frac{27b^2}{4a^3 - 27b^2}\right) \text{C} - \frac{17.20}{21.24}\left(\frac{27b^2}{4a^3 - 27b^2}\right)$$

D +, &c.

Where the signs are to be taken as in the latter part of the preceding case.

EXAMPLES.

1 Given $x^3 + 6x = 2$, to find the value of x.

Here $a = 6$, and $b = 2$, whence

$$\frac{27b^2}{27b^2 + 4a^3} = \frac{27 \times 4}{27 \times 4 + 4 \times 216} = \frac{1}{1+8} = \frac{1}{9}; \text{ and}$$

$$\frac{2b}{\sqrt[3]{[2(27b^2 + 4a^3)]}} = \frac{4}{\sqrt[3]{[2(4 \times 27 + 4 \times 216)]}} = \frac{4}{2\sqrt[3]{(27 + 8 \times 27)}} = \frac{4}{6\sqrt[3]{9}} = \frac{2\sqrt[3]{81}}{27} = \frac{\sqrt[3]{648}}{27}.$$ Consequently, by formula 1, we shall have

1	1.0000000 (A)
$\frac{2.5}{6.9} \times \frac{1}{9}$ A	.0205761 (B)
$\frac{8.11}{12.15} \times \frac{1}{9}$ B	.0011177 (C)
$\frac{14.17}{18.21} \times \frac{1}{9}$ C	.0000782 (D)

---

also have $\left(\frac{b}{2s}\right)^2 = \frac{27b^2}{27b^2 - 4a^3} = -\frac{27b^2}{4a^3 - 27b^2}$, &c., and consequently, $s^{\frac{1}{3}} \sqrt{-3} = \sqrt[6]{\left(\frac{4a^3 - 27b^2}{4}\right)}$

Hence, if these values be substituted for their equals in the above series, the result will give the above expressions for the two remaining roots of the equation.

$\frac{20.23}{24.27}\times\frac{1}{9}$ D	.0000062 (E)
$\frac{26.29}{30.33}\times\frac{1}{9}$ E	.0000005 (F)
	1.0217787
Log. 1.0217787	0.0093570
Log. $\sqrt[3]{648}$	0.9371916
Colog. 27	8.5686362
No. 3274801	$-$ 1.5151848.

Therefore $x = .3274801$

2. Given $x^3 - 9x = 12$, to find the real value of x.

Here $a = 9$ and $b = 12$;

whence $2\sqrt[3]{\frac{12}{2}} = 2\sqrt[3]{6}$ and $\frac{27b^2 - 4a^3}{27b^2} = \frac{27\times144-4\times27^2}{27\times144}$

$= \frac{144 - 108}{144} = \frac{36}{144} = \frac{1}{4}$.

Consequently, by formula 2, we shall have

1	1.0000000 (A)
$-\frac{2}{3.6}\times\frac{1}{4}$ (A)	$-$.0277778 (B)
$-\frac{5.8}{9.12}\times\frac{1}{4}$ (B)	$-$.0025720 (C)
$-\frac{11.14}{15.18}\times\frac{1}{4}$ (C)	$-$.0003667 (D)
$-\frac{17.20}{21.24}\times\frac{1}{4}$ (D)	$-$.0000619 (E)
$-\frac{23.26}{27.30}\times\frac{1}{4}$ (E)	$-$.0000114 (F)
$-\frac{29.32}{33.36}\times\frac{1}{4}$ (F)	$-$.0000022 (G)
Sum	$-$.0307920
Comp.	.9692080
Log. 969208	$-$1.9864137

Log. $2\sqrt[3]{6}$, or Log. $\sqrt[3]{48}$	0.5604137
No. 3.522334	.5478274

therefore $x = 3.522334$.

3. Given $x^3 - 12x = 16$, to find the three values of x.

Here $a = 12$ and $b = 15$;

Whence $2\sqrt[3]{\frac{b}{2}} = 2\sqrt[3]{\frac{15}{2}} = \sqrt[3]{60}$ and $\frac{4a^3 - 27b^2}{27b^2} = \frac{4.12^3 - 27.15^2}{27.15^2} = \frac{256 - 225}{225} = \frac{31}{225}$.

Consequently, by formula 3, we shall have

1	$+1.0000000$ (A)
$+\frac{2}{3.6} \times \frac{31}{225}$ A	$+0.0153086$ (B)
$-\frac{5.\ 8}{9.12} \times \frac{31}{225}$ B	-0.0007812 (C)
$+\frac{11.14}{15.18} \times \frac{31}{225}$ C	$+0.0000614$ (D)
$-\frac{17.20}{21.24} \times \frac{31}{225}$ D	-0.0000057 (E)
$-\frac{23.26}{27.30} \times \frac{31}{225}$ E	$+0.0000006$ (F)
Sum of + Terms	$+1.0153706$
Sum of − Terms	$-.0007869$
Difference	1.0145837
Log. 1.014837	.0062880
Log. $\sqrt[3]{60}$	.5927171
No. 3.971962	.5990051

Therefore the affirmative value of x or first root, $r =$ 3.971962.

Again, $\frac{\sqrt{(4a^3 - 27b^2)}}{9\sqrt[3]{2b^2}} = \frac{\sqrt{837}}{9\sqrt[3]{450}} = \frac{\sqrt{(9 \times 93)}}{9\sqrt[3]{450}} = \frac{\sqrt{93}}{3\sqrt[3]{450}}$

and $\frac{4a^3 - 27b^2}{27b^2} = \frac{31}{225}$.

Hence,

$+1$	1.0000000 (A)
$-\frac{2.5}{6.9} \times \frac{31}{225}$	$-.0255144$ (B)

$+\frac{8.11}{12.15}\times\frac{31}{225}$ B	+.0017186 (C)
$-\frac{14.17}{18.21}\times\frac{31}{225}$ C	—.0001490 (D)
$+\frac{20.23}{24.27}\times\frac{31}{225}$ D	+.0000145 (E)
$-\frac{26.29}{30.33}\times\frac{31}{225}$ E	—.0000014 (F)
Sum	.9760683
Log. 9760683	—1.9894802
Log. $\sqrt{93}$	0.9842415
Colog. $\sqrt[3]{450}$	9.1155958
Colog. 3	9.5228787
No. .4099445	—1.6121964
Also $-\frac{r}{2}$	—1.9859810
Last No	±0.4099445
Result	—1.5760365
Or	—2.3959255

Whence the three roots or values of x are 3.971962 — 1.5760365, and —2.3959255.

4. Given $x^3 - 6x = 2$, to find the three values of x.

Here $\frac{-2b}{\sqrt[3]{[2(4a^3-27b^2)]}} = \frac{-4}{\sqrt[3]{[2(4.6^3-27.4)]}} =$

$\frac{-4}{3\sqrt[3]{[2(4.8-4)]}} = \frac{-4}{6\sqrt[3]{7}} = \frac{-2}{3\sqrt[3]{7}} = -\frac{2\sqrt{49}}{21}$, and $\frac{27b^2}{4a^3-27b^2}$

$= \frac{4.27}{4.6^3-27.4} = \frac{1}{8-1} = \frac{1}{7}$.

Hence, by the formula 4, we shall have

1	1.0000000 (A)
$-\frac{2.5}{6.9}\times\frac{1}{7}$ A	-- 0264550 (B)
$+\frac{8.11}{12.15}\times\frac{1}{7}$ B	+.0018476 (C)
$-\frac{14.17}{18.21}\times\frac{1}{7}$ C	—.0001662 (D)

$+\frac{20.23}{24.27}\times\frac{1}{7}$ D	+.0000168 (E)
$-\frac{26.29}{30.33}\times\frac{1}{7}$ E	—.0000018 (F)
Sum	+.9752414
Log. .9752414	—.9891120
Log. 2	0.3010300
L. $\sqrt[3]{49}$	0.5633987
Colog. 21	8.6777807
No. 329870	—1.5313214

Therefore one of the negative roots or values of x, is $-.339870=-r$.

Again, $\sqrt[6]{\frac{4a^3-27b^2}{4}}=\sqrt[6]{\frac{4.6^3-27.4}{4}}=\sqrt[6]{(6^3-27)}=\sqrt[6]{189}$, and $\frac{27b^2}{4a^3-27b^2}=\frac{1}{7}$.

Hence,

1	1.0000000 (A)
$+\frac{2}{3.6}\times\frac{1}{7}$ A	+0.0158730 (B)
$-\frac{5.8}{9.12}\times\frac{1}{7}$ B	—.0008398 (C)
$+\frac{11.14}{15.18}\times\frac{1}{7}$ C	+.0000684 (D)
$-\frac{17.20}{21.24}\times\frac{1}{7}$ D	—.0000066 (E)
$+\frac{23.26}{27.30}\times\frac{1}{7}$ E	+.0000002 (F)
Sum	1.0150952
Log. 1.0150952	.0065070
Log. $\sqrt[6]{189}$	.3794103
No. 2.431741	.3859173
Therefore $\frac{r}{2}$	+.169935

Last number	± 2.431741
Result	$+2.601676$
Or	-2.261806

And consequently, $+2.601676$, -2.261806, and $-.339870$, are the three roots required.

EXAMPLES FOR PRACTICE.

1. Given $x^3 + 9x = 30$, to find the root of the equation, or the value of x. Ans. $x = 2.180849$.
2. Given $x^3 - 2x = 5$, to find the root of the equation, or the value of x. Ans. $x = 2.0945515$
3. Given $x^3 - 3x = 3$, to find the root of the equation, or the value of x. Ans. 2.103803
4. Given $x^3 - 27x = 36$, to find the three roots or values of x. Ans 5.765722, $-$ 4.320684, and $-$ 1.445038.
5. Given $x^3 - 48x^2 = -200$, to find the root of the equation, or the value of x. Ans. 47.9128.
6. Given $x^3 - 22x = 24$, to find the root of the equation, or the value of x. Ans. 5.162277

OF BIQUADRATIC EQUATIONS.

A *biquadratic equation*, as before observed, is one that rises to the fourth power, or which is of the general form —

$$x^4 + ax^3 + bx^2 + cx + d = 0.$$

The root of which may be determined by means of the following formula; substituting the numbers of the given equation, with their proper signs, in the places of the literal coefficients, a, b, c, d.

RULE 1.*—Find the value of z in the cubic equation $z^3 + \left(\frac{1}{4}ac - \frac{1}{12}b^2 - d\right) z = \frac{1}{108}b^3 + \frac{1}{8}(c^2 + da^2) - \frac{b}{24}(ac + 8d)$

---

* *This method* is that given by Simpson, p. 120 of his *Algebra*, which consists in supposing the given biquadratic to be formed by taking the difference of two complete squares, being the same in principle as that of *Ferrari*.

Thus, let the proposed equation be of the form $x^4 + ax^3 + bx^2 + cx + d = 0$ (1), having all its terms complete; and assume $(x^2 + \frac{1}{2}ax + p)^2 - (qx + r)^2 = x^4 + ax^3 + bx^2 + cx + d$.

Then, if $x^2 + \frac{1}{2}ax + p$ and $qx + r$ be actually involved, we shall have

$$x^4 + ax^3 + \left.\begin{array}{l} 2p \\ +\frac{1}{4}a^2 \\ -q^2 \end{array}\right| x^2 + \left.\begin{array}{l} ap \\ \\ -2qr \end{array}\right| x + \left.\begin{array}{l} p^2 \\ -r^2 \end{array}\right\} = x^4 + ax^3 + bx^2 + cx + d.$$

And consequently, by equating the homologous terms, there will arise

by one of the former rules; and let the root, thus determined, be denoted by r. Then find the two values of x in each of the following quadratic equations :—

$$x^2 + (\frac{1}{2}a + \sqrt{\left\{\frac{1}{4}a^2 + 2(r - \frac{1}{3}b)\right\}})x = -(r + \frac{1}{6}b) + \sqrt{\left\{(r + \frac{1}{6}b)^2 - d\right\}},$$

$$x^2 + (\frac{1}{2}a - \sqrt{\left\{\frac{1}{4}a^2 + 2(r - \frac{1}{3}b)\right\}})x = -(r + \frac{1}{6}b) - \sqrt{\left\{(r + \frac{1}{6}b)^2 - d\right\}},$$

and they will be the four roots of the biquadratic required.

---

1. $2p + \frac{1}{4}a^2 - q^2 = b$		$2p + \frac{1}{4}a^2 - b = q^2$
2. $ap - 2qr = c$	or	$ap - c = 2qr$
3. $p^2 - r^2 = d$		$p^2 - d = r^2$

where, since the product of the first and last of the absolute terms of these equations is evidently equal to $\frac{1}{4}$ of the square of the second, we shall have

$$2p^3 + (\tfrac{1}{4}a^2 - b)p^2 - 2dp - d(\tfrac{1}{4}a^2 - b) = \tfrac{1}{4}(a^2p^2 - 2acp + c^2).$$

Or, by bringing the unknown quantities to the lefthand side, and the known to the right, and then dividing by 2,

$$p^3 - \frac{b}{2}p^2 + \tfrac{1}{4}(ac - 4d)p = \tfrac{1}{8}(c^2 + a^2d) - \tfrac{1}{2}bd \ldots . (2.)$$

From which last equation p can be determined by the rules before given for cubics.

And since, from the preceding equations, it appears that

$$q = \sqrt{(2p + \tfrac{1}{4}a^2 - b)} \text{ and } r = \frac{ap - c}{2q}, \text{ or } \sqrt{(p^2 - d)},$$

it is evident that the several values of z can be obtained from the quantities thus found.

For, because $x^4 + ax^3 + bx^2 + cx + d$, or its equal $(x^2 + \frac{1}{2}ax + p)^2 - (qx + r)^2 = 0$, it is plain that $(x^2 + \frac{1}{2}ax + p)^2 = (qx + r)^2$. And, therefore, by extracting the roots of each side of this equation, there will arise

$$x^2 + \tfrac{1}{2}ax + p = qx + r; \text{ or } x^2 + (\tfrac{1}{2}a - q)x = r - p.$$

Whence, by substituting the above values of p, q, and r, for their equals, and transposing the terms, we shall have

$$x^2 + \left\{\tfrac{1}{2}a \mp \sqrt{(2p + \tfrac{1}{4}a^2 - b)}\right\}x + p \mp \sqrt{(p^2 - d)} = 0,$$

for the case where $ap - c$ is positive; and

$$x^2 + \left\{\tfrac{1}{2}a \mp \sqrt{2p + \tfrac{1}{4}a^2 - b)}\right\}x + p \pm \sqrt{(p^2 - d)} = 0,$$

for the case where $ap - c$ is negative: which two quadratics give the four roots of the proposed equation.

And by putting $p = z + \frac{b}{6}$, in the reducing equation (2), in order to destroy its second term, the several steps of the investigation may be made to agree with the expressions given in the above rule.

EXAMPLES.

1. Given the equation $x^4 - 10x^3 + 35x^2 - 50x + 24 = 0$, to find its roots.

Here $a = -10$, $b = 35$, $c = -50$, and $d = 24$.

Whence, by substituting these numbers in the cubic equation,

$z^3 + (\frac{1}{4}ac - \frac{1}{12}b^2 - d)z = \frac{1}{108}b^3 + \frac{1}{8}(c^2 + da^2) - \frac{b}{24}(ac + 8d)$, we shall have the following reduced equation,

$$z^3 - \frac{13}{12}z = \frac{35}{108},$$

which, being resolved according to the rule before laid down for that purpose, gives

$$z = \frac{1}{6}\left\{\sqrt[3]{(35 + 18\sqrt{-3})} + \sqrt[3]{(35 - 18\sqrt{-3})}\right\}.$$

But, by the rule for binomial surds, given in the former part of the work,

$\sqrt[3]{(35 + 18\sqrt{-3})} = \frac{7}{2} + \frac{1}{2}\sqrt{-3}$, and $\sqrt[3]{(35 - 18\sqrt{-3})}$

$$= 7 - \frac{1}{2}\sqrt{-3}$$

Wherefore $z = \frac{1}{6}\left\{\frac{7}{2} + \frac{1}{2}\sqrt{-3} + \frac{7}{2} - \frac{1}{2}\sqrt{-3}\right\} = \frac{7}{6}$.

And if this number be substituted for r, -10 for a, 35 for b, and 24 for d, in the two quadratic equations,

$$x^2 + (\frac{1}{2}a + \sqrt{\left\{\frac{1}{4}a^2 + 2(r - \frac{1}{3}b)\right\}})x = -(r + \frac{1}{6}b) + \sqrt{\left\{(r + \frac{1}{6}b)^2 - d\right\}},$$

$$x^2 + (\frac{1}{2}a - \sqrt{\left\{\frac{1}{4}a^2 + 2(r - \frac{1}{3}b)\right\}})x = -(r + \frac{1}{6}b) - \sqrt{\left\{(r + \frac{1}{6}b)^2 - d\right\}},$$

they will become, after reducing them to their most simple terms,

$$x^2 - 3x = -2, \text{ and } x^2 - 7x = -12:$$

from the first of which $x = \frac{3}{2} \pm \sqrt{\frac{1}{4}} = \frac{3}{2} \pm \frac{1}{2} = 2$, or 1, and

from the second, $x = \frac{7}{2} \pm \sqrt{\frac{1}{4}} = \frac{7}{2} \pm \frac{1}{2} = 4$ or 3;

Whence the four roots of the given equation are 1, 2, 3, and 4.

Or, when its second term is taken away, it will be of the form

$$x^4 + bx^2 + cx + d = 0,$$

to which it can always be reduced; and in that case, its solution may be obtained by the following rule:—

RULE 2.—Find the value of z in the cubic equation

$$z^3 - (\frac{1}{12}b^2 + d)z = \frac{1}{108}b^3 + \frac{1}{8}c^2 - \frac{1}{3}bd,$$

and let the root thus determined be denoted by r,

Then find the two values of x in each of the following quadratic equations:—

$$x^2 + [\sqrt{\{2(r - \frac{1}{3}b)\}}]x = -(r + \frac{1}{6}b) + \sqrt{\{(r + \frac{1}{6}b)^2 - d\}}$$

$$x^2 - [\sqrt{\{2(r - \frac{1}{3}b)\}}]x = -(r + \frac{1}{6}b) - \sqrt{\{(r + \frac{1}{6}b)^2 - d\}},$$

and they will be the four roots of the biquadratic equation required.*

---

* The method of solving biquadratic equations was first discovered by Louis Ferrari, a disciple of the celebrated Cardan before mentioned; but the above rule is derived from that given by Descartes, in his *Geometry*, published in 1637, the truth of which may be shown as follows:—

Let the given or proposed equation be

$$x^4 + ax^2 + bx + c = 0,$$

and conceive it to be produced by the multiplication of the two quadratics

$$x^2 + px + q = 0; \text{ and } x^2 + rx + s = 0.$$

Then, since these equations, as well as the given one, are each $= 0$, there will arise, by taking their product,

$$x^4 + (p + r)x^3 + (s + q + pr)x^2 + (ps + qr)x + qs = x^4 + ax^2 + bx + c.$$

And consequently, by equating the homologous terms of this last equation, we shall have the four following equations,

$$p + r = 0; \ s + q + pr = a; \ ps + qr = b; \ qs = c;$$

$$\text{Or, } r = -p; \ s + q = a + p^2; \ s - q = \frac{b}{p}; \ qs = c.$$

Whence, subtracting the square of the third of these from that of the second, and then changing the sides of the equation, we shall have

$$a^2 + 2ap^2 + p^4 - \frac{b^2}{p^2} = 4qs, \text{ or } 4c; \text{ or } p^6 + 2ap^4 + (a^2 - 4c)p^2 = b^2.$$

Where the value of p may be found by the rule before given for cubic equations.

Hence, also, since $s + q = a + p^2$, and $s - q = \frac{b}{p}$, there will arise, by addition and subtraction,

$$s = \frac{1}{2}a + \frac{1}{2}p^2 + \frac{b}{2p}; \ q = \frac{1}{2}a + \frac{1}{2}p^2 - \frac{b}{2p};$$

where p being known, s and q are likewise known.

And, consequently, by extracting the roots of the two assumed quad-

Or the four roots of the given equation, in this last case, will be as follows:—

$$x = -\tfrac{1}{2}\sqrt{\left\{2\left(r - \frac{1}{3}b\right)\right\}} + \sqrt{\left\{-\frac{r}{2} + \frac{b}{3} + \sqrt{\left[\left(r + \frac{1}{6}b\right)^2 - d\right]}\right\}}$$

$$x = -\tfrac{1}{2}\sqrt{\left\{2\left(r - \frac{1}{3}b\right)\right\}} - \sqrt{\left\{-\frac{r}{2} - \frac{b}{3} + \sqrt{\left[\left(r + \frac{1}{6}b\right)^2 - d\right]}\right\}}$$

$$x = +\tfrac{1}{2}\sqrt{\left\{2\left(r - \frac{1}{3}b\right)\right\}} + \sqrt{\left\{-\frac{r}{2} - \frac{b}{3} - \sqrt{\left[\left(r + \frac{1}{6}b\right)^2 - d\right]}\right\}}$$

$$x = +\tfrac{1}{2}\sqrt{\left\{2\left(r - \frac{1}{3}b\right)\right\}} - \sqrt{\left\{-\frac{r}{2} - \frac{b}{3} - \sqrt{\left[\left(r + \frac{1}{6}b\right)^2 - d\right]}\right\}}$$

2. Given $x^4 + 12x - 17 = 0$, to find the four roots of the equation.

Here $a = 0$, $b = 0$, $c = 12$, and $d = -17$;

Whence, by substituting these numbers in the cubic equation,

$$z^3 - \left(\frac{1}{12}b^2 + d\right)z = \frac{1}{108}b^3 + \frac{1}{8}c^2 - \frac{1}{3}bd,$$

we shall have, after simplifying the results,

$$z^3 + 17z = 18,$$

Where it is evident, by inspection, that $z = 1$.

And if this number be substituted for r, 0 for b, and -17 for d in the two quadratic equations in the above rule, their solution will give

$$x = -\tfrac{1}{2}\sqrt{2} + \sqrt{(-\tfrac{1}{2} + \sqrt{18})} = -\tfrac{1}{2}\sqrt{2} + \sqrt{(-\tfrac{1}{2} + 3\sqrt{2})}$$
$$x = -\tfrac{1}{2}\sqrt{2} - \sqrt{(-\tfrac{1}{2} + \sqrt{18})} = -\tfrac{1}{2}\sqrt{2} - \sqrt{(-\tfrac{1}{2} + 3\sqrt{2})}$$
$$x = +\tfrac{1}{2}\sqrt{2} + \sqrt{(-\tfrac{1}{2} - \sqrt{18})} = +\tfrac{1}{2}\sqrt{2} + \sqrt{(-\tfrac{1}{2} - 3\sqrt{2})}$$
$$x = +\tfrac{1}{2}\sqrt{2} - \sqrt{(-\tfrac{1}{2} - \sqrt{18})} = +\tfrac{1}{2}\sqrt{2} - \sqrt{(-\tfrac{1}{2} - 3\sqrt{2})}$$

Which are the four roots of the proposed equation; the first two being real, and the last two imaginary.

RULE* 3.—The roots of any biquadratic equation of the form $x^4 + ax^2 + bx + c = 0$, may also be determined by the

---

ratics, $x^2 + px + q = 0$, and $x^2 + rx + s = 0$, or its equal $x^2 - px + s = 0$ we shall have

$$x = -\tfrac{1}{2}p \pm \sqrt{(\tfrac{1}{4}p^2 - q)};\ x = \tfrac{1}{2}p \pm \sqrt{(\tfrac{1}{4}p^2 - s)};$$

which expression, when taken in + and —, give the four roots of the proposed biquadratic as was required.

* This method, which differs considerably from either of the former, consists in supposing the root of the given equation,

$$x^4 + ax^2 + bx + c = 0\ (1),$$

to be of the following trinomial surd form

$$x = \sqrt{p} + \sqrt{q} + \sqrt{r};$$

where p, q, r, denote the roots of the cubic equation

$$y^3 + fy^2 + gy = h\ (2),$$

of which the coefficients f, g, and the absolute term h, are the unknown quantities that are to be determined.

following general formulæ first given by EULER; which are remarkable for their elegance and simplicity.

Find the three roots of the cubic equation $z^3 + 2az^2 + (a^2 - 4c)z = b^2$, by one of the former rules before given for this purpose; and let them be denoted by r', r'', and r'''

Then, agreeably to the theory of equations before given, we shall have $p+q+r=-f$; $pq+pr+qr=g$; $pqr=h$. And by squaring each side of the formula expressing the value of x,

$$x^2 = p+q+r+2\sqrt{pq}+2\sqrt{pr}+2\sqrt{qr}.$$

Or, by substituting f for its equal $-(p+q+r)$, and bringing the term, so obtained, to the other side of the equation

$$x^2+f=2\sqrt{pq}+2\sqrt{pr}+2\sqrt{qr}.$$

Also, by again squaring each side of this last expression, we shall have $x^4+2fx^2+f^2=4pq+4pr+4qr+8\sqrt{p^2qr}+8\sqrt{q^2pr}+8\sqrt{r^2pq}$.

Or, substituting $4g$ for its equal $4pq+4pr+4qr$, and bringing the term to the other side as before,

$$x^4+2fx^2+f^2-4g=8\sqrt{pqr}(\sqrt{p}+\sqrt{q}+\sqrt{r}).$$

But since, from what has been above laid down, we have

$$\sqrt{p}+\sqrt{q}+\sqrt{r}=x, \text{ and } \sqrt{pqr}=\sqrt{h},$$

if these be put for their equals in the last equation, it will become, by this substitution,

$$x^4+2fx^2-8h^{\frac{1}{2}}x+f^2-4g=0.$$

Whence, comparing these coefficients with those of the given equation, there will arise

$$2f=a;\ -8\sqrt{h}=b;\ f^2-4g=c;\ \text{or,}$$

$$f=\frac{a}{2};\ h=\frac{b^2}{64};\ g=\frac{a^2}{16}-\frac{c}{4}$$

And consequently, by substituting these values in the assumed cubic equation (2), we shall have

$$y^3+\tfrac{1}{2}ay^2+\frac{1}{16}(a^2-4c)y=\frac{b^2}{64}\ (3),$$

the three roots of which last equation, when substituted for p, q, and r, in the formula $x=\sqrt{p}+\sqrt{r}+\sqrt{q}$, will give, by taking each term of the expression both in $+$ and $-$, all the four values of x.

Or, in order to render this result more commodious in practice, by freeing it from fractions, let $y=\frac{1}{4}z$. Then, by substitution and reduction, we shall have the corresponding equation

$$z^3+2az^2+(a^2-4c)z=b^2,\ (4),$$

the three roots of which are each, evidently, four times those of the former. Hence, using this instead of equation (3), and denoting its roots by r', r'', r''', the last mentioned formula, taking each of its terms in $+$ and $-$, as before, will give the values of x, as in the above expressions.

Note.—If we were to take all the possible changes of the signs, in this case, which the terms of the assumed formula admit of, it would appear that x should have eight different values; but it is to be observed, that, according to the first part of the above investigation, the product $\sqrt{p}\times\sqrt{q}\times\sqrt{r}=\sqrt{h}$, or $\frac{1}{8}b$; and consequently, that when b is positive, either all the three radicals must be taken in $+$, or two in $-$, and one in $+$; and when b is negative, they must either be all $-$, or two $+$ and one $-$; which considerations reduce the number of roots to four.

Then, we shall have,

When b is positive,	When b is negative,
$x = \dfrac{-\sqrt{r'} - \sqrt{r''} - \sqrt{r'''}}{2}$	$x = \dfrac{+\sqrt{r'} + \sqrt{r''} + \sqrt{r'''}}{2}$
$x = \dfrac{-\sqrt{r'} + \sqrt{r''} + \sqrt{r'''}}{2}$	$x = \dfrac{+\sqrt{r'} - \sqrt{r''} - \sqrt{r''}}{2}$
$x = \dfrac{+\sqrt{r'} - \sqrt{r''} + \sqrt{r'''}}{2}$	$x = \dfrac{-\sqrt{r'} + \sqrt{r''} - \sqrt{r''}}{2}$
$x = \dfrac{+\sqrt{r'} + \sqrt{r''} - \sqrt{r'''}}{2}$	$x = \dfrac{-\sqrt{r'} - \sqrt{r''} + \sqrt{r'''}}{2}$

Note.—If the three roots r', r'', r''', of the auxiliary cubic equation be all real and positive, the four roots of the proposed equation will also be real; and if one of these roots be positive and the other two imaginary, or both of them negative and equal to each other, two of the roots of the given equation will be real, and two imaginary; which are the only cases that produce real results.

3. Given $x^4 - 25x^2 + 60x - 36 = 0$, to find the four roots of the equation.

Here $a = -25$, $b = 60$, and $c = -36$;

Whence, by substituting these values for their equals in the cubic equation above given, we shall have $z^3 - 2 \times 25 z^2 + (25^2 + 4 \times 36) z = 60^2$, or $z^3 - 50z^2 + 769 z = 3600$: the three roots of which last equation, as found by trial, or by one of the former rules, are 9, 16, and 25, respectively; whence

$x = \frac{1}{2}(-\sqrt{9} - \sqrt{16} - \sqrt{25}) = \frac{1}{2}(-3 - 4 - 5) = -6$
$x = \frac{1}{2}(-\sqrt{9} + \sqrt{16} + \sqrt{25}) = \frac{1}{2}(-3 + 4 + 5) = +3$
$x = \frac{1}{2}(+\sqrt{9} - \sqrt{16} + \sqrt{25}) = \frac{1}{2}(+3 - 4 + 5) = +2$
$x = \frac{1}{2}(+\sqrt{9} + \sqrt{16} - \sqrt{25}) = \frac{1}{2}(+3 + 4 - 5) = +1$

And consequently the four roots of the proposed equation are 1, 2, 3, and -6.

EXAMPLES FOR PRACTICE.

1. Given $x^4 - 55x^2 - 30x + 504 = 0$, to find the four roots, or values of x. Ans. 3, 7, -4, and -6.

2. Given $x^4 + 2x^3 - 7x^2 - 8x = -12$, to find the four roots, or values of x. Ans. 1, 2, -3, and -2.

3. Given $x^4 - 8x^3 + 14x^2 + 4x = 8$, to find the four roots, or values of x. Ans. $\begin{cases} 3 + \sqrt{5},\ 3 - \sqrt{5} \\ 1 + \sqrt{3},\ 1 - \sqrt{3}. \end{cases}$

4. Given $x^4 - 17x^2 - 20x - 6 = 0$, to find the four roots, or values of x. $\begin{cases} 2 + \sqrt{7},\quad 2 - \sqrt{7} \\ -2 + \sqrt{2},\ -2 - \sqrt{2}. \end{cases}$

5. Given $x^4 - 3x^2 - 4x = 3$, to find the four roots, or values of x. Ans. $\begin{cases} \frac{1}{2} + \frac{1}{2}\sqrt{\ 13}, & \frac{1}{2} - \frac{1}{2}\sqrt{\ 13} \\ -\frac{1}{2} + \frac{1}{2}\sqrt{-3}, & -\frac{1}{2} - \frac{1}{2}\sqrt{-\ 3}. \end{cases}$

6. Given $x^4 - 19x^3 + 132x^2 - 302x + 200 = 0$, to find the four roots, or values of x. Ans. $\begin{cases} 4.27768, & .80955 \\ +6.956377, & \pm\sqrt{(-9.3686)}. \end{cases}$

7. Given $x^4 - 27x^3 + 162x^2 + 356x - 1200 = 0$, to find the four roots, or values of x. Ans. $\begin{cases} 2.05608, & -3.00000 \\ 13.15306, & 14.79086 \end{cases}$

8. Given $x^4 - 12x^2 + 12x - 3 = 0$, to find the four roots, or values of x. Ans. $\begin{cases} .606018, & -3.907378 \\ 2.858083, & .443277. \end{cases}$

9. Given $x^4 - 36x^2 + 72x - 36 = 0$, to find the four roots, or values of x. Ans. $\begin{cases} 0.8729836, & 1.2679494 \\ 4.7320506, & -6.8729836. \end{cases}$

10. Given $x^4 - 12x^3 + 47x^2 - 72x + 36 = 0$, to find the roots, or values of x. Ans. 1, 2, 3, and 6.

11. Given $x^4 + 24x^3 - 114x^2 - 24x + 1 = 0$, to find the roots, or values of x. Ans. $\begin{cases} +\sqrt{197} - 14, & 2 + \sqrt{5} \\ -\sqrt{197} - 14, & 2 - \sqrt{5}. \end{cases}$

12. Given $x^4 - 6x^3 - 58x^2 - 114x - 11 = 0$, to find the roots, or values of x.

Ans. $\pm\frac{5}{2}\sqrt{3} + \frac{3}{2} \pm \sqrt{(17 \pm \frac{21}{2}\ \sqrt{3})}$.

OF THE RESOLUTION OF EQUATIONS BY APPROXIMATION.

Equations of the fifth power, and those of higher dimensions, cannot be resolved by any rule or algebraic formula that has yet been discovered; except in some particular cases where certain relations subsist between the coefficients of their several terms, or when the roots are rational; and, for that reason, can be easily found by means of a few trials.

In these cases, therefore, recourse must be had to some of the usual methods of approximation; among which, that commonly employed is the following, which is universally applicable to all kinds of numeral equations, whatever may be the number of their dimensions, and, though not strictly accurate, will give the value of the root sought to any required degree of exactness.

Rule.—Find by trials, a number nearly equal to the root sought, which call r; and let z be made to denote the difference between this assumed root, and the true root x.

Then, instead of x, in the given equation, substitute its

equal $r \pm z$, and there will arise a new equation involving only z and known quantities.

Reject all the terms of this equation in which z is of two or more dimensions; and the approximate value of z may then be determined by means of a simple equation.

And if the value, thus found, be added to, or subtracted from, that of r, according as r was assumed too little or too great, it will give a near value of the root required.

But as this approximation will seldom be sufficiently exact, the operation must be repeated, by substituting the number thus found for r in the abridged equation exhibiting the value of z; when a second correction of z will be obtained, which, being added to, or subtracted from, r, will give a nearer value of the root than the former.

And by again substituting this last number for r, in the abovementioned equation, and repeating the same process as often as may be thought necessary, a value of x may be found to any degree of accuracy required.

Note.—The decimal part of the root, as found both by this and the next rule, will, in general, about double itself at each operation; and therefore it would be useless, as well as troublesome, to use a much greater number of figures than those in the several substitutions for the values of r.*

EXAMPLES.

1. Given $x^3 + x^2 + x = 90$, to find the value of x by approximation.

Here the root, as found by a few trials, is nearly equal to 4.

Let, therefore, $4 = r$, and $r + z = x$.

$$\text{Then} \left| \begin{array}{l} x^3 = r^3 + 3r^2z + 3rz^2 + z^3 \\ x^2 = r^2 + 2rz + z^2 \\ x = r + z \end{array} \right| = 90.$$

And by rejecting the terms z^3, $3rz^2$, and z^2, as small in comparison with z, we shall have

$$r^3 + r^2 + r + 3r^2z + 2rz + z, = 90;$$

* It may here be observed, that if any of the roots of an equation be whole numbers, they may be determined by substituting 1, 2, 3, 4, &c., successively, both in *plus* and in *minus*, for the unknown quantity, till a result is obtained equal to that in the question; when those that are found to succeed, will be the roots required.

Or, since the last term of any equation is always equal to the continued product of all its roots, the number of these trials may be generally diminished, by finding all the divisors of that term, and then substituting them both in *plus* and *minus*, as before, for the unknown quantity, when those that give the proper result will be the rational roots sought; but if none of them are found to succeed, it may be concluded that the equation cannot be resolved by this method; the roots, in that case, being either irrational or imaginary.

Whence $z = \frac{90 - r^3 - r^2 - r}{3r^2 + 2r + 1} = \frac{90 - 64 - 16 - 4}{48 + 8 + 1} = \frac{6}{57} = .10.$

And consequently, $x = 4.1$, *nearly*.

Again, if 4.1 be substituted in the place of r, in the last equation, we shall have

$$z = \frac{90 - r^3 - r^2 - r}{3r^2 + 2r + 1} = \frac{90 - 68.921 - 16.81 - 4.1}{50.43 + 8.2 + 1} = .00283;$$

And consequently, $x = 4.1 + .00283 = 4.10283$ for a *second approximation.*

And if the first four figures, 4,102, of this number be again substituted for r, in the same equation, a still nearer value of the root will be obtained; and so on, as far as may be thought necessary.

2. Given $x^2 + 20x = 100$, to find the value of x by approximation. Ans. $x = 4.1421356$.

3. Given $x^3 + 9x^2 + 4x = 80$, to find the value of x by approximation. Ans. $x = 2.4721359$.

4. Given $x^4 - 38x^3 + 210x^2 + 538x + 289 = 0$, to find the value of x by approximation. Ans. $x = 30.53565375$.

5. Given $x^5 + 6x^4 - 10^3 - 112x^2 - 207x + 110 = 0$, to find the value of x by approximation. Ans. 4.46410161.

The roots of equations, of all orders, can also be determined, to any degree of exactness, by means of the following easy rule of double position; which, though it has not been generally employed for this purpose, will be found in some respects superior to the former, as it can be applied, at once to any unreduced equation consisting of surds, or compound quantities, as readily as if it had been brought to its usual form

RULE 2.—Find, by trial, two numbers as near the true root as possible, and substitute them in the given equation instead of the unknown quantity, noting the results that are obtained from each.

Then, as the difference of these results is to the difference of the two assumed numbers, so is the difference between the true result, given by the question, and either of the former, to the correction of the number belonging to the result used; which correction being added to that number when it is too little, or subtracted from it when it is too great, will give the root required *nearly*.

And if the number thus determined, and the nearest of the two former, or any other that appears to be more accurate, be now taken as the assumed roots, and the operation be repeated as before, a new value of the unknown quantity will be obtained still more correct than the first; and so on, proceeding in this manner as far as may be judged necessary.*

EXAMPLES.

1. Given $x^3 + x^2 + x = 100$, to find an approximate value of

Here it is soon found by a few trials, that the value of x lies between 4 and 5.

Hence, by taking these as the two assumed numbers, the operation will stand as follows:—

First Sup.		Second Sup.
4	x	5
16	x^2	25
64	x^3	125
84	Results	155

Therefore

155	. . 5 . .	100		
84	. . 4 . .	84		
71	: 1 ::	16	:	.225

And consequently $x = 4 + .225 = 4.225$, *nearly*.

---

* The above rule for Double Position, which is much more simple and commodious than the one commonly employed for this purpose, is the same as that which was first given at p. 311 of the octavo edition of my Arithmetic, published in 1810.

To this we may further add, that when one of the roots of an equation has been found, either by this method or the former, the rest may be determined as follows:—

Bring all the terms to the lefthand side of the equation, and divide the whole expression, so formed, by the difference between the unknown quantity (x) and the root first found; and the resulting equation will then be depressed by a degree lower than the given one.

Find a root of this equation, by approximation, as in the first instance, and the number so obtained will be a second root of the original equation.

Then by means of this root, and the unknown quantity, depress the second equation a degree lower, and thence find a third root; and so on, till the equation is reduced to a quadratic; when the two roots of this, together with the former, will be the roots of the equation required.

Thus in the equation $x^3 - 15x^2 + 63x = 50$, the first root is found by approximation to be 1.02804. Hence,

$$x - 1.02804)\, x^3 - 15x^2 + 63x - 50\, (x^2 - 13.97195x + 48.63627 = 0.$$

And the two roots of the quadratic equation, $x^2 - 13{,}97195x = -48.63627$, found in the usual way, are 6.57653 and 7.39543.

So that the three roots of the given cubic equation $x^3 - 15x^2 + 63x = 50$, are 1.02804, 6.57653, and 7.39543, their sum being $= 15$, the coefficient of the second term of the equation, as it ought to be when they are right.

Again, if 4.2 and 4.3 be taken as the two assumed numbers the operation will stand thus :—

First Sup.		*Second Sup.*
4.2	. . x . .	4.3
17.64	. . x^2 . .	18.49
74.088	. . x^3 . .	79.507
95.928	Results	102.297

Therefore

102.297	. .	4.3	. .	102.297		
95.928	. .	4.2	. .	100		
6.369	:	.1	: :	2.297	:	.036.

And consequently, $x = 4.3 - .036 = 4.264$, *nearly*.

Again, let 4.264 and 4.265 be the two assumed numbers; then

First Sup.		*Second Sup.*
4.264	x . .	4.265
18.181696 .	. x^2 .	. 18.190225
77.526752 .	. x^3 .	. 77.581310
99.972448	Results	100.036535

Therefore,

100.036535		4.265		100	
99.972448		4.264		99.972448	
.064087	:	.001	: :	.027552	: 0004299

And consequently,

$x = 4.264 + .0004299 = 4.2644299$, *very nearly*.

2. Given $(\frac{1}{5}x^2 - 15)^2 + x\sqrt{x} = 90$, to find an approximate value of x.

Here, by a few trials, it will be soon found, that the value of x lies between 10 and 11; which let, therefore, be the two assumed numbers, agreeably to the directions given in the rule.

Then,

First Sup.		*Second Sup.*
25 .	. $(\frac{1}{5}x^2-15)^2$ . .	84.64
31.622 .	$x\sqrt{x}$ . .	36.482
56.622	Results	121.122

Hence

121.122	. .	11	. .	121.122		
56.622	. .	10	. .	90		
64.5	:	1	: :	31.122	:	.482.

And consequently, $x = 11 - .482 = 10.518$.

Again, let 10.5 and 10.6 be the two assumed numbers

Then,

First Sup.		*Second Sup.*
49.7025 . .	$(\frac{1}{5}x^2 - 15)^2$	. . 55,830784
34.0239 . .	$x\sqrt{x}$	. . 34.511099
83.7264	Results	90.341883

Hence,

90.341883 . .	10.6 . .	90.341883
83.7264 . .	10.5 . .	90
6.615483 . .	.1 : :	.341883 : .0051679

And consequently,

$x = 10.6 - .0051679 = 10.5948321$, *very nearly.*

EXAMPLES FOR PRACTICE.

1. Given $x^3 + 10x^2 + 5x = 2600$, to find a near approximate value of x. Ans. $x = 11.00673$.

2. Given $2x^4 - 16x^3 + 40x^2 - 30x + 1 = 0$, to find a near value of x. Ans. $x = 1.284724$.

3. Given $x^5 + 2x^4 + 3x^3 + 4x^2 + 5x = 54321$, to find the value of x. Ans. 8.414455.

4. Given $\sqrt[3]{(7x^3 + 4x^2)} + \sqrt{(20x^2 - 10x)} = 28$, to find the value of x. Ans. 4.510661.

5. Given $\sqrt{[144x^2 - (x^2 + 20)^2]} + \sqrt{[196x^2 - (x^2 + 24)^2]} = 114$, to find the value of x. Ans. 7.123883.

OF EXPONENTIAL EQUATIONS.

An exponential quantity is that which is to be raised to some unknown power, or which has a variable quantity for its index; as,

$$a^x,\ a^{\frac{1}{x}},\ x^x, \text{ or } x^{\frac{1}{x}}, \ \&c.$$

And an exponential equation is that which is formed between any expression of this kind and some other quantity whose value is known; as,

$$a^x = b,\ x^x = a,\ \&c.$$

Where it is to be observed, that the first of these equations, when converted into logarithms, is the same as

$x \log. a = b$, or $x = \dfrac{\log. b}{\log. a}$; and the second equation $x^x = a$ is the same as $x \log. x = \log. a$.

In the latter of which cases, the value of the unknown

quantity x may be determined, to any degree of exactness, by the method of double position, as follows :—

RULE.—Find, by trial, as in the rule before laid down, two numbers as near the number sought as possible, and substitute them in the given equation

$$x \log. x = \log. a,$$

instead of the unknown quantity, noting the results obtained from each.

Then, as the difference of these results is to the difference of the two assumed numbers, so is the difference between the true result, given in the question, and either of the former, to the correction of the number belonging to the result used; which correction, being added to that number when it is too little, or subtracted from it when it is too great, will give the root required, *nearly*.

And if the number thus determined, and the nearest of the two former, or any other that appears to be nearer, be taken as the assumed roots, and the operation be repeated as before, a new value of the unknown quantity will be obtained still more correct than the first; and so on, proceeding in this manner as far as may be thought necessary.

EXAMPLES.

1. Given $x^x = 100$ to find an approximate value of x.

Here, by the above formula, we have

$$x \log. x = \log. 100 = 2.$$

And since x is readily found, by a few trials, to be nearly in the middle between 3 and 4, but rather nearer the latter than the former, let 3.5 and 3.6 be taken for the two assumed numbers.

Then log. 3.5 = .5440680, which, being multiplied by 3.5, gives 1.904238 = first result;

And log. 3.6 = .5563025, which, being multiplied by 3.6, gives 2.002689 for the second result.

Whence,

2.002689 . . 3.6 . . 2.002689
1.904238 . . 3.5 . . 2.

.098451 : .1 : : .002689 : .00273

for the first correction; which, taken from 3.6, leaves $x = 3.59727$, *nearly*.

And as this value is found, by trial, to be rather too small, let 3.59727 and 3.59728 be taken as the two assumed numbers

Then log. 3.59728 = 0.555974243134677 to 15 places.

The log. 3.59727 = 0.555973035847267 to 15 places,

which logarithms, multiplied by their respective numbers, give the following products :—

1.999995025343512 } both true to the last figure.
1.999985122662298 }

Therefore, the errors are 4974656488
and 14877337702
and the difference of errors 9902681214

Now, since only 6 additional figures are to be obtained, we may omit the three last figures in these errors; and state thus: as difference of errors 9902681 : difference of sup. 1 : : error 4974656 : the correction 502354, which, united to 3,59728, gives us the true value of $x = 3.597285\text{0}2354$.*

2. Given $x^x = 2000$, to find an approximate value of x.
Ans. $x = 4.82782263$.

3. Given $(6x)^x = 96$, to find the approximate value of x.
Ans. $x = 1.8826432$.

4. Given $x^x = 123456789$, to find the value of x.
Ans. 8.6400268.

5. Given $x^x - x = (2x - x^x)^{\frac{1}{x}}$, to find the value of x.
Ans. $x = 1.747933$.

OF THE BINOMIAL THEOREM.

The binomial theorem is a general algebraical expression, or formula, by which any power, or root of a given quantity, consisting of two terms, is expanded into a series; the form of which, as it was first proposed by Newton, being as follows :—

$$(P + PQ)^{\frac{m}{n}} = P^{\frac{m}{n}}[1 + \frac{m}{n}Q + \frac{m}{n}\left(\frac{m-n}{2n}\right)Q^2 + \frac{m}{n}\left(\frac{m-n}{2n}\right)\left(\frac{m-2n}{3n}\right)Q^3, \&c.]$$

Or,

$$(P + PQ)^{\frac{m}{n}} = P^{\frac{m}{n}} + \frac{m}{n}AQ + \frac{m-n}{2n}BQ + \frac{m-2n}{3n}CQ + \frac{m-3n}{4n}DQ, \&c.$$

When P is the first term of the binomial, Q the second

---

* The correct answer to this question has been first given by *Doctor Adrain*, in his edition of Hutton's Mathematics, who plainly proves that Hutton's answer, which is the same as Bonnycastle's, is incorrect. *See Hutton's Mathematics*, Vol. I. p. 263, *N. Y. Edition*.—ED.

term divided by the first, $\frac{m}{n}$ the index of the power, or root, and A, B, C, &c., the terms immediately preceding those in which they are first found, including their signs + or —.

Which theorem may be readily applied to any particular case, by substituting the numbers, or letters, in the given example, for P, Q, m, and n, in either of the above formulæ, and then finding the result according to the rule.*

EXAMPLES.

1. It is required to convert $(a^2+x)^{\frac{1}{2}}$ into an infinite series.

Here $P=a^2$, $Q=\frac{x}{a^2}$, $\frac{m}{n}=\frac{1}{2}$, or $m=1$, and $n=2$;

whence

$$P^{\frac{m}{n}}=(a^2)^{\frac{m}{n}}=(a^2)^{\frac{1}{2}}=a=A,$$

$$\frac{m}{n}AQ=\frac{1}{2}\times\frac{a}{1}\times\frac{x}{a^2}=\frac{x}{2a}=B,$$

---

* This celebrated theorem, which is of the most extensive use in algebra and various other branches of analysis, may be otherwise expressed as follows:—

$$(a+x)^{\frac{m}{n}}=a^{\frac{m}{n}}[1+\frac{m}{n}\left(\frac{x}{a}\right)+\frac{m}{n}\cdot\frac{m-n}{2n}\left(\frac{x}{a}\right)^2+\frac{m}{n}\cdot\frac{m-n}{2n}\cdot\frac{m-2n}{3n}\left(\frac{x}{a}\right)^3,$$
&c.]

$$\text{Or, } (a+x)^{\frac{m}{n}}=$$

$$a^{\frac{m}{n}}[1+\frac{m}{n}\left(\frac{x}{a+x}\right)+\frac{m}{n}\cdot\frac{m+n}{2n}\left(\frac{x}{a+x}\right)^2+\frac{m}{n}\cdot\frac{m+n}{2n}\cdot\frac{m+2n}{3n}\left(\frac{x}{a+x}\right)^3,$$
&c.]

$$\text{Or, } (a+x)^{\frac{m}{n}}=$$

$$2a^{\frac{m}{n}}[1+\frac{m}{n}\left(\frac{a-x}{a+x}\right)+\frac{m}{n}\cdot\frac{m+n}{2n}\left(\frac{a-x}{a+x}\right)^2-\frac{m}{n}\cdot\frac{m+n}{2n}\cdot\frac{m+2n}{3n}\left(\frac{a+x}{a+x}\right)^3],$$
&c.

It may here also be observed, that if m be made to represent any whole, or fractional number, whether positive or negative, the first of these expressions may be exhibited in a more simple form,

$$(a+x)^m=a^m+ma^{m-1}x+\frac{m(m-1)}{1.2}a^{m-2}x^2+\frac{m(m-1)(m-2)}{1.2.3}a^{m-3}x^3$$

$$\cdots\cdots\cdots\frac{m(m-1)(m-2)\ldots\ldots[m-(n-1)]\,a^{m-n}x^m}{1.2.3.4\quad\ldots\ldots\quad n}$$

Where the last term is called the *general term of the series*, because if 1, 2, 3, 4, &c., be substituted successively for n, it will give all the rest.

14*

$$\frac{m-n}{2n}\text{BQ} = \frac{1-2}{4} \times \frac{x}{2a} \times \frac{x}{a^2} = -\frac{x^2}{2.4a^3} = \text{C},$$

$$\frac{m-2n}{3n}\text{CQ} = \frac{1-4}{6} \times -\frac{x^2}{2.4a^3} \times \frac{x}{a^2} = \frac{3x^3}{2.4.6a^5} = \text{D},$$

$$\frac{m-3n}{4n}\text{DQ} = \frac{1-6}{8} + \frac{3x^3}{2.4.6a^5} \times \frac{x}{a^2} = -\frac{3.5x^4}{2.4.6.8a^7} = \text{E},$$

$$\frac{m-4n}{5n}\text{EQ} = \frac{1-8}{10} \times -\frac{3.5x^4}{2.4.6.8a^7} \times \frac{x}{a^2} = \frac{3.5.7x^5}{2.4.6.8.10a^9} = \text{F},$$

&c. &c. &c.

Therefore $(a^2+x)^{\frac{1}{2}} =$

$$x + \frac{x}{2a} - \frac{x^2}{2.4a^3} + \frac{3x^3}{2.4.6a^5} - \frac{3.5x^4}{2.4.6.8a^7} + \frac{3.5.7x^5}{2.4.6.8.10a^9} -, \text{ \&c.}$$

Where the law of formation of the several terms of the series is sufficiently evident.

2. It is required to convert $\frac{1}{(a+b)^2}$, or its equal $(a+b)^{-2}$, into an infinite series.

Here $\text{P} = a$, $\text{Q} = \frac{b}{a}$, and $\frac{m}{n} = -2$, or $m = -2$, and $n = 1$; whence

$$\text{P}^{\frac{m}{n}} = (a)^{\frac{m}{n}} = a^{-2} = \frac{1}{a^2} = \text{A},$$

$$\frac{m}{n}\text{AQ} = -\frac{2}{1} \times \frac{1}{a^2} \times \frac{b}{a} = -\frac{2b}{a^3} = \text{B},$$

$$\frac{m-n}{2n}\text{BQ} = \frac{-2-1}{2} \times -\frac{2b}{a^3} \times \frac{b}{a} = \frac{3b^2}{a^4} = \text{C},$$

$$\frac{m-2n}{3n}\text{CQ} = \frac{-2-2}{3} \times \frac{3b^2}{a^4} \times \frac{b}{a} = -\frac{4b^3}{a^5} = \text{D},$$

$$\frac{m-3n}{4n}\text{DQ} = \frac{-2-3}{4} \times -\frac{4b^3}{a^5} \times \frac{b}{a} = \frac{5b^4}{a^6} = \text{E},$$

&c. &c. &c.

Consequently, $\frac{1}{(a+b)^2} = \frac{1}{a^2} - \frac{2b}{a^3} + \frac{3b^2}{a^4} - \frac{4b^3}{a^5} + \frac{5b^4}{a^6}$, &c.

3. It is required to convert $\frac{a^2}{(a^2-x)^{\frac{1}{2}}}$, or its equal $a^2 (a^2-x)^{-\frac{1}{2}}$, into an infinite series.

Here

$\text{P} = a^2$, $\text{Q} = -\frac{x}{a^2}$, and $\frac{m}{n} = \frac{-1}{2}$, or $m = -1$, and $n = 2$;

whence

$$\mathrm{P}^{\frac{m}{n}} = (a^2)^{\frac{m}{n}} = (a^2)^{-\frac{1}{2}} = \frac{1}{a} = \mathrm{A},$$

$$\frac{m}{n}\,\mathrm{AQ} = -\frac{1}{2} \times \frac{1}{a} \times -\frac{x}{a^2} = \frac{x}{2a^3} = \mathrm{B},$$

$$\frac{m-n}{2n}\,\mathrm{BQ} = \frac{-1-2}{4} \times \frac{x}{2a^3} \times -\frac{x}{a^2} = \frac{3x^2}{2.4a^5} = \mathrm{C},$$

$$\frac{m-2n}{3n}\,\mathrm{CQ} = \frac{-1-4}{6} \times \frac{3x^2}{2.4a^5} \times = \frac{x}{a^2} = \frac{3.5x^3}{2.4.6a^7} = \mathrm{D},$$

$$\frac{m-3n}{4n}\,\mathrm{DQ} = \frac{-1-6}{8} \times \frac{3.5x^3}{2.4.6a^7} \times -\frac{x}{a^2} = \frac{3.5.7x^4}{2.4.6.8a^9} = \mathrm{E},$$

&c. &c. &c.

Therefore,

$$\frac{1}{(a^2-x)^{\frac{1}{2}}} = \frac{1}{a} + \frac{1}{2}\left(\frac{x}{a^3}\right) + \frac{3}{2.4}\left(\frac{x^2}{a^5}\right) + \frac{3.5}{2.4.6}\left(\frac{x^3}{a^7}\right) + \frac{3.5.7}{2.4.6.8}\left(\frac{x^4}{a^9}\right) +, \text{ \&c.}$$

And

$$\frac{a^2}{(a^2-x)^{\frac{1}{2}}} = a + \frac{1}{2}\left(\frac{x}{a}\right) + \frac{2}{2.4}\left(\frac{x^2}{a^3}\right) + \frac{3.5}{2.4.6}\left(\frac{x^3}{a^5}\right) + \frac{3.5.7}{2.4.6.8}\left(\frac{x^4}{a^7}\right) +, \text{ \&c.}$$

4. It is required to convert $\sqrt[3]{9}$, or its equal $(8+1)^{\frac{1}{3}}$, into an infinite series.

Here $\mathrm{P} = 8$, $\mathrm{Q} = \frac{1}{8}$, and $\frac{m}{n} = \frac{1}{3}$, or $m = 1$ and $n = 3$;

Whence,

$$\mathrm{P}^{\frac{m}{n}} = (8)^{\frac{m}{3n}} = 8^{\frac{1}{2}} = 2 = \mathrm{A},$$

$$\frac{m}{n}\,\mathrm{AQ} = \frac{1}{3} \times \frac{2}{1} \times \frac{1}{2^3} = \frac{1}{3.2^2} = \mathrm{B},$$

$$\frac{m-n}{2n}\,\mathrm{BQ} = \frac{1-3}{6} \times \frac{1}{3.2^2} \times \frac{1}{2^3} = -\frac{1}{3.6.2^4} = \mathrm{C},$$

$$\frac{m-2n}{3n}\,\mathrm{CQ} = \frac{1-6}{9} \times -\frac{1}{3.6.2^4} \times \frac{1}{2^3} = \frac{5}{3.6.9.2^7} = \mathrm{D},$$

$$\frac{m-3n}{4n}\,\mathrm{DQ} = \frac{1-9}{12} \times \frac{5}{3.6.9.2^7} \times \frac{1}{2^3} = -\frac{5.8}{3.6.9.12.2^{10}} = \mathrm{E},$$

$$\frac{m-4n}{5n}\,\mathrm{EQ} = \frac{1-12}{15} \times -\frac{5.8}{3.6.9.12.2^{10}} \times \frac{1}{2^3} = \frac{5.8.11}{3.6.9.12.15.2^{13}}$$

&c. &c. &c.

Therefore, $\sqrt[3]{9} =$

$$2 + \frac{1}{3.2^2} - \frac{1}{3.6.2^4} + \frac{5}{3.6.9.2^7} - \frac{5.8}{3.6.9.12.2^{10}} + \frac{5.8.11}{3.6.9.12.15.2^{13}},$$

&c.

5. It is required to convert $\sqrt{2}$, or its equal $\sqrt{(1+1)}$, into an infinite series.

$$\text{Ans. } 1 + \frac{1}{2} - \frac{1}{2.4} + \frac{3}{2.4.6} - \frac{3.5}{2.4.6.8} + \frac{3.5.7}{2.4.6.8.10}, \text{ \&c.}$$

6. It is required to convert $\sqrt[5]{7}$, or its equal $(8-1)^{\frac{1}{3}}$, into an infinite series.

$$\text{Ans. } 2 - \frac{1}{3.2^2} - \frac{1}{3.6.2^4} - \frac{5}{3.6.9.2^7} - \frac{5.8}{3.6.9.12.2^{10}} -, \text{ \&c.}$$

7. It is required to convert $\sqrt[5]{240}$, or its equal $(243-3)^{\frac{1}{5}}$, into an infinite series.

$$\text{Ans. } 3 - \frac{1}{5.3^3} - \frac{4}{5.10.3^7} - \frac{4.9}{5.10.15.3^{11}} - \frac{4.9.14}{5.10.15.20.3^{15}}, \text{ \&c.}$$

8. It is required to convert $(a \pm x)^{\frac{1}{2}}$ into an infinite series.

$$\text{Ans. } a^{\frac{1}{2}}\left\{1 \pm \frac{x}{2a} - \frac{x^2}{2.4a^2} \pm \frac{3x^3}{2.4.6a^3} - \frac{3.5x^4}{2.4.6.8a^4} \pm \text{ \&c.}\right\}$$

9. It is required to convert $(a \pm b)^{\frac{1}{3}}$ into an infinite series.

$$\text{Ans. } a^{\frac{1}{3}}\left\{1 \pm \frac{b}{3a} - \frac{2b^2}{3.6a^2} \pm \frac{2.5b^3}{3.6.9a^3} - \frac{2.5.8b^4}{3.6.9.12a^4} \pm, \text{ \&c.}\right\}$$

10. It is required to convert $(a-b)^{\frac{1}{4}}$ into an infinite series.

$$\text{Ans. } a^{\frac{1}{4}}\left\{1 - \frac{b}{4a} - \frac{3b^2}{4.8a^2} - \frac{3.7b^3}{4.8.12a^3} - \frac{3.7.11b^4}{4.8.12.16a^4} -, \text{ \&c.}\right\}$$

11. It is required to convert $(a+x)^{\frac{2}{3}}$ into an infinite series.

$$\text{Ans. } a^{\frac{2}{3}}\left\{1 + \frac{x}{3a} - \frac{x^2}{9a^2} + \frac{4x^3}{9^2a^3} - \frac{4.7x^4}{9^2.12a^4} + \frac{4.7.10x^5}{9^2.12.15a^5} -, \text{\&c.}\right\}$$

12. It is required to convert $(1-x)^{\frac{2}{5}}$ into an infinite series.

$$\text{Ans. } 1 - \frac{2x}{5} - \frac{2.3x^2}{5.10} - \frac{2.3.8x^3}{5.10.15} - \frac{2.3.8.13x^4}{5.10.15.20} -, \text{ \&c.}$$

13. It is required to convert $\frac{1}{(a \pm x)^{\frac{1}{2}}}$, or its equal $(a \pm x)^{-\frac{1}{2}}$ into an infinite series.

$$\text{Ans. } \frac{1}{a^{\frac{1}{2}}}\left\{1 \mp \frac{x}{2a} + \frac{3x^2}{2.4a^2} \mp \frac{3.5x^3}{2.4.6a^3} + \frac{3.5.7x^4}{2.4.6.8a^4} \mp, \text{ \&c.}\right\}$$

14. It is required to convert $\frac{a}{(a \pm x)^{\frac{1}{3}}}$, or its equal $(a \pm x)^{-\frac{1}{3}}$, into an infinite series.

Ans. $a^{\frac{2}{3}}\left\{1 \mp \frac{x}{3a} + \frac{4x^2}{3.6a^2} \mp \frac{4.7x^3}{3.6.9a^3} + \frac{4.7.10x^4}{3.6.9.12a^4} \mp, \text{\&c.}\right\}$

15. It is required to convert $\frac{1}{(1 + x)^{\frac{1}{5}}}$, or its equal $(1 + x)^{-\frac{1}{5}}$, into an infinite series.

Ans. $1 - \frac{x}{5} + \frac{6x^2}{5.10} - \frac{6.11x^3}{5.10.15} + \frac{6.11.16x^4}{5.10.15.20} -, \text{\&c.}$

16. It is required to convert $\left(\frac{a + x}{a - x}\right)^{\frac{1}{2}}$, or its equal $(a + x)(a^2 - x^2)^{-\frac{1}{2}}$, into an infinite series.

Ans. $1 + \frac{x}{a} + \frac{x^2}{2a^2} + \frac{x^3}{2a^3} + \frac{3x^4}{8a^4} + \frac{3x^5}{8a^5} + \frac{5x^6}{16a^6} + \frac{5x^7}{16a^7}$, &c.

OF THE INDETERMINATE ANALYSIS.

In the common rules of Algebra, such questions are usually proposed as require some certain or definite answer; in which case it is necessary that there should be as many independent equations, expressing their conditions, as there are unknown quantities to be determined; or otherwise the problem would not be limited.

But in other branches of the science, questions frequently arise that involve a greater number of unknown quantities than there are equations to express them; in which instances they are called indeterminate or unlimited problems, being such as usually admit of an indefinite number of solutions; although, when the question is proposed in integers, and the answers are required only in whole positive numbers, they are, in some cases, confined within certain limits, and in others the problem may become impossible.

Problem 1.—To find the integral values of the unknown quantities x and y in the equation

$$ax - by = \pm c, \text{ or } ax + by = c.$$

Where a and b are supposed to be given whole numbers, which admit of no common divisor, except when it is also a divisor of c.

RULE 1.—Let wh denote a whole, or integral number; and reduce the equation to the form

$$x = \frac{by \pm c}{a} wh, \text{ or } x = \frac{c - by}{a} wh.$$

2. Throw all whole numbers out of that of these two expressions, to which the question belongs, so that the numbers d and e, in the remaining parts, may be each less than a; then

$$\frac{dy \pm e}{a} = wh, \text{ or } \frac{e - dy}{a} = wh.$$

3. Take such a multiple of one of these last formulæ, corresponding with that abovementioned, as will make the coefficient of y nearly equal to a, and throw the whole numbers out of it as before.

Or, find the sum or difference of $\frac{ay}{a}$, and the expression above used, or any multiple of it that comes near $\frac{ay}{a}$, and the result, in either of these cases, will still be $= wh$, a whole number.

4. Proceed in the same manner with this last result; and so on, till the coefficient of y becomes $= 1$, and the remainder $=$ some number r; then

$$\frac{y \pm r}{a} = wh, = p, \text{ and } y = ap \mp r.$$

Where p may be 0, or any integral number whatever that makes y positive; and, as the value of y is now known, that of x may be found from the given equation, when the equation is possible.*

NOTE.—Any indeterminate equation of the form

$$ax - by = \pm c,$$

in which a and b are prime to each other, is always possible, and will admit of an infinite number of answers in whole numbers.

But if the proposed equation be of the form

$$ax + by = c,$$

the number of answers will always be limited; and, in some cases, the question is impossible; both of which circum-

* This rule is founded on the obvious principle, that the sum, difference, or product of any two whole numbers, is a whole number; and that if a number divides the whole of any other number and a part of it, it will also divide the remaining part.

stances may be readily discovered from the mode of solution above given.*

EXAMPLES.

1. Given $19x - 14y = 11$, to find x and y in whole numbers.

Here $x = \frac{14y + 11}{19} = wh.$, and also $\frac{19y}{19} = wh.$

Whence, by subtraction, $\frac{19y}{19} - \frac{14y + 11}{19} = \frac{5y - 11}{19} = wh.$

Also, $\frac{5y - 11}{19} \times 4 = \frac{20y - 44}{19} = y - 2 + \frac{y - 6}{19} = wh.$

And, by rejecting $y - 2$, which is a whole number,

$$\frac{y - 6}{19} = wh. = p.$$

Whence we have $y = 19p + 6$.

And $x = \frac{14y + 11}{19} = \frac{14(19p + 6) + 11}{19} = \frac{266p + 95}{19} = 14p + 5$.

Whence, if p be taken $= 0$, we shall have $x = 5$ and $y = 6$, for their least values; the number of solutions being obviously indefinite.

2. Given $2x + 3y = 25$, to determine x and y in whole positive numbers.

$$\text{Here } x = \frac{25 - 3y}{2} = 12 - y + \frac{1 - y}{2}.$$

Hence, since x must be a whole number, it follows that $\frac{1 - y}{2}$ must also be a whole number.

---

* That the coefficients a and b, when these two formulæ are possible, should have no common divisor, which is not at the same time a divisor of c, is evident; for if $a = md$, and $b = me$, we shall have $ax \pm by = mdx \pm mey = c$; and consequently, $dx + ey = \frac{c}{m}$. But d, e, x, y, being supposed to be whole numbers, $\frac{c}{m}$ must also be a whole number, which it cannot be except when m is a divisor of c.

Hence, if it were required to pay 100*l.* in guineas and moidores only, the question would be impossible; since, in the equation $21x + 27y = 2000$, which represents the conditions of the problem, the coefficients, 21 and 27, are each divisible by 3, whilst the absolute term 2000 is not divisible by it. See my *Treatise on Algebra*, for the method of resolving questions of this kind by means of *Continued Fractions*.

Let, therefore, $\frac{1-y}{2} = wh. = p$;

Then $1 - y = 2p$, or $y = 1 - 2p$.

and since

$$x = 12 - y + \frac{1-y}{2} = 12 - (1 - 2p) + p = 12 + 3p - 1,$$

We shall have $x = 11 + 3p$, and $y = 1 - 2p$;

Where p may be any whole number whatever, that will render the values of x and y in these two equations positive.

But it is evident, from the value of y, that p must be either 0 or negative; and consequently, from that of x, that it must be $0, -1, -2$, or -3.

Whence, if $p = 0$, $p = -1$, $p = -2$, $p = -3$,

Then $\begin{cases} x = 11, & x = 8, & x = 5, & x = 2, \\ y = 1, & y = 3, & y = 5, & y = 7. \end{cases}$

Which are all the answers in whole positive numbers that the question admits of.

3. Given $3x = 8y - 16$, to find the values of x and y in whole numbers.

Here $x = \frac{8y - 16}{3} = 2y - 5 + \frac{2y - 1}{3} = wh.$; or $\frac{2y-1}{3} = wh.$

Also, $\frac{2y - 1}{3} \times 2 = \frac{4y - 2}{3} = y + \frac{y - 2}{3} = wh.$

Or, by rejecting y, which is a whole number, there will remain $\frac{y-2}{3} = wh. = p$.

Therefore, $y = 3p + 2$,

And $x = \frac{8y - 16}{3} = \frac{8(3p + 2) - 16}{3} = \frac{24p}{3} = 8p$.

Where, if p be put $= 1$, we shall have $x = 8$ and $y = 5$ for their least values; the number of answers being, as in the first question, indefinite.

4. Given $21x + 17y = 2000$, to find all the possible values of x and y in whole numbers.

Here $x = \frac{2000 - 17y}{21} = 95 + \frac{5 - 17y}{21} = wh.$;

Or, omitting the 95, $\frac{5 - 17y}{21} = wh.$;

Consequently, by addition, $\frac{21y}{21} + \frac{5 - 17y}{21} = \frac{4y + 5}{21} = wh.$;

Also, $\frac{4y + 5}{21} \times 5 = \frac{20y + 25}{21} = 1 + \frac{4 + 20y}{21} = wh$;

Or, by rejecting the whole number 1, $\frac{4+20y}{21} = wh.$;

And, by subtraction, $\frac{21y}{21} - \frac{4+20y}{21} = \frac{y-4}{21} = wh. = p$;

$$\text{Whence } y = 21p + 4,$$

And $x = \frac{2000 - 17y}{21} = \frac{2000 - 17(21p+4)}{21} = 92 - 17p.$

Where, if p be put $= 0$, we shall have the least value of $y = 4$, and the corresponding or greatest value of $x = 92$.

And the rest of the answers will be found by adding 21 continually to the least value of y, and subtracting 17 from the greatest value of x; which being done, we shall obtain the six following results:—

$x = 92$	75	58	41	24	7
$y = 4$	25	46	67	88	109

These being all the solutions, in whole numbers, that the question admits of.

Note 1.—When there are three or more unknown quantities, and only one equation by which they can be determined, as

$$ax + by + cz = d,$$

it will be proper first to find the limit of the quantity that has the greatest coefficient, and then to ascertain the different values of the former, from 1 up to that extent, as in the following question:—

5. Given $3x + 5y + 7z = 100$, to find all the different values of x, y, and z in whole numbers.*

Here each of the least integer values of x and y are 1, by the question; whence it follows, that

$$z = \frac{100 - 5 - 3}{7} = \frac{100 - 8}{7} = \frac{92}{7} = 13\tfrac{1}{7}.$$

Consequently, z cannot be greater than 13, which is also the limit of the number of answers; though they may be considerably less.

---

* If any indeterminate equation, of the kind above given, has one or more of its coefficients, as c, negative, the equation may be put under the form

$$ax + by = d + cz,$$

in which case it is evident that an indefinite number of values may be given to the second side of the equation by means of the indefinite quantity z; and consequently, also, to x and y in the first. And if the coefficients a, b, c, in any such equation, have a common divisor, while d has not, the question, as in the first case, becomes impossible.

By proceeding, therefore, as in the former rule, we shall have

$$x = \frac{100 - 5y - 7z}{3} = 33 - y - 2z + \frac{1 - 2y - z}{3} = wh.$$

And, by rejecting $33 - y - 2z$,

$$\frac{1 - 2y - z}{3} = wh.; \text{ or } \frac{3y}{3} + \frac{1 - 2y - z}{3} = \frac{y + 1 - z}{3} = wh.;$$

$$\text{Whence } \frac{y + 1 - z}{3} = p,$$

$$\text{And } y = 3p + z - 1.$$

And consequently, putting $p = 0$, we shall have the least value of $y = z - 1$; where z may be any number, from 1 up to 13, that will answer the conditions of the question.

When, therefore, $z = 2$, we have $y = 1$,

$$\text{And } x = \frac{100 - 19}{3} = 27.$$

Hence, by taking $z = 2, 3, 4, 5$, &c., the corresponding values of x and y, together with those of z, will be found to be as below:—

$z =$ 2	3	4	5	6	7	8
$y =$ 1	2	3	4	5	6	7
$x =$ 27	23	19	15	11	7	3

Which are all the integral values of x, y, and z, that can be obtained from the given equation.

Note 2.—If there be three unknown quantities, and only two equations for determining them, as

$$ax + by + cz = d, \text{ and } ex + fy + gz = h,$$

exterminate one of these quantities in the usual way, and find the values of the other two from the resulting equation, as before.

Then, if the values, thus found, be separately substituted in either of the given equations, the corresponding values of the remaining quantities will likewise be determined: thus:—

6. Let there be given $x - 2y + z = 5$, and $2x + y - z = 7$, to find the values of x, y, and z.

Here, by multiplying the first of these equations by 2, and subtracting from the second the product, we shall have

$$3z - 5y = 3, \text{ or } z = \frac{3 + 5y}{3} = 1 + y + \frac{2y}{3} = wh;$$

$$\text{And consequently } \frac{2y}{3}, \text{ or } \frac{3y}{3} - \frac{2y}{3} = \frac{y}{3} = wh. = p.$$

$$\text{Whence } y = 3p.$$

And, by taking $p = 1, 2, 3, 4$, &c., we shall have $y = 3, 6, 9, 12, 15$, &c., and $z = 6, 11, 16, 21, 26$, &c.

But from the first of the two given equations

$$x = 5 + 2y - z;$$

whence, by substituting the above values for y and z, the results will give $x = 5, 6, 7, 8, 9$, &c.

And therefore the first six values of x, y, and z, are as below:—

$x = 5$	6	7	8	9	10
$y = 3$	6	9	12	15	18
$z = 6$	11	16	21	26	31

Where the law by which they can be continued is sufficiently obvious.

EXAMPLES FOR PRACTICE.

1. Given $3x = 8y - 16$, to find the least values of x and y in whole numbers. Ans. $x = 8$, $y = 5$.

2. Given $14x = 5y + 7$, to find the least values of x and y in whole numbers. Ans. $x = 3$, $y = 7$.

3. Given $27x = 1600 - 16y$, to find the least values of x and y in whole numbers. Ans. $x = 48$, $y = 19$.

4. It is required to divide 100 into two such parts, that one of them may be divisible by 7, and the other by 11.
Ans. The only parts are 56 and 44.

5. Given $9x + 13y = 2000$, to find the greatest value of x, and the least value of y in whole numbers.
Ans. $x = 215$, $y = 5$.

6. Given $11x + 5y = 254$, to find all the possible values of x and y in whole numbers.
Ans. $x = 19, 14, 9, 4$; $y = 9, 20, 31, 42$.

7. Given $17x + 19y + 21z = 400$, to find all the answers in whole numbers which the question admits of.
Ans. 10 different answers.

8. Given $5x + 7y + 11z = 224$, to find all the possible values x, y, and z, in whole positive numbers.
Ans. The number of answers is 59.

9. It is required to find in how many different ways it is possible to pay 20*l*, in half-guineas and half-crowns, without using any other sort of coin. Ans. 7 different ways.

10. I owe my friend a shilling, and have nothing about me but guineas, and he has nothing but louis-d'ors; how must I contrive to acquit myself of the debt, the louis being valued at 17*s*. apiece, and the guineas at 21*s*.?
Ans. I must give him 13 guineas, and he must give me 16 louis.

11. How many gallons of British spirits, at 12*s*., 15*s*., and 18*s*. a gallon, must a rectifier of compounds take to make a mixture of 1000 gallons that shall be worth 17*s*. a gallon?

Ans. $111\frac{1}{9}$ at 12*s*.; $111\frac{1}{9}$ at 15*s*.; and $777\frac{7}{9}$ at 18*s*.

PROBLEM 2.—To find such a whole number as, being divided by other given numbers, shall leave given remainders.

RULE 1.—Call the number that is to be determined x, the numbers by which it is to be divided a, b, c, &c., and the given remainders f, g, h, &c.

2. Subtract each of the remainders from x, and divide the differences by a, b, c, &c., and there will arise

$$\frac{x-f}{a}, \frac{x-g}{b}, \frac{x-h}{c}, \text{ \&c.,} = \text{whole numbers.}$$

3. Put the first of these fractions $\frac{x-f}{a} = p$, and substitute the value of x, as found in terms of p, from this equation, in the place of x, in the second fraction.

4. Find the least value of p in this second fraction, by the last problem, which put $= r$, and substitute the value of x, as found in terms of r, in the place of x in the third fraction.

Find, in like manner, the least value of r, in this third fraction, which put $= s$. and substitute the value of x, as found in the terms of s, in the fourth fraction as before.

Proceed in the same way with the next following fraction, and so on to the last; when the value of x, thus determined, will give the whole number required.

EXAMPLES.

1. It is required to find the least whole number, which, being divided by 17, shall leave a remainder of 7, and when divided by 26, shall leave a remainder of 13.

Let $x =$ the number required.

Then $\frac{x-7}{17}$ and $\frac{x-13}{26} =$ whole numbers.

And, putting $\frac{x-7}{17} = p$, we shall have $x = 17p + 7$.

Which value of x, being substituted in the second fraction, gives $\frac{17p+7-13}{26} = \frac{17p-6}{26} = wh.$

But it is obvious that $\frac{26p}{26}$ is also $= wh.$

And consequently, $\frac{26p}{26} - \frac{17p-6}{26} = \frac{9p+6}{26} = wh.$

Or $\frac{9p+6}{26} \times 3 = \frac{27p+18}{26} = p + \frac{p+18}{26} = wh.$

Where, by rejecting p, there remains $\frac{p+18}{26}$ $wh. = r.$

Therefore, $p = 26r - 18$;

Hence, if r be taken, $= 1$, we shall have $p = 8$.

And consequently, $x = 17p + 7 = 17 \times 8 + 7 = 143$, the number sought.

2. It is required to find the least whole number, which, being divided by 11, 19, and 29, shall leave the remainders 3, 5, 10 respectively.

Let $x =$ the number required.

Then $\frac{x-3}{11}$, $\frac{x-5}{19}$, and $\frac{x-10}{29}$ = whole numbers.

And, putting $\frac{x-3}{11} = p$, we shall have $x = 11p + 3$.

Which value of x, being substituted in the second fraction, gives $\frac{11p-2}{19} = wh.$

Or $\frac{11p-2}{19} \times 2 = \frac{22p-4}{19} = p + \frac{3p-4}{19} = wh.$

And, by rejecting p, there will remain $\frac{3p-4}{19} = wh.$

Also, by multiplication, $\frac{3p-4}{19} \times 6 = \frac{18p-24}{19} = \frac{18p-5}{19}$ $- 1 = wh.$;

Or, by rejecting the 1, $\frac{18p-5}{19} = wh.$

But $\frac{19p}{19}$ is likewise $= wh.$

Whence $\frac{19p}{19} - \frac{18p-5}{19} = \frac{p+5}{19} = wh.$, which put $= r.$

Then we shall have

$p = 19r - 5$, and $x = 11(19r - 5) + 3 = 209r - 52.$

And if this value be substituted for x in the third fraction, there will arise

$$\frac{209r-62}{29} = 7r - 2 + \frac{6r-4}{29} = wh.;$$

Or, by neglecting $7r - 2$, we shall have the remaining part of the expression $\frac{6r-4}{29} = wh.$;

But, by multiplication,

$$\frac{6r-4}{29} \times 5 = \frac{30r-20}{29} = r + \frac{r-20}{29} = wh.$$

Or, by rejecting r, there will remain $\frac{r-20}{29} = wh.$, which put $= s$.

Then $r = 29s + 20$; where, by taking $s = 0$, we shall have $r = 20$.

And consequently,

$$x = 209r - 52 = 209 \times 20 - 52 = 4128,$$

the number required.

3. To find a number, which, being divided by 6, shall leave the remainder 2, and when divided by 13, shall leave the remainder 3. Ans. 68.

4. It is required to find a number, which, being divided by 7, shall leave 5 for a remainder, and if divided by 9, the remainder shall be 2. Ans. 47, 110, &c.

5. It is required to find the least whole number, which, being divided by 39, shall leave the remainder 16, and when divided by 56, the remainder shall be 27. Ans. 1147.

6. It is required to find the least whole number, which, being divided by 7, 8, and 9, respectively, shall leave the remainders 5, 7, and 8. Ans. 215.

7. It is required to find the least whole number, which, being divided by each of the nine digits, 1, 2, 3, 4, 5, 6, 7, 8, 9, shall leave no remainders. Ans. 2520.

8. A person receiving a box of oranges observed, that when he told them out by 2, 3, 4, 5, and 6 at a time, he had none remaining; but when he told them out by 7 at a time, there remained 5; how many oranges were there in the box?

Ans. 180.

OF THE DIOPHANTINE ANALYSIS.

This branch of Algebra, which is so called from its inventor, Diophantus, a Greek mathematician of Alexandria in Egypt, who flourished in or about the third century after Christ, relates chiefly to the finding of square and cube numbers, or to the rendering certain compound expressions free from surds: the method of doing which is by making such substitutions for the unknown quantity, as will reduce the resulting equation to a simple one, and then finding the value of that quantity in terms of the rest.

It is to be observed, however, that questions of this kind do not always admit of answers in rational numbers, and that when they are resolvable in this way, no rule can be given, that will apply in all the cases that may occur; but as far as respects a particular class of these problems relating to squares, they may generally be determined by means of some of the rules derived from the following formula :—

PROBLEM 1.—To find such values of x as will make $\sqrt{(ax^2 + bx + c)}$ rational, or $ax^2 + bx + c =$ a square.*

RULE 1.—When the first term of the formula is wanting, or $a = 0$, put the side of the square sought $= n$; then $bx + c = n^2$.

And consequently, by transposing c, and dividing by the coefficient b, we shall have $x = \frac{n^2 - c}{b}$; where n may be any number taken at pleasure.

2. When the last term is wanting, or $c = 0$, put the side of the square sought $= nx$, or, for the sake of greater generality $= \frac{mx}{n}$; then, in this case, we shall have $ax^2 + bx = \frac{m^2x^2}{n^2}$.

And consequently, by multiplying by n^2, and dividing by x, there will arise $an^2x + bn^2 = m^2x$, and $x = \frac{bn^2}{m^2 - an^2}$, where m and n, both in this and the following cases, may be any whole numbers whatever, that will give positive answers.†

3. When the coefficient a, of the first term, is a square number, put it $= d^2$, and assume the side of the square sought $= dx + \frac{m}{n}$; then, $d^2x^2 + bx + c = d^2x^2 + \frac{2dm}{n}x + \frac{m^2}{n^2}$

And consequently, by cancelling d^2x^2, and multiplying by

---

* The coefficients a, b, of the unknown quantities, as well as the absolute term c, are here supposed to be all integers; for if they were fractions, they could be readily reduced to a common square denominator; which, being afterwards rejected, will not alter the nature of the question; since any square number, when multiplied or divided by a square number, is still a square.

† The unknown quantity x, in this case, can always be found in integers when b is positive; and, in Case 4, next following, its integral value can always be determined, whether b be positive or negative. See Vol. II. of Bonnycastle's Treatise on Algebra, Art. (H.)

n^2, we shall have $bn^2x + cn^2 = 2dmnx + m^2$, and $x = \frac{m^2 - cn^2}{bn^2 - 2dmn}$.

4. When the last term c is a square number, put it $= e^2$, and assume the side of the square sought $= \frac{mx}{n} + e$; then $ax^2 + bx + e^2 = \frac{m^2x^2}{n^2} + \frac{2em}{n}x + e^2$. And consequently, by cancelling e^2, and dividing by x, we shall have $ax + b = \frac{m^2x}{n^2} + \frac{2em}{n}$ and $x = \frac{bn^2 - 2emn}{m^2 - an^2}$.

5. When the given formula, or general expression,

$$ax^2 + bx + c$$

can be divided into two factors of the form $fx + g$ and $hx + k$, which it always can when $b^2 - 4ac$ is a square, let there be taken $(fx + g) \times (hx + k) = \frac{m^2}{n^2}(fx + g)^2$; then, by reduction, we shall have $x = \frac{gm^2 - kn^2}{hn^2 - fm^2}$; where it may be observed, that if the square root of $b^2 - 4ac$, when rational, be put $= \delta$, the two factors abovementioned, will be

$$ax + \frac{b - \delta}{2}, \text{ and } x + \frac{b + \delta}{2a},^*$$

And, consequently, by substituting them in the place of the former, we shall have,

$$x = \frac{am^2(b - \delta) - n^2(b + \delta)}{2a(n^2 - am^2)}.$$

6. When the formula, last mentioned, can be separated into two parts, one of which is a square, and the other the product of two factors, its solution may be obtained by putting the sum of the square and the product so formed, equal to the square of the sum of its roots, and $\frac{m}{n}$ times one of the

---

* These factors are found by putting the given formula $ax^2 + bx + c = 0$, and then determining its roots; which, by the rule for quadratics, are $x = -\frac{b}{2a} + \frac{1}{2a}\sqrt{(b^2 - 4ac)}$, and $x = -\frac{b}{2a} - \frac{1}{2a}\sqrt{(b^2 - 4ac)}$. Whence, if $b^2 - 4ac$ be a square, of which the root is δ, we shall have a, $x + \frac{b}{2a} - \frac{\delta}{2a}$, and $x + \frac{b}{2a} + \frac{\delta}{2a}$, for the divisors of $ax^2 + bx + c$, or $ax + \frac{b - \delta}{2}$, and $x + \frac{b + \delta}{2a}$, for its two factors, as in the above rule.

factors, and then finding the values of x as in the former instances.

7. These being all the cases of the general formula that are resolvable by any direct rule, it only remains to observe, that, either in these, or other instances of a different kind, if we can find, by trials, any one simple value of the unknown quantity which satisfies the condition of the question, an expression may be derived from this that will furnish as many other values of it as we please.

Thus, let p, in the given formula $ax^2 + bx + c$, be a value of x so found, and make $ap^2 + bp + c = q^2$.

Then, by putting $x = y + p$, we shall have $ax^2 + bx + c = a(y+p)^2 + b(y+p) + c = ay^2 + (2ap+b)y + ap^2 + bp + c$, or $ax^2 + bx + c = ay^2 + (2ap+b)y + q^2$.

From which latter expression, the values of y, and consequently those of x, may be found as in Case 4.

Or, because $c = q^2 - bp - ap^2$, if this value be substituted for c, in the original formula $ax^2 + bx + c$, it will become $a(x^2 - p^2) + b(x - p) + q^2$, or

$$q^2 + (x-p) \times (ax + ap + b) = \text{a square};$$

which last expression can be resolved by Case 6.

It may here, also, be farther observed, that by putting the given formula, $ax^2 + bx + c = \frac{y^2}{4}$, and taking $x = \frac{z-b}{2a}$; we shall have, by substituting this value for x in the former of these expressions, and then multiplying by $4a$, and transposing the terms $ay^2 + (b^2x - 4ac) = z^2$; or putting, for the sake of greater simplicity, $b^2 - 4ac = b'$, this last expression may then be exhibited under the form $ay^2 + b' = z^2$, where it is obvious, that if $ay^2 + (b^2 - 4ac)$, or its equal $ay^2 + b'$, it can be made a square, $ax^2 + bx + c$ will also be a square.

And as the proposed formula can always be reduced to one of this kind, which consists only of two terms, the possibility or impossibility of resolving the question, in this state of it, can be more easily perceived.*

---

* It may here be observed, that an infinite number of expressions, of the kind $ay^2 + (b^2 - ac)$, or $ay^2 + b' = z^2$, here mentioned, are wholly irresolvable; among which we may reckon

$$2y^2 \pm 3,\ 5y^2 \pm 6,\ 7y^2 \pm 5, \text{ \&c.}$$

none of which can ever become squares, whatever number, either whole or fractional, be substituted for y; although there are a variety of instances in which the value of y may be found, even in integers, so as to render the formula $ay^2 + b = z^2$.

For a further detail of which circumstances, as well as for other particulars relating to this part of the subject, see the second volume of *Euler's Algebra*, or the second volume of *Bonnycastle's Algebra*.

EXAMPLES.

1. It is required to find a number, such that if it be multiplied by 5 and then added to 19, the result shall be a square.

Let $x=$ the required number: then, as in Case 1, $5x+19=n^2$, or $x=\frac{n^2-19}{5}$; where it is evident that n may be any number whatever greater than $\sqrt{19}$.

Whence, if n be taken $=5, 6, 7$, respectively, we shall have $x=\frac{25-19}{5}=1\frac{1}{5}$, or $\frac{36-19}{5}=3\frac{2}{5}$, or $\frac{49-19}{5}=6$; the latter of which is the least value of x, in whole numbers, that will answer the conditions of the question; and consequently $5x+19=5\times 6+19=30+19=49$, a square number as was required.

2. It is required to find an integral number, such that it shall be both a triangular number and a square.

It is here to be observed, that all triangular numbers are of the form $\frac{x^2+x}{2}$; and therefore the question is reduced to the making $\frac{x^2+x}{2}$, or its equal $\frac{2x^2+2x}{4}$, a square.

Where, since the divisor 4 is a square number, it is the same as if it were required to make $2x^2+2x$ a square.

Let, therefore, $2x^2+2x=\left(\frac{mx}{n}\right)^2=\frac{m^2x^2}{n^2}$, agreeably to the method laid down in Case 2.

Then, by dividing by x, and multiplying the result by n^2, the equation will become $2n^2x+2n^2=m^2x$, or $(m^2-2n^2)x=2n^2$; and consequently, $x=\frac{2n^2}{m^2-2n^2}$; where, if n be taken $=2$, and $m=3$, we shall have $x=8$, and $\frac{x^2+x}{2}=\frac{64+8}{2}=\frac{72}{2}=36$, which is the least integral triangular number that is at the same time a square.

3. It is required to find the least integral number, such that if 4 times its square be added to 29, the result shall be a square.

Here it is evident, that this is the same as to make $4x^2+29$ a square.

And, as the first term in the expression is a square, let

$4x^2 + 29 = (2x + \frac{m}{n})^2 = 4x^2 + \frac{4m}{n}x + \frac{m^2}{n^2}$; agreeably to Case 3.

Then, $\frac{4m}{n}x + \frac{m^2}{n^2} = 29$, or $\frac{4m}{n}x = 29 - \frac{m^2}{n^2}$; and consequently, $x = \frac{29n^2 - m^2}{4mn}$; where, if m and n be each taken $=1$, we shall have $x = \frac{29-1}{4} = 7$, and $4x^2 + 29 = 4 \times 49 + 29 = 225 = (15)^2$, which is a square number, as was required.

4. It is required to find such a value of x as will make $7x^2 - 5x + 1$ a square.

Here the last term 1 being a square, let there be taken, according to Case 4,

$$7x^2 - 5x + 1 = (\frac{m}{n}x - 1)^2 = \frac{m^2}{n^2}x^2 - \frac{2m}{n}x + 1.$$

Then, by rejecting the 1 on each side of the equation, and dividing by x, we shall have $7x - 5 = \frac{m^2}{n^2}x - \frac{2m}{n}$, and consequently, $x = \frac{2mn - 5n^2}{m^2 - 7n^2}$; where, if m and n be each taken $= 1$, the result will give $x = \frac{2-5}{1-7} = \frac{3}{6} = \frac{1}{2}$, or by taking $n = 3$, and $m = 8$, we shall have $x = \frac{48-45}{64-63} = 3$, which makes $7 \times 3^2 - 5 \times 3 + 1 = 49 = 7^2$, as required.

5. It is required to find such a value of x as will make $8x^2 + 14x + 6$ a square.

Here, by comparing this expression with the general formula, $ax^2 + bx + c$, we shall have $a = 8$, $b = 14$, and $c = 6$.

And, as neither a nor c, in the present instance, are squares, but $b^2 - 4ac = 196 - 192 = 4$ is a square, the given expression can be resolved, by Case 5, into the two following factors, $8x + 6$ and $x + 1$.

Let, therefore, $8x^2 + 14x + 6 = (8x + 6)(x + 1) = \frac{m^2}{n^2}(x + 1)^2$, agreeably to the rule there laid down.

Then there will arise, by dividing each side by $x + 1$, $8x + 6 = \frac{m^2}{n^2}(x + 1)$.

And consequently, by multiplication and reduction, we shall

have, in this case, $x = \frac{m^2 - 6n^2}{8n^2 - m^2}$; where it appears that, in order to obtain a rational answer, $\frac{m^2}{n^2}$ must be less than 8 and greater than 6.

Whence, by taking $m = 5$, and $n = 2$, we shall have $x = \frac{25 - 24}{32 - 25} = \frac{1}{7}$, which makes $\frac{8}{49} + \frac{14}{7} + 6 = \frac{400}{49} = \left(\frac{20}{7}\right)^2$, as required.

6. It is required to find such a value of x as will make $2x^2 - 2$ a square.

Here, by comparing this with the general formula $ax^2 + bx + c$, as before, we shall have $a = 2$, $b = 0$, and $c = -2$.

And, as neither a nor c are squares, but $b^2 - 4ac = -4ac = -4(2 \times -2) = 16$ is a square, the root of which is 4, the given expression can be resolved, by Case 5, into the two factors $2x - 2$, and $x + 1$, or $2(x - 1)$, and $(x + 1)$, which is evident, indeed, in this case, from inspection.

Let, therefore, $2x^2 - 2 = 2(x - 1) \times (x + 1) = \frac{m^2}{n^2}(x + 1)^2$, agreeably to the rule; and there will arise, by division, $2x - 2 = \frac{m^2}{n^2}(x + 1)$. And consequently, by multiplication, and reducing the result, we shall have $x = \frac{2n^2 + m^2}{2n^2 - m^2}$; where, by taking $n = 1$, and $m = 1$, we shall have $x = 3$, and $2x^2 - 2 = 18 - 2 = 16 = (4)^2$; or, taking $n = 2$ and $m = 3$, the result will give $x = -17$.

But as x enters the problem only in its second power, $+17$ may be taken instead of -17; since either of them give $2x^2 - 2 = 576 = (24)^2$.

7. It is required to find such a value of x as will make $5x^2 + 36x + 7$ a square.

Here, by comparing the expression with the general formula, we shall have $a = 5$, $b = 36$, and $c = 7$.

And as neither a nor c are squares, but $b^2 - 4ac = 1296 - 140 = 1156 = (34)^2$ is a square, it can be resolved, as in the last example, into the two factors, $5x + 1$, and $x + 7$.

Whence, putting $5x^2 + 36x + 7 = (5x + 1) \times (x + 7) = \frac{m^2}{n^2}(x + 7)^2$, there will arise, by dividing, by $x + 7$, $5x + 1 = \frac{m^2}{n^2}(x + 7)$.

And consequently, by multiplication, and reducing the resulting expression, we shall have $x = \frac{7m^2 - n^2}{5n^2 - m^2}$; where, taking $m = 2$, and $n = 1$, the substitution will give $x = \frac{7 \times 4 - 1}{5 \times 1 - 4} = 27$, which makes $5 \times (27)^2 + 36 \times 27 + 7 = 4624 = (68)^2$, as required.

8. It is required to find such a value of x as will make $6x^2 + 13x + 10$ a square.

Here, by comparing the given expression with the general formula $ax^2 + bx + c$, we have $a = 6$, $b = 13$, and $c = 10$. And as neither a, c, nor $b^2 - 4ac$, are squares, the question, if possible, can only be resolved by the method pointed out in Case 6.

In order, therefore, to try it in this way, let the first simple square 4, be subtracted from it, and there will remain, in that case, $6x^2 + 13x + 6$.

Then since $(13)^2 - 4(6 \times 6) = 169 - 144 = 25$ is now a square, this part of the formula can be resolved, by Case 5, into the two factors

$$3x + 2, \text{ and } 2x + 3.$$

Whence, by assuming, according to the rule, $6x^2 + 13x + 10 = 4 + (3x + 2) \times (2x + 3) = \left\{2 + \frac{m}{n}(3x + 2)\right\}^2 = 4 + \frac{4m}{n}(3x + 2) + \frac{m^2}{n^2}(3x + 2)^2$, we shall have, by cancelling the 4 on each side, and dividing by $3x + 2$, $2x + 3 = \frac{4m}{n} + \frac{m^2}{n^2}(3x + 2)$.

And consequently, by multiplying by n^2, and transposing the terms, we have $2n^2x - 3m^2x = 4mn + 2m^2 - 3n^2$, or $x = \frac{4mn + 2m^2 - 3n^2}{2n^2 - 3m^2}$.

Where, putting $m = 2$, and $n = 3$, the result will give $x = \frac{24 + 8 - 27}{18 - 12} = \frac{5}{6}$; or, if m be taken $= 13$, and $n = 17$, we shall have $x = \frac{4 \times 17 \times 13 + 2 \times (13)^2 - 3 \times (17)^2}{2(17)^2 - 3(13)^2} = \frac{355}{71} = 5$,

Which makes $6 \times (5)^2 + 13 \times 5 + 10 = 225 = (15)^2$, as required.

9. It is required to find such a value of x as will make $13x^2 + 15x + 7$ a square.

Here, by comparing this with the general formula, as before, we have $a = 13$, $b = 15$, and $c = 7$. And as neither a, b, nor $b^2 - 4ac$, are squares, the answer to the question, if it be resolvable, can only be obtained by Case 6. In order, therefore, to try it in that way, let $(1 - x)^2$ or $1 - 2x + x^2$ be subtracted from the given expression, and there will remain $12x^2 + 17x + 6$.

And as $(17)^2 - 4(6 \times 12)$, which is $= 1$, is now a square, this part of the formula can be resolved by Case 5, into the two factors $4x + 3$ and $3x + 2$. Whence, assuming $13x^2 + 15x + 7 = (1 - x)^2 + (4x + 3) \times (3x + 2) = \left\{(1 - x) + \frac{m}{n}(3x + 2)\right\}^2 = (1 - x)^2 + \frac{2m}{n}(1 - x) \times (3x + 2) + \frac{m^2}{n^2}(3x + 2)^2$, we shall have, by cancelling $(1 - x)^2$, and dividing by $3x + 2$, $4x + 3 = \frac{2m}{n}(1 - x) + \frac{m^2}{n^2}(3x + 2)$; and consequently, by multiplying by n^2, and transposing the terms, there will arise $4n^2x + 2mnx - 3m^2x = 2mn + 2m^2 - 3n^2$, or $x = \frac{2mn + 2m^2 - 3n^2}{4n^2 + 2mn - 3m^2}$.

Where, putting m and n each $= 1$, we shall have $x = \frac{2 + 2 - 3}{4 + 2 - 3} = \frac{1}{3}$, which makes $\frac{13}{9} + \frac{15}{3} + 7 = \frac{13}{9} + \frac{45}{9} + \frac{63}{9} = \frac{121}{9} = \left(\frac{11}{3}\right)^2$, as required.

10. It is required to find such a value of x as will make $7x^2 + 2$ a square.

Here it is easy to perceive that neither of the former rules will apply.

But as the expression evidently becomes a square when $x = 1$, let, therefore, $x = 1 + y$, according to Case 7, and we shall have

$$7x^2 + 2 = 9 + 14y + 7y^2;$$

Or, putting $9 + 14y + 7y^2 = (3 + \frac{m}{n}y)^2$, according to the rule, and squaring the righthand side, $9 + 14y + 7y^2 = 9 + \frac{6m}{n}y + \frac{m^2}{n^2}y^2$.

Hence, rejecting the 9s and dividing the remaining terms by y, we have $7n^2y + 14n^2 = 6mn + m^2y$; and consequently,

$y = \frac{6mn - 14n^2}{7n^2 - m^2}$, and $x = 1 + \frac{6mn - 14n^2}{7n^2 - m^2}$; where it is evident that m and n may be any positive or negative numbers whatever.

If, for instance, m and n be each taken $= 1$, we shall have $y = -\frac{4}{3}$ and $x = -\frac{1}{3}$. Or, since the second power of x only enters the formula, we may take, as in a former instance, $x = \frac{1}{3}$, which value makes $7x^2 + 2 = \frac{7}{9} + 2 = \frac{7}{9} + \frac{18}{9} = \frac{25}{9}$ a square.

Or, if $m = 3$ and $n = -1$, we shall have $x = 17$, and $7x^2 + 2 = 7 \times (17)^2 + 2 = 2025 = (45)^2$, a square as before.

And by proceeding in this manner, we may obtain as many other values of x as we please.

Problem 2.—To find such values of x as will make $\sqrt{(ax^3 + bx^2 + cx + d)}$ rational, or $ax^3 + bx^2 + cx + d =$ a square. This problem is much more limited and difficult to be resolved than the former; as there are but a few cases of it that admit of answers in rational numbers, and in these the rules for obtaining them are of a very confined nature, being mostly such as are subject to certain limitations, or that admit only of a few simple answers, which, in the instances here mentioned, may be found as follows:—

Rule 1.—When the third and fourth terms of the formula are wanting, or c and d are each $= 0$, put the side of the square sought $= nx$, than $ax^3 + bx^2 = n^2x^2$.

And consequently, by dividing each side of the equation by x^2, we shall have $ax + b = n^2$, or $x = \frac{n^2 - b}{a}$, where n may be any integral or fractional number whatever.

2. When the last term d is a square, put it $= e^2$, and assume the side of the required square $= e + \frac{c}{2e}x$, and the proposed formula is $e^2 + cx + bx^2 + ax^3 = e^2 + cx + \frac{c^2}{4e^2}x^2$.

Whence, by expunging the terms $e^2 + cx$, which are common, and dividing by x^2, we shall have $4ae^2x + 4be^2 = c^2$; and consequently, $x = \frac{c^2 - 4be^2}{4ae^2}$.

Or if, the same case, there be put $e + \frac{c}{2e}x + \frac{4be^2 - c^2}{8e^3}x^2$ for the side of the required square, we shall have, by squaring,

$e^2 + cx + bx^2 + ax^3 = e^2 + cx + bx^2 + \frac{e(4be^2 - c^2)}{8e^4}x^3 + \frac{(4be^2 - c^2)^2}{64e^6}x^4$. And, as the first three terms $(e^2 + cx + bx^2)$ are now common, there will arise, by expunging them, and then multiplying them by $64e^6$, $64ae^6x^3 = 8ce^2(4be^2 - c^2)x^3 + (4be^2 - c^2)^2x^4$.

Whence, by dividing each side of this last equation by x^3, and reducing the result, we shall have

$$x = \frac{64ae^6 - 8ce^2(4be^2 - c^2)}{(4be^2 - c^2)^2},$$

which last method gives a new value of x, different from that before obtained.

It must be observed, however, that each of these forms fail when the second and third terms of the given formula are wanting, or b and c each $= 0$.*

3. When neither of the above rules can be applied to the question, the formula can be resolved, by first finding, by trial, as in the former problem, some value of the unknown quantity that makes the given expression a square: in which case other values of it may be determined from this, when they are possible, as follows :—

Thus, let p be a value of x so found, and make

$$ap^3 + bp^2 + cp + d = q^2;$$

Then, by putting $x = y + p$, we shall have $ap^3 + bp^2 + cp + d = a(y + p)^3 + b(y + p)^2 + c(y + p) + d = ay^3 + (3ap + b)y^2 + (3ap^2 + 2ap + c)y + ap^3 + bp^2 + cp + d$, or $ax^3 + bx^2 + cx + d = ay^3 + (3ap + b)y^2 + (3ap^2 + 2ap + c)y + q^2$.

From which latter form, the value of y, and consequently that of x, may be found by either of the methods given in Case 2.

It may also be further remarked, that if the given formula,

---

* In the first of these methods, the assumed root, $e + \frac{c}{2e}x$, is determined by first taking it in the form $e + nx$, and then equating the second term of it, when squared, with the corresponding term of the original formula; when it will be found that $n = \frac{c}{2e}$.

In like manner the assumed root $e + \frac{c}{2e}x + \frac{4be^2 - c^2}{8e^3}x^2$, in the second method, is determined by first taking it in the form $e + nx + mx^2$, and then equating the second and third terms of it, when squared, with the corresponding terms of the given formula, when it will be found that $n = \frac{c}{2e}$ and $m = \frac{4be^2 - c^2}{8e^3}$.

in any case of this kind, can be resolved into factors, such that one of them shall be a square, it will be sufficient to make the remaining factor a square, in order to render the whole expression so; since a square, multiplied or divided by a square, is still a square.*

EXAMPLES.

1. It is required to find such a value of x as will make $11x^3 + 3x^2$ a square.

Let the given expression $11x^3 + 3x^2 = n^2x^2$: agreeably to Case 1.

Then, by dividing by x^2, we shall have $11x + 3 = n^2$; and consequently, $x = \frac{n^2 - 3}{11}$; where n may be any number, positive or negative, that is greater than $\sqrt{3}$.

Taking, therefore, $n = 2, 3, 4, 5$, &c., respectively, we shall have, in this case, $x = \frac{1}{11}, \frac{6}{11}, \frac{13}{11}$, or 2, the last of which is the least integral answer that the question admits of.

2. It is required to find such values of x as will make $x^3 - 2x^2 + 2x + 1$ a square.

Here the last term 1, being a square, let $1 + 2x - 2x^2 + x^3 - (1 + x)^2 = 1 + 2x + x^2$, agreeably to the first part of Case 2.

Then, since the first two terms, on both sides of the equation, destroy each other, we shall have $x^3 - 2x^2 = x^2$, or $x^3 = 3x^2$, and consequently $x = 3$; which, by substitution, makes $1 + 2x - 2x^2 + x^3 = 1 + 6 - 18 + 27 = 16$ a square, as required.

Again, by putting $x = y + 3$, according to Case 3, we shall

* The method of determining the factors of which any formula is composed, when it can be done, is to put the given expression $= 0$, and then find the roots r, r', &c., of the equation so formed; each of which will give a factor $x - r$, $x - r'$, and these are generally easily discovered, as we here seek only the rational roots, which are always divisors of the absolute term, or of that which does not contain x.

Thus the formula $x^3 - x^2 - x + 1$ is resolvable into the factors $(1-x)\times(1+x)\times(1-x)$, or $(1-x)^2\times(1+x)$; and by putting $1+x = n^2$, we have $x = n^2 - 1$; where, if n be taken equal to any number whatever, $x^3 - x^2 - x + 1$ will be a square; though, by any other mode of solution, it would be difficult to find even two or three values of x.

It may here also be observed, there are but few questions in this problem that can be determined in whole numbers. Several of them, likewise, admit only of one answer, and others are totally irresolvable, either in integers or fractions. Thus, if it were required to make $x^3 + 1$ a square, the only positive value of x that renders this possible, is 2; and the making of $3x^2 - 1$ a square, is impossible.

have $1+2x-2x^2+x^3=1+2(y+3)-2(y+3)^2+(y+3)^3=16+17y+7y^2+y^3$.

And consequently, by making $16+17y+7y^2+y^3=(4+\frac{17}{8}y)^2=16+17y+\frac{289}{64}y^2$, agreeably to the first part of Case 2, by cancelling $16+17y$, there will arise $7y^2+y^3=\frac{289}{64}y^2$, or $y+7=\frac{289}{64}$.

Whence $y=\frac{289}{64}-7=\frac{289-448}{64}=-\frac{159}{64}$, and $x=3-\frac{159}{64}=\frac{192-159}{64}=\frac{33}{64}$, for another value of x.
Which number, being substituted in the original formula, makes $1+2x-2x^2+x^3=\frac{429025}{262144}=\left(\frac{655}{512}\right)^2$, a square, as before.

3. It is required to find such values of x as will make $3x^3-5x^2+6x+4$ a square.

Here, 4 being a square, let $4+6x-5x^2+3x^3=(2+\frac{3}{4}x)^2=4+6x+\frac{9}{4}x^2$, as in the first part of Case 2.

Then, since the first two terms on each side of the equation destroy each other, we shall have $3x^3-5x^2=\frac{9}{4}x^2$, or $3x-5=\frac{9}{4}$, and consequently, in this case, $x=\frac{5+\frac{9}{4}}{3}=\frac{29}{12}$.

Whence $(2+\frac{3}{2}\times\frac{29}{12})^2=(2+\frac{29}{8})^2=(\frac{45}{8})^2$ a square, as was required.

Or, by the second method of the same Case, let $4+6x-5x^2+3x^3=(2+\frac{3}{2}x-\frac{29}{16}x^2)^2=4+6x-5x^2-\frac{87}{16}x^3+\frac{841}{256}x^4$; then, as the first three terms on each side of this equation destroy each other, we shall have $\frac{841}{256}x^4-\frac{87}{16}x^3=3x^3$, or $\frac{841}{256}x-\frac{87}{16}=3$, or $841x-1392=768$; and consequently, $x=\frac{1392+768}{841}=\frac{2160}{841}$, which is another value of x, that, being substituted in the original formula, will make it a square.

4. It is required to find such values of x as will make x^3+3 a square.

Here, it is evident that the expression is a square when $x=1$. Let, therefore, $x=1+y$, and we shall have $3+x^3=4+3y+3y^2+y^3$.

And, as the first part of this is a square, make, according to the first part of Case 2, $4+3y+3y^2+y^3=(2+\frac{3}{4}y)^2=4+3y+\frac{9}{16}y^2$. Then, because the first two terms on each side of the equation destroy each other, we shall have $y^3+3y^2=\frac{9}{16}y^2$, or $y+3=\frac{9}{16}$.

Whence $y=\frac{9}{16}-3=\frac{9-48}{16}=-\frac{39}{16}$, and $x=1-\frac{39}{16}=\frac{16-39}{16}=-\frac{23}{16}$; which is a second value of x.

Again, let $4+3y+3y^2+y^3=(2+\frac{3}{4}y+\frac{39}{64}y^2)^2=4+3y+3y^2+\frac{117}{128}y^3+\frac{1521}{4096}y^4$, according to the second part of Case 2.

Then, as the first three terms on each side of the equation destroy each other, we shall have $\frac{1521}{4096}y^4+\frac{117}{128}y^3=y^3$, or $\frac{1521}{4096}y+\frac{117}{128}=1$.

Whence, also, $y=\frac{352}{1521}$, and $x=1+\frac{352}{1521}=\frac{1873}{1521}$, which is a third value of x.

And by proceeding in the same way with either of these new values of x as with the first, other values of it may be obtained; but the resulting fraction will become continually more complicated in each operation.

Problem 3.—To find such values of x as will make $\sqrt{(ax^4+bx^3+cx^2+dx+e)}$ rational, or $ax^4+bx^3+cx^2+dx+e=$ a square.

The resolution of expressions of this kind, in which the indeterminate, or unknown quantity, rises to the fourth power, is the utmost limit of the researches that have hitherto been made on the formula affected by the sign of the square root; and in this problem, as well as in that last given, there are only a few particular cases that admit of answers in rational numbers; the rest being either impossible, or such as afford one or two simple solutions; which may generally be found as follows:*—

RULE 1.—When the last term e, of the given formula, is a square, put it $=f^2$, and make $f^2+dx+cx^2+bx^3+ax^4=(f+\frac{d}{2f}x+\frac{4cf^2-d^2}{8f^3}x^2)^2=f^2+dx+cx^2+\frac{d(4cf^2-d^2)}{8f^4}x^3+\frac{(4cf^2-d^2)}{64f^6}x^4$.

Then, by expunging the first three terms, which are common to each side of the equation, there will remain $bx^3+ax^4=\frac{d(4cf^2-d^2)}{8f^4}x^3+\frac{(4cf^2-d^2)^2}{64f^6}x^4$. And consequently, by dividing by x^3, and reducing the result, we shall have $x=\frac{64bf^6-8df^2(4cf^2-d^2)}{(4cf^2-d^2)^2-64af^6}$; which form fails when the coefficients c and d, or b and d, are each $=0$.

2. When the coefficient a, of the first term of the formula, is a square, put it $=g^2$, and make $g^2x^4+bx^3+cx^2+dx+e=(gx^2+\frac{b}{2g}x+\frac{4cg^2-b^2}{8g^3})^2=g^4x^4+bx^3+cx^2+\frac{b(4cg^2-b^2)^2}{8g^4}x+\frac{(4cg^2-b^2)^2}{64g^6}$.

Then, $dx+e=\frac{b(4cg^2-b^2)}{8g^4}x+\frac{(4cg^2-b^2)^2}{64g^6}$; and consequently, $x=\frac{(4cg^2-b^2)^2-64eg^6}{64dg^6-8bg^2(4cg^2-b^2)}$; which form likewise fails under similar circumstances with the former.

3. When the first and last terms of the formula are both squares, put $a=g^2$, and $e=f^2$, and make $f^2+dx+cx^2+bx^3+g^2x^4=(f+\frac{d}{2f}x+gx^2)^2=f^2+dx+(2fg+\frac{d^2}{4f^2})^2+\frac{dg}{f}x^3+g^2x^4$. Then, $cx^2+bx^3=(2fg+\frac{d^2}{4f_2})x^2+\frac{dg}{f}x^3$.

And consequently, $x=\frac{f^2(2fg-c)+\frac{1}{4}d^2}{f(bf-dg)}$.

Or, because g enters the given formula only in its second

* As an instance of what is above said, it may be observed, that the only value of x that renders the formula $2x^4-3x^2+2$ a square, is 1; and the formula x^4-x^2+1 can never be a square, except when $x=+1$, or -1.

power, it may be taken either negatively or positively; and consequently, we shall also have $x = \frac{\frac{1}{4}d^2 - f^2(2fg + c)}{f(bf + dg)}$

So that this mode of solution furnishes two different answers.

Also, if there be taken for another supposition $f^2 + dx + cx^2 + bx^3 + g^2x^4 = (f + \frac{b}{2g}x + gx^2)^2 = f^2 + \frac{bf}{g}x + (2fg + \frac{b^2}{4g^2})x^2 + bx^3 + g^2x^4$, hence, by cancelling, $dx + cx^2 = \frac{bf}{g}x + (2fg + \frac{b^2}{4g^2})x^2$; and consequently, $x = \frac{g(bg - bf)}{\frac{1}{4}b^2 + g^2(2fg - c)}$.

And because f enters the given formula only in the second power, it may be taken either negatively or positively; and consequently, we shall also have $x = \frac{g(bg + bf)}{\frac{1}{4}b^2 - g^2(2fg + c)}$.

So that this solution likewise furnishes two values for x, which are each different from the former.

But these forms all fail under similar circumstances with those of the second Case.

4. When neither the first nor the last terms are squares, the formula cannot be resolved in any other way than by first endeavouring to discover, by trials, some simple value of the unknown quantity that will answer the conditions of the question, and then finding other values of it according to the methods pointed out in the last two problems.

Thus, let p be a value of x so found, and make $ap^4 + bp^3 + cp^2 + dp + e = q^2$.

Then, by putting $x = y + p$, we shall have $ap^4 + bp^3 + cp^2 + dp + e = a(y + p)^4 + b(y + p)^3 + c(y + p)^2 + d(y + p) + e = ay^4 + (ap + b)y^3 + (6ap^2 + 3bp + c)y^2 + (4ap^3 + 3bp^2 + 2cp + d)y + ap^4 + bp^3 + cp^2 + dp + e$, or $ax^4 + bx^3 + cx^2 + dx + e = ay^4 + (ap + b)y^3 + (6ap^2 + 3bp + c)y^2 + (4ap^3 + 3bp^2 + 2cp + d)y + q^2$. From which last formula, the value of y, and consequently that of x, may be found by Case 1.

EXAMPLES.

1. It is required to find such a value of x as will make $1 - 2x + 3x^2 - 4x^3 + 5x^4$ a square.

Here, the first term 1, being a square, let $1 - 2x + 3x^2 - 4x^3 + 5x^4 = (1 - x + x^2)^2 = 1 - 2x + 3x^2 - 2x^3 + x^4$, agreeably to the method in Case 1.

Then we shall have $5x^4 - 4x^3 = x^4 - 2x^3$.

And consequently, $5x - 4 = x - 2$; whence $x = \frac{2}{4} = \frac{1}{2}$.

And consequently, $1-2x+3x^2-4x^3+5x^4=1-1+\frac{3}{4}-\frac{1}{2}+\frac{5}{16}=\frac{9}{16}$, which is a square number, as was required.

2. It is required to find such a value of x as will make $4x^4-2x^3-x^2+3x-2$ a square.

Here, the first term being a square, let $4x^4-2x^3-x^2+3x-2=(2x^2-\frac{1}{2}x-\frac{5}{16})^2=4x^4-2x^3-2x^2+\frac{5}{16}x+\frac{25}{256}$, according to the method in Case 2.

Then we shall have $3x-2=\frac{5}{16}x+\frac{25}{256}$, or $3x-\frac{5}{16}x=2+\frac{25}{256}$. Whence $768x-80x=512+25$; and consequently, $x=\frac{512+25}{768-80}=\frac{537}{688}$

Or, if we put $x=\frac{1}{y}$, the formula in that case will become $\frac{4}{y}-\frac{2}{y^3}-\frac{1}{y^2}+\frac{3}{y}-2$.

And, therefore, multiplying this by y^4, which is a square, it will be $4-2y-y^2+3y^3-2y^4$. Where the first term being now a square, if the expression, so transformed, be resolved by Case 1, we shall have $y=\frac{688}{537}$, and $x=\frac{1}{y}=\frac{537}{688}$, as before.

3. It is required to find such values of x as will make $1+3x+7x^2-2x^3+4x^4$ a square.

Here, both the first and last terms being squares, let $1+3x+7x^2-2x^3+4x^4=(1+\frac{3}{2}x+2x^2)^2=1+3x+\frac{25}{4}x^2+6x^3+4x^4$, according to the method in Case 3.

Then, we shall have $6x^3+\frac{25}{4}x^2=7x^2-2x^3$; or $6x+2x=7-\frac{25}{4}$; and, by reduction, $x=\frac{3}{32}$.

And if we put the same formula, $1+3x+7x^2-2x^3+4x^4=(1+\frac{3}{2}x-2x^2)^2=1+3x-\frac{7}{4}x^2-6x^3+4x^4$, we shall have, by cancelling, $7x^2-2x^3=-\frac{7}{4}x^2-6x^3$; whence $6x-2x=-\frac{7}{4}-7=-\frac{35}{4}$; or $x=-\frac{35}{16}$.

And, in a similar manner, other values of x may be found, by employing the method of substitution pointed out in the latter part of Case 3.

4. It is required to find such values of x as will make $2x^4 - 1$ a square.

Here, 1 being an obvious value of x, let, according to Case 4, $x = 1 + y$.

Then $2x^4 - 1 = 2(1 + y)^4 - 1 = 2(1 + 4y + 6y^2 + 4y^3 + y^4) - 1 = 1 + 8y + 12y^2 + 8y^3 + 2y^4$. And since the first term of this last expression is now a square, we shall have, by Case 1, $1 + 8y + 12y^2 + 8y^3 + 2y^4 = (1 + 4y - 2y^2)^2 = 1 + 8y + 12y^2 - 16y^3 + 4y^4$.

Whence, as the three first terms of the two members of this equation destroy each other, there will remain $4y^4 - 16y^3 = 2y^4 + 8y^3$; or $y = 12$; and consequently $x = 1 + y = 13$; which value being substituted for x, makes $2x^4 - 1 = 57121 = (239)^2$, as required. And if 13 be now taken as the known value of x, and the operation be repeated as before, we shall obtain, for another value of x, the complicated fraction $\frac{10607469769}{2447192159}$.

PROBLEM 4.—To find such values of x as will make $\sqrt[3]{(ax^3 + bx^2 + cx + d)}$ rational, or $ax^3 + bx^2 + cx + d =$ a cube. This formula, like the two latter of those relating to squares, cannot be resolved by any direct method, except in the cases where the first or last terms of the expression are cubes; it being necessary, in all the rest, that some simple number answering the conditions of the question, should be first found by trial before we can hope to obtain others, but when this can be done, the problem, in each of the cases here mentioned, may be resolved as follows :—

RULE 1.—When the last term d of the given formula is a cube, put it $= e^3$, and make $e^3 + cx + bx^2 + ax^3 = (e + \frac{c}{3e^2}x)^3 = e^3 + cx + \frac{c^2}{3e^3}x^2 + \frac{c^3}{27e^6}x^3$.

Then, by expunging the two first terms on each side of the equation, which are common, there will remain $ax^3 + bx^2 = \frac{c^3}{27e^6}x^3 + \frac{c^2}{3e^3}x^2$; whence, by division and reduction, we shall have $27ae^6x + 27be^6 = c^3x + 9c^2e^3$, and consequently $x = \frac{9e^3(3be - c^2)}{c^3 - 27ae^6}$; which form fails when the coefficients b and c, or a and c, are each equal 0.

2. When the coefficients a of the first term is a cube, put it $=f^3$, and make $f^3x^3+bx^2+cx+d=(fx+\frac{b}{3f^2})^3=f^3x^3+bx^2+\frac{b^2}{3f^3}x+\frac{b^3}{27f^6}$.

Then, by expunging the two first terms on each side of the equation, as before, there will remain $cx+d=\frac{b^2}{3f^3}x+\frac{b^3}{27f^6}$; whence, by multiplying by $27f^6$ we shall have $27f^6cx+27df^6=9b^2f^3x+b^3$, and consequently, $x=\frac{b^3-27df^6}{9f^3(3cf^3-b^2)}$; which form likewise fails when b and c, or b and d, are each $=0$.

3. When the first and last terms are both cubes, put $a=f^3$ and $d=e^3$, and make $e^3+cx+bx^2+f^3x^3=(e+fx)^3=e^3+3fe^2x+3f^2ex^2+f^3x^3$.

Then $cx+bx^2=3fe^2x+3f^2ex^2$;

Whence we shall have $bx-3f^2ex=3fe^2-c$, and consequently, $x=\frac{3fe^2-c}{b-3f^2e}$; which formula may also be resolved by either of the two first cases.

4. When neither the first nor the last terms are cubes, let p be a value of x, found by inspection or by trials, and make $ap^3+bp^2+cp+d=q^3$.

Then, by putting $x=y+p$, we shall have $ap^3+bp^2+cp+d=a(y+p)^3+b(y+p)^2+c(y+p)+d=ay^3+(3ap+b)y^2+(3ap^2+2bp+c)y+ap^3+bp^2+cp+d$, or $ax^3+bx^2+cx+d=ay^3+(3ap+b)y^2+(3ap^2+2bp+c)y+q^3$.

From which latter form, the value of y, and consequently that of x, may be found as in Case 1.

EXAMPLES.

1. It is required to find such a value of x as will make x^2+x+1 a cube.

Here, the last term being a cube, let the root of the cube sought $=1+\frac{1}{3}x$ according to Case 1.

Then, by cubing, we shall have $1+x+x^2=1+x+\frac{1}{3}x^2+\frac{1}{27}x^3$;

And since the two first terms on each side of this equation destroy each other, there will remain $x^2=\frac{1}{3}x^2+\frac{1}{27}x^3$.

Whence, dividing by x^2, we shall have $\frac{1}{27}x+\frac{1}{3}=1$, or $x+9=27$; and consequently, $x=27-9=18$; which number, by substitution, makes $1+x+x^2=1+18+324=343$ $=7^3$ a cube number, as was required.

And if we now take this value of x, and proceed according to the method employed in Case 4, we shall obtain $x=-\frac{137826}{50653}$; which last number will also lead, in like manner, to other new values.

2. It is required to find such a value of x as will make x^3+3x^2+133 a cube.

Here, the first term being a cube, let its root $=1+x$, according to Case 2.

Then, by cubing, we shall have $133+3x^2+x^3=(1+x)^3$ $=1+3x+3x^2+x^3$.

And since the two last terms of this equation destroy each other, there will remain $1+3x=133$, or $3x=133-1=132$; whence $x=\frac{132}{3}=44$, and $x^3+3x^2+133=91125$ $=(45)^3$, a cube number, as was required.

And if 45 be now taken as a known value of x, other values of it may be found, as in the last example.

3. It is required to find such a value of x as will make $8+28x+89x^2-125x^3$ a cube.

Here, let the root sought $=2-5x$, according to Case 3.

Then, by cubing, we shall have $8+28x+89x^2-125x^3$ $=(2-5x)^3=8-60x+150x^2-125x^3$.

And since the first and last terms of this equation destroy each other, there will remain $28x+89x^2=-60x+150x^2$.

Whence, by dividing by x, and transposing the terms, we shall have $150x-89x=28+60$, or $61x=88$; and consequently $x=\frac{88}{61}$.

And as this formula can also be resolved either by the first or second Case, other values of x may be obtained, that will equally answer the conditions of the question.

4. It is required to find such a value of x as will make $2x^3-3x+7$ a cube.

Here, -1 being a value of x that is readily found, by inspection, let $x=y-1$, agreeably to Case 4.

Then, by substitution, we shall have $2x^3-3x+7=$ $2(y-1)^3-3(y-1)+7=2y^3-6y^2+3y+8$.

And as the last term of this expression is a cube, let $8+3y-6y^2+2y^3=(2+\frac{1}{4}y)^3=8+3y+\frac{3}{8}y^2+\frac{1}{64}y^3$, according to Case 1. Then, by expunging the equal terms on each side, there will remain $2y^3 - 6y^2 = \frac{3}{8}y^2 + \frac{1}{64}y^3$.

Whence, dividing by y^2, and reducing the terms, we shall have $128y - 384 = y + 24$, or $127y = 408$; and consequently, $y = \frac{408}{127}$, and $x = \frac{408}{127} - 1 = \frac{281}{127}$.

Which number, by substitution, makes $2x^3 - 3x + 7 = \frac{2 \times (281)^3}{(127)^3} - 3 \times \frac{281}{127} + 7 = \frac{45118016}{2048383} = \left(\frac{356}{127}\right)^3$, as required. And, by taking this last as a new value of x, others may be determined by the same method.

Problem 5.—*Of the resolutions of double and triple equalities*

When a single formula, containing one or more unknown quantities, is to be transformed into a perfect power, such as a square or a cube, this is called, in the Diophantine Analysis, a simple equality; and when two formulæ, containing the same unknown quantity or quantities, are to be each transformed to some perfect power, it is then called a double equality, and so on; the methods of resolving which, in such cases as admit of any direct rule, are as follows:—

Rule 1.—In the case where the unknown quantity does not exceed the first degree, as in the double equality,

$$ax + b = \square, \text{ and } cx + d = \square,$$

let the first of these formulæ $ax + b = z^2$, and the second $cx + d = w^2$.

Then, by equating the two values of x, as found from these equations, we shall have $cz^2 + ad - bc = aw^2$, or $acz^2 + a(ad - bc) = a^2w^2$.

And since the quantity on the righthand side of this equation is now a square, it only remains to find such a value of z as will make, when the question is resolvable, $acz^2 + a(ad - bc) = \square$; which being done, according to the method pointed out in Problem 1, we shall have $x = \frac{z^2 - b}{a}$

2. When the unknown quantity does not exceed the second degree, and is found in each of the terms of the two formulæ; as in the double equality

$$ax^2 + bx = \square, \text{ and } cx^2 + dx = \square.$$

Let $x = \frac{1}{y}$; then, by substitution, and multiplying each of the resulting expressions by y^2, we shall have

$$a + by = \square, \text{ and } c + dy + \square,$$

from which last formulæ, the value of y, when the question is possible, and consequently that of x, may be determined as in Case 1.

But if it were required to make the two general expressions

$$ax^2 + bx + c = \square, \text{ and } dx^2 + ex + f = \square,$$

the solution could only be obtained in a few particular cases, as the resulting equality would rise to the fourth power.

In the case of a triple equality, where it is required to make

$$ax + by = \square, \; cx + dy = \square, \text{ and } ex + fy = \square,$$

let the first of them $ax + by = u^2$, the second $cx + dy = v^2$, and the third $ex + fy = w^2$.

Then, by first eliminating x in each of these equations, and afterwards y in the two resulting equations, we shall have $(af - be)v^2 - (cf - de)\,u^2 = (ad - bc)\,w^2$;

or, putting $v = uz$, and reducing the terms, the result will give the simple equality $\dfrac{af - be}{ad - bc}z^2 - \dfrac{cf - de}{ad - bc} = \dfrac{w^2}{u^2}$; where the righthand member being a square, it only remains to find a value of z that will make the lefthand member a square; which, when possible, may be done by Problem 1.

Hence, having z, we have, as above, $v = uz$; and the first two equations will give $x = \dfrac{d - bz^2}{ad - bc}u^2$ and $y = \dfrac{az^2 - c}{ad - bc}u^2$, where u may be any whole or fractional number whatever.

But if the three formulæ, here proposed, contained only one variable quantity, the simple equality, to which it would be necessary to reduce them, would rise, as in the last case, to the fourth power, and be equally limited with respect to its solution.

4. In other cases of this kind, all that can be done is to find successively by the former rules, several answers, when one is known; and if neither this nor any of the abovementioned modes of solution are found to succeed, the problem under consideration can only be determined by adopting some artifice of substitution that will fulfil one or more of the required conditions, and then resolving the remaining formulæ, when they are possible, by the methods already delivered for that purpose; but as no general precepts can be given, for obtaining the solution in this way, the proper mode of proceeding, in such cases, must chiefly depend upon the skill and sagacity of the learner.

EXAMPLES.

1. It is required to find a number x, such that $x + 128$ and $x + 192$ shall be both squares.

Here, according to Case 1, let $x + 128 = w^2$, and $x + 192 = z^2$.

Then, by eliminating x, and equating the result, we shall have $w^2 - 128 = z^2 - 192$, or $w^2 + 64 = z^2$.

And as the quantity on the righthand side of the equation is now a square, it only remains to make $w^2 + 64$ a square.

For which purpose, put its root $= w + n$; then $w^2 + 64 = w^2 + 2nw + n^2$, or $2nw + n^2 = 64$; and consequently, $w = \frac{64 - n^2}{2n}$; where, taking n, which is arbitrary, $= 2$, we shall have $w = \frac{64 - 4}{4} = \frac{60}{4} = 15$; and consequently $x = w^2 - 128 = 15^2 - 128 = 225 - 128 = 97$, the answer.

2. It is required to find a number x, such that $x^2 + x$ and $x^2 - x$ shall be both squares.

Here, according to Case 2, of the last Problem, let $x = \frac{1}{y}$; then we shall have to make $\frac{1}{y^2} + \frac{1}{y}$, and $\frac{1}{y^2} - \frac{1}{y}$ squares; or, by reduction, $\frac{1}{y^2}(1 + y) = \square$, and $\frac{1}{y^2}(1 - y) = \square$.

Or, since a square number, when divided by a square number, is still a square, it is the same as to make

$$1 + y = \square, \text{ and } 1 - y = \square;$$

For this purpose, therefore, let $1 + y = z^2$, or $y = z^2 - 1$; then $1 - y = 2 - z^2$; which is also to be made a square.

But as neither the first nor the last terms of this formula are squares, we must, in order to succeed, find some simple number that will answer the condition required; which, it is evident from inspection, will be the case when $z = 1$.

Let, therefore, $z = 1 - w$, agreeably to Problem 1, Case 7, and we shall have $1 - y = 2 - z^2 = 2 - (1 - w)^2 = 1 + 2w - w^2$; or $y = w^2 - 2w$;
Or, putting $1 - nw$ for the root of the former of these expressions, there will arise, by squaring, $1 + 2w - w^2 = 1 - 2nw + n^2w^2$.

Whence, expunging the 1 on each side, and dividing by w, we shall have $2 - w = -2n + n^2w$; and consequently

$$w = \frac{2n^2 + 2}{n^2 + 1}, \text{ and } x = \frac{1}{y} = \frac{1}{w^2 - 2w} = \frac{(n^2 + 1)^2}{4n - 4n^3},$$

where, in order to render the value of x positive, n may be taken equal to any proper fraction whatever.

Or if, for the sake of greater generality, $\frac{m}{n}$ be substituted for n, we shall have

$$x = \frac{(m^2 + n^2)^2}{4mn(n^2 - m^2)},$$

where m and n may now be taken equal to any integral numbers whatever, provided n be made greater than m.

If, for instance, $n = 2$ and $m = 1$, we shall have $x = \frac{25}{24}$; and if $n = 3$ and $m = 2$, $x = \frac{169}{120}$; and so on, for any other number.

3. It is required to find three whole numbers in arithmetical progression, such, that the sum of every two of them should be a square.

Let x, $x + y$, and $x + 2y$, be the three numbers sought; and put $2x + y = u^2$, $2x + 2y = v^2$, and $2x + 3y = w^2$, agreeably to Case 3.

Then, by eliminating x and y from each of these equations, we shall have $v^2 - u^2 = w^2 - v^2$, or $2v^2 - u^2 = w^2$.

And if we now put $v = uz$, there will arise $2u^2z^2 - u^2 = w^2$; or, by dividing by u^2, $2z^2 - 1 = \frac{w^2}{u^2}$; where the righthand member being a square, it only remains to make $2z^2 - 1$ a square, which it evidently is when $z = 1$.

But as this value would be found not to answer the conditions of the question, let $z = 1 - p$; then $2z^2 - 1 = 2(1 - p)^2 - 1 = 1 - 4p + 2p^2$.

And consequently, if this last expression be put $= (1 - np)^2$, we shall have, by squaring, $1 - 4p + 2p^2 = 1 - 2np + n^2p^2$, or $-4 + 2p = -2n + n^2p$; whence

$$p = \frac{2n - 4}{n^2 - 2} \text{ and } z = 1 - \frac{2n - 4}{n^2 - 2} = \frac{n^2 - 2n + 2}{n^2 - 2}.$$

Or if, for the sake of greater generality, $\frac{m}{n}$ be substituted for n in the last expression, we shall have

$$z = \frac{m^2 - 2mn + 2n^2}{m^2 - 2n^2}.$$

And since, by the two first equations, $y = v - u^2 = u^2z^2 - u^2 = (z^2 - 1)u^2$, and $x = \frac{1}{2}(u^2 - y) = \frac{1}{2}(2 - z^2)u^2$, it is evident that z must be some number greater than 1, and less than $\sqrt{2}$

If, therefore, $m = 9$ and $n = 5$, we shall have

$$z = \frac{81 - 90 + 50}{81 - 50} = \frac{41}{31},\ x = \frac{241}{31^2} \times \frac{u^2}{2},\ \text{and } y = \frac{720}{31^2} \times u^2.$$

Or, taking $u = 2 \times 31$, $x = 482$, and $y = 2880$, we have $x = 482$, $x + y = 3362$, and $x + 2y = 6242$, which are the numbers required.

4. It is required to divide a given square number into two such parts that each of them shall be a square.*

Let $a^2 =$ given square number, and x^2 and $a^2 - x^2$ its two parts. Then, since x^2 is a square, it only remains to make $a^2 - x^2$ a square.

For which purpose let its root $= nx - a$, and we shall have $a^2 - x^2 = n^2x^2 - 2anx + a^2$, or $-x^2 = n^2x^2 - 2anx$; whence, by reduction, $x = \dfrac{2an}{n^2 + 1}$, the root of the first part,

and $nx - a = \dfrac{2an^2}{n^2 + 1} - a = \dfrac{an^2 - a}{n^2 + 1}$ the root of the second.

Therefore $\left(\dfrac{2an}{n^2 + 1}\right)^2$ and $\left(\dfrac{an^2 - a}{n^2 + 1}\right)^2$ are the parts required; where a and n may be any numbers taken at pleasure, provided n be greater than 1.

5. It is required to divide a given number, consisting of two known square numbers, into two other square numbers.

Let $a^2 + b^2$ be the given numbers, and x^2, y^2, the two required numbers, whose sum, $x^2 + y^2$, is to be equal to $a^2 + b^2$.

Then it is evident, that if x be either greater or less than a, y will be accordingly less or greater than b. Let, therefore, $x = a + mz$, and $y = b - nz$, and we shall have $a^2 + 2amz + m^2z^2 + b^2 - 2bnz + n^2z^2 = a^2 + b^2$.

Or, by transposition and rejecting the terms which are common to each side of the equation, $m^2z^2 + n^2z^2 = 2bnz -$

---

* To this we may add the following useful property:—

If s and r be any two unequal numbers, of which r is the greater, it can then be readily shown, from the nature of the problem, that

$$2rs,\ s^2 - r^2,\ \text{and } s^2 + r^2.$$

will be the perpendicular, base, and hypothenuse of a right-angled triangle.

From which expressions, two square numbers may be found, whose sum or difference shall be square numbers; for $(2rs)^2 + (s^2 - r^2)^2 = (s^2 + r^2)^2$, and $(s^2 + r^2)^2 - (2rs)^2 = (s^2 - r^2)^2$, or $(s^2 + r^2)^2 - (s^2 - r^2)^2 = (2rs)^2$; where s and r may be any numbers whatever, provided r be greater than s.

$2amz$, or $m^2z + n^2z = 2bn - 2am$; whence

$$z = \frac{2bn - 2am}{m^2 + n^2}, x = \frac{2bmn + a(n^2 - m^2)}{m^2 + n^2}, y = \frac{2amn + b(m^2 - n^2)}{m^2 + n^2}$$

where m and n may be any numbers, taken at pleasure, provided their assumed values be such as will render the values of x, y, and z, in the above expressions, all positive.

6. It is required to find two square numbers, such that their difference shall be equal to a given number.

Let $d =$ the given difference; which resolve into two factors a, b, making a the greater and b the less.

Then, putting $x =$ the side of the less square, and $x + b =$ side of the greater, we shall have $(x + b)^2 - x^2 = x^2 + 2bx + b^2 - x^2 = d = (ab)$, or $2bx + b^2 = d = (ab)$.

Whence, dividing each side of this equation, by b, we shall have $x = \frac{a - b}{2} =$ the side of the less square sought, and

$x + b = \frac{a - b}{2} + b = \frac{a + b}{2} =$ the side of the greater.

If, for instance, $d = 60$, take $a \times b = 30 \times 2$, and we shall have $x = \frac{30 - 2}{2} = 14$, and $x + 2 = \frac{30 + 2}{2} = 16$, or $16^2 - 14^2 = 256 - 196 = 60$, the given difference.

7. As an instance of the great use of resolving formula of this kind into factors, let it be proposed, in addition to what has been before said, to find two numbers, x and y, such that the difference of their squares, $x^2 - y^2$, shall be an integral square.

Here the factors of $x^2 - y^2$, being $x + y$ and $x - y$, we shall have $(x + y) \times (x - y) = x^2 - y^2$. And since this product is to be a square, it will evidently become so, by making each of its factors a square, or the same multiple of a square.

Let there be taken, therefore, for this purpose,

$$x + y = mr^2, x - y = ms^2.$$

Then, by the question, we shall have $(x + y) \times (x - y)$ or its equal $x^2 - y^2 = m^2r^2s^2$; which is evidently a square, whatever may be the value of $m, r. s.$

But by addition and subtraction, the above equations give, when properly reduced,

$$x = \frac{m(r^2 + s^2)}{2}, y = \frac{m(r^2 - s^2)}{2}$$

where, as above said, m, r, and s, may be assumed at pleasure. Thus, if we take $m = 2$, we shall have $x = r^2 + s^2$, and $y = r^2 = s^2$, which expressions will obviously give integral values of x and y, if r and s be taken = any integral numbers.

8. It is required to find two numbers, such that, if either of them be added to the square of the other, the sums shall be squares.

Let x and y be the numbers sought; and consequently x^2+y and y^2+x the expressions that are to be transformed into squares. Then, if $r-x$ be assumed for the side of the first square, we shall have $x^2+y=r^2-2rx+x^2$, or $y=r^2-2rx$; and consequently, $x=\frac{r^2-y}{2r}$.

And if $s+y$ be taken for the side of the second square, we shall have $y^2+\frac{r^2-y}{2r}=s^2+2sy+y^2$; or, by reducing the equation, $r^2-y=4rsy+2rs^2$, and consequently, by reduction, $y=\frac{r^2-2rs^2}{4rs+1}$, and $x=\frac{2r^2s+s^2}{4rs+1}$; where r and s may be any numbers, taken at pleasure, provided r be greater than $2s^2$.

9. It is required to find two numbers, such that their sum and difference shall be both squares.

Let x and x^2-x be the two numbers sought; then, since their sum is evidently a square, it only remains to make their difference, x^2-2x, a square.

For this purpose, therefore, put the root $=x-r$, and we shall have $x^2-2x=x^2-2rx+r^2$;

Or, by transposition, and cancelling x^2 on each side of the equation, $2rx-2x=r^2$: whence

$$x=\frac{r^2}{2r-2}, \text{ and } x^2-2x=\tfrac{1}{4}\left(\frac{r^2}{r-1}\right)^2-\frac{r^2}{r-1};$$

where r may be any number, taken at pleasure, provided it be greater than 2.

10. It is required to find three numbers, such that not only the sum of all three of them, but also the sum of every two, shall be a square number.

Let $4x$, x^2-4x, and $2x+1$, be the three numbers sought; then, $4x+(x^2-4x)=x^2$, $(x^2-4x)+(2x+1)=x^2-2x+1$, and $4x+(x^2-4x)+(2x+1)=x^2+2x+1$, being all squares, it only remains to make $4x+(2x+1)$, or its equal, $6x+1$, a square. For which purpose, let $6x+1=n^2$, and we shall have, by transposition and division, $x=\frac{n^2-1}{6}$;

whence $\frac{4n^2-4}{6}$, $\frac{(n^2-1)^2}{36}-\frac{4n^2-4}{6}$, and $\frac{2n^2-2}{6}+1$, or their equals $\frac{2n^2-2}{3}$, $\frac{n^4-26n^2+25}{36}$, and $\frac{n^2+2}{3}$, are the numbers required.

Where n may be any number, taken at pleasure, provided it be greater than 5.

QUESTIONS FOR PRACTICE.

1. It is required to find a number x, such that $x+1$ and $x-1$ shall be both squares. Ans. $x=\frac{5}{4}$.

2. It is required to find a number x, such that $x+4$ and $x+7$ shall be both squares. Ans. $\frac{57}{16}$.

3. It is required to find a number x, such that $10+x$ and $10-x$ shall be both squares. Ans. $x=6$.

4. It is required to find a number x, such that x^2+1 and $x+1$ shall be both squares. Ans. $\frac{40}{9}$.

5. It is required to find three integral square numbers, such that the sum of every two of them shall be squares.

Ans. $(528)^2$, $(5796)^2$, and $(6325)^2$.

6. It is required to find two numbers, x and y, such that x^2+y and y^2+x shall be both squares.

Ans. $x=\frac{3}{20}$, and $y=\frac{7}{10}$.

7. It is required to find three integral square numbers that shall be in harmonical proportion. Ans. 25, 49, and 1225.

8. It is required to find three integral cube numbers, x^3, y^3, and z^3, whose sum may be equal to a cube.

Ans. $x=3$, $y=4$, $z=5$.

9. It is required to divide a given square number (100) into two such parts, that each of them may be a square number.

Ans. 64, and 36.

10. It is required to find two numbers, such that their difference may be equal to the difference of their squares, and that the sum of their squares shall be a square number.

Ans. $\frac{4}{7}$ and $\frac{3}{7}$.

11. To find two numbers, such that if each of them be added to their product, the same shall be both squares.

Ans. $\frac{2}{3}$ and $\frac{5}{3}$.

12. To find three square numbers in arithmetical progression. Ans. 1, 25, and 49.

13. To find three numbers in arithmetical progression, such that the sum of every two of them shall be a square number.

Ans. $120\frac{1}{2}$, $840\frac{1}{2}$, and $1560\frac{1}{2}$.

14. To find three numbers, such that if to the square of

each, the sum of the other two be added, the three sums shall be all squares. Ans. 1, $\frac{8}{3}$, and $\frac{16}{3}$.

15. To find two numbers in proportion as 8 is to 15, and such that the sum of their squares shall be a square number. Ans. 576 and 1080.

16. To find two numbers, such that if the square of each be added to their product, the sums shall be both squares. Ans. 9 and 16.

17. To find two whole numbers such, that the sum or difference of their squares, when diminished by unity, shall be a square. Ans. 8 and 9.

18. It is required to resolve 4225, which is the square of 65, into two other integral squares. Ans. 2704 and 1521.

19. To find three numbers in geometrical proportion, such that each of them, when increased by a given number (19), shall be square numbers. Ans. 81, $\frac{5}{4}$, and $\frac{25}{1296}$.

20. To find two numbers, such that if their product be added to the sum of their squares, the result shall be a square number. Ans. 5 and 3, 8 and 7, 16 and 5, &c.

21. To find three whole numbers such, that if to the square of each the product of the other two be added, the three sums shall be all squares. Ans. 9, 73, and 328.

22. To find three square numbers, such that their sum, when added to each of their three sides, shall be all square numbers. Ans. $\frac{4418}{62920}$, $\frac{13254}{62920}$, and $\frac{19818}{62920}$ = roots required.

23. To find three numbers in geometrical progression, such, that if the mean be added to each of the extremes, the sums, in both cases, shall be squares. Ans. 5, 20, and 80.

24. To find two numbers such, that not only each of them, but also their sum and their difference, when increased by unity, shall be all square numbers. Ans. 3024 and 5624.

25. To find three numbers such, that whether their sum be added to, or subtracted from, the square of each of them, the numbers thence arising shall be all squares. Ans. $\frac{406}{96}$, $\frac{518}{96}$, and $\frac{791}{96}$.

26. To find three square numbers such, that the sum of their squares shall also be a square number. Ans. 9, 16, and $\frac{144}{25}$.

27. To find three square numbers such, that the difference of every two of them shall be a square number. Ans. 485809, 34225, and 23409.

28. To divide any given cube number (8), into three other cube numbers. Ans. 1, $\frac{64}{27}$, and $\frac{125}{27}$.

29. To find three square numbers such, that the difference

between every two of them and the third shall be a square number. Ans. 149^2, 241^2, and 269^2.

30. To find three cube numbers such, that if from each of them a given number (1) be subtracted, the sum of the remainders shall be a square number.

Ans. $\frac{4913}{3375}$, $\frac{21952}{3375}$, and 8

OF THE SUMMATION AND INTERPOLATION OF INFINITE SERIES.

THE doctrine of Infinite Series is a subject which has engaged the attention of the greatest mathematicians, both of ancient and modern times; and when taken in its whole extent, is, perhaps, one of the most abstruse and difficult branches of abstract mathematics.

To find the sum of a series, the number of the terms of which is inexhaustible, or infinite, has been regarded by some as a paradox, or a thing impossible to be done; but this difficulty will be easily removed, by considering that every finite magnitude whatever is divisible *in infinitum*, or consists of an indefinite number of parts, the aggregate, or sum of which, is equal to the quantity first proposed.

A number actually infinite, is, indeed, a plain contradiction to all our ideas; for any number that we can possibly conceive, or of which we have any notion, must always be determinate and finite; so that a greater may still be assigned, and a greater after this; and so on, without a possibility of ever coming to an end of the increase or addition.

This inexhaustibility, therefore, in the nature of numbers, is all that we can distinctly comprehend by their infinity: for though we can easily conceive that a finite quantity may become greater and greater without end, yet we are not, by that means, enabled to form any notion of the *ultimatum*, or last magnitude, which is incapable of farther augmentation.

Hence we cannot apply to an infinite series the common notion of a sum, or of a collection of several particular numbers, which are joined and added together, one after another; as this supposes that each of the numbers composing that sum, is known and determined. But as every series generally observes some regular law, and continually approaches towards a term, or limit, we can easily conceive it to be a whole of its own kind, and that it must have a certain real value, whether that value be determinable or not.

Thus, in many series, a number is assignable, beyond which no number of its terms can ever reach, or, indeed, be ever perfectly equal to it; but yet may approach towards it in such a manner as to differ from it by less than any quantity that can be named. So that we may justly call this the value or sum of the series; not as being a number found by the common method of addition, but such a limitation of the value of the series, taken in all its infinite capacity, that, if it were possible to add all the terms together, one after another, the sum would be equal to that number.

In other series, on the contrary, the aggregate, or value of the several terms, taken collectively, has no limitation; which state of it may be expressed by saying, that the sum of the series is infinitely great; or, that it has no determinate or assignable value, but may be carried on to such a length, that its sum shall exceed any given number whatever.

Thus, as an illustration of the first of these cases, it may be observed, that if r be the ratio, g the greatest term, and l the least, of any decreasing geometric series, the sum, according to the common rule, will be $(rg - l) \div (r - 1)$: and if we suppose the less extreme l to be diminished till it becomes $= 0$, the sum of the whole series will be $rg \div (r - 1)$: for it is demonstrable that the sum of no assignable number of terms of the series can ever be equal to that quotient; and yet no number less than it will ever be equal to the value of the series.

Whatever consequences, therefore, follow from the supposition of $rg \div (r - 1)$ being the true and adequate value of the series taken in all its infinite capacity, as if all the parts were actually determined, and added together, no assignable error can possibly arise from them, in any operation or demonstration, where the sum is used in that sense; because, if it should be said that the series exceeds that value, it can be proved, that this excess must be less than any assignable difference; which is, in effect, no difference at all; whence the supposed error cannot exist, and consequently $rg \div (r - 1)$ may be looked upon as expressing the true value of the series, continued to infinity.

We are, also, farther satisfied of the reasonableness of this doctrine, by finding, in fact, that a finite quantity is frequently convertible into an infinite series, as appears in the case of circulating decimals. Thus, two thirds expressed decimally is $\frac{2}{3} = .66666$, &c., $= \frac{6}{10} + \frac{6}{100} + \frac{6}{1000} + \frac{6}{10000} +$, &c., continued *ad infinitum*. But this is a geometric series, the first term of which is $\frac{6}{10}$, and the ratio $\frac{1}{10}$; and therefore the sum of all its terms, continued to infinity, will evidently be equal to $\frac{2}{3}$, or the number from which it was originally derived. And the same may be shown of many other series, and of all circulating decimals in general.

With respect to the processes by which the summation of various kinds of infinite series are usually obtained, one of the principal is by the method of differences pointed out and illustrated in Prob. 4, next following.

Another method is that first employed by James and John Bernoulli, which consists in resolving the given series into several others of which the summation is known; or by subtracting from an assumed series, when put = s, the same series, deprived of some of its first terms; in which case a new series will arise, the sum of which will be known.

A third method, which is that of Demoivre, consists in putting the sum of the series = s, and multiplying each side of the equation by some binomial or trinomial expression, which involves the powers of the unknown quantity x, and certain known coefficients; then taking x, after the process is performed, of such a value that the assumed binomial, &c., shall become = 0, and transposing some of the first terms, a series will arise, the sum of which will be known as before.

Each of which methods, modified so as to render it more commodious in practice, together with several other artifices for the same purpose, will be found sufficiently elucidated in the miscellaneous questions succeeding the following problems:—

Problem 1.—Any series being given to find its several orders of differences.

Rule 1.—Take the first term from the second, the second from the third, the third from the fourth, &c., and the remainders will form a new series, called the *first order of differences*.

2. Take the first term of this last series from the second, the second from the third, the third from the fourth, &c., and the remainders will form another new series, called the *second order of differences*.

3. Proceed, in the same manner, for the third, fourth, fifth, &c., order of differences; and so on till they terminate, or are carried as far as may be thought necessary.*

EXAMPLES.

1. Required the several orders of differences of the series $1, 2^2, 3^2, 4^2, 5^2, 6^2$, &c.

* When the several terms of the series continually increase, the differences will all be positive; but when they decrease, the differences will be negative and positive alternately.

1, 4, 9, 16, 25, 36, &c.
3, 5, 7, 9, 11, &c., 1st diff.
2, 2, 2, 2, &c., 2d diff.
0, 0, 0, &c., 3d diff.

2. Required the different orders of differences of the series 1, 2^3, 3^3, 4^3, 5^3, 6^3, &c.

1, 8, 27, 64, 125, 216, &c.
7, 19, 37, 61, 91, &c., 1st diff.
12, 18, 24, 30, &c., 2d diff.
6, 6, 6, &c., 3d diff.
0, 0, &c., 4th diff.

3. Required the several orders of differences of the series 1, 3, 6, 10, 15, 21, &c.

Ans. 1st, 2, 3, 4, 5, &c.; 2d, 1, 1, 1, &c.

4. Required the several orders of differences of the series 1, 6, 20, 50, 105, 196, &c.

Ans. 1st, 5, 14, 30, 55, 91, &c.; 2d, 9, 16, 25, 36, &c.; 3d, 7, 9, 11, &c.; 4th, 2, 2, &c.

5. Required the several orders of differences of the series $\frac{1}{2}, \frac{1}{4}, \frac{1}{8}, \frac{1}{16}, \frac{1}{32}$, &c.

Ans. 1st, $-\frac{1}{4}, -\frac{1}{8}, -\frac{1}{16}-$, &c.; 2d, $\frac{1}{8}, \frac{1}{16}, \frac{1}{32}$, &c.; 3d, $-\frac{1}{16}, -\frac{1}{32}$, &c.; 4th, $\frac{1}{32}$, &c.

PROBLEM 2.—Any series a, b, c, d, e, &c., being given to find the first term of the nth order of differences.

RULE.—Let δ stand for the first term of the nth differences.

Then will $a - nb + n \cdot \frac{n-1}{2} c - n \cdot \frac{n-1}{2} \cdot \frac{n-2}{3} d + n \cdot \frac{n-1}{2} \cdot \frac{n-2}{3} \cdot \frac{n-3}{4} e$, &c., to $n+1$ terms $= \delta$, when n is an even number.

And $-a + nb - n \cdot \frac{n-1}{2} c + n \cdot \frac{n-1}{2} \cdot \frac{n-2}{3} d - n \cdot \frac{n-1}{2} \cdot \frac{n-2}{3} \cdot \frac{n-3}{3} e$, &c., to $n+1$ terms $= \delta$, when n is an odd number.*

* When the terms of the several orders of differences happen to be very great, it will be more convenient to take the logarithms of the quantity concerned whose differences will be smaller: and when the operation is finished, the quantity answering to the last logarithm may be easily found.

EXAMPLES.

1. Required the first term of the third order of differences of the series 1, 5, 15, 35, 70, &c.

Here a, b, c, d, e, &c., $= 1, 5, 15, 35, 70$, &c., and $n = 3$.

Whence $-a + nb - n.\frac{n-1}{2}c + n.\frac{n-1}{2}.\frac{n-2}{3}d = -a + 3b - 3c + d = -1 + 15 - 45 + 35 = 4 =$ the first term required.

2. Required the first term of the fourth order of differences of the series 1, 8, 27, 64, 125, &c.

Here a, b, c, d, e, &c., $= 1, 8, 27, 64, 125$, &c., and $n = 4$.

Whence $a - nb + n.\frac{n-1}{2}c - n.\frac{n-1}{2}.\frac{n-2}{3}d + n.\frac{n-1}{2}.\frac{n-2}{3}.\frac{n-3}{4}e = a - 4b + 6c - 4d + e = 1 - 32 + 162 - 256 + 125 = 0$; so that the first term of the fourth order is 0.

3. Required the first term of the eighth order of differences of the series 1, 3, 9, 27, 81, &c.* Ans. 256.

4. Required the first term of the fifth order of differences of the series $1, \frac{1}{2}, \frac{1}{4}, \frac{1}{8}, \frac{1}{16}, \frac{1}{32}, \frac{1}{64}$, &c. Ans. $-\frac{1}{32}$

PROBLEM 3.—To find the nth term of the series a, b, c, d, e, &c., when the differences of any order become at last equal to each other.

RULE.—Let d', d'', d''', d^{iv}, &c., be the first of the several orders of differences, found as in the last problem.

Then will $a + \frac{n-1}{1}d' + \frac{n-1}{1}.\frac{n-2}{2}d'' + \frac{n-1}{1}.\frac{n-2}{2}.\frac{n-3}{3}d''' + \frac{n-1}{1}.\frac{n-2}{2}.\frac{n-3}{3}.\frac{n-4}{4}d^{iv}$, &c., $=$ nth term required.

EXAMPLES.

1. It is required to find the twelfth term of the series, 2, 6, 12, 20, 30, &c.

2, 6, 12, 20, 30, &c.
4, 6, 8, 10, &c.
2, 2, 2, &c.
0, 0, &c.

* The labour in questions of this kind may be often abridged, by putting ciphers for some of the terms at the beginning of the series; by which means several of the differences will be equal to 0, and the answer, on that account, obtained in fewer terms.

Here 4 and 2 are the first terms of their differences.

Let, therefore, $4 = d'$, $2 = d''$, and $n = 12$.

Then $a + \frac{n-1}{1} d' + \frac{n-1}{1} . \frac{n-2}{2} d'' = 2 + 11d' + 55d'' = 2 + 44 + 110 = 156 = 12\text{th}$ term, or the answer required.

2. Required the twentieth term of the series 1, 3, 6, 10, 15, 21, &c.

1, 3, 6, 10, 15, 21, &c.
2, 3, 4, 5, 6, &c.
1, 1, 1, 1, &c.
0, 0, 0, &c.

Here 2 and 1 are the first terms of the differences.

Let, therefore, $2 = d'$, $1 = d''$, and $n = 20$.

Then $a + \frac{n-1}{1} d' + \frac{n-1}{1} . \frac{n-2}{2} d' = 1 + 19d' + 171d'' = 1 + 38 + 171 = 210 = 20\text{th}$ term required.

3. Required the fifteenth term of the series 1, 4, 9, 16, 25, 36, &c. Ans. 225.

4. Required the twentieth term of the series 1, 8, 27, 64, 125, &c. Ans. 8000.

5. Required the thirtieth term of the series $1, \frac{1}{3}, \frac{1}{6}, \frac{1}{10}, \frac{1}{15}, \frac{1}{21}$, &c. Ans. $\frac{1}{465}$.

PROBLEM 4.*—To find the sum of n terms of the series a, b, c, d, e, &c., when the differences of any order become at last equal to each other.

RULE.—Let d', d'', d''', d^{iv}, &c., be the first of the several orders of differences.

Then will $na + n\frac{n-1}{2} d' + n . \frac{n-1}{2} . \frac{n-2}{3} d'' + n . \frac{n-1}{2} . \frac{n-2}{3} . \frac{n-3}{4} d''' + n . \frac{n-1}{2} . \frac{n-2}{3} . \frac{n-3}{4} \frac{n-4}{5} d^{iv}$, &c., = to the sum of n terms of the series.

---

* When the differences in this or the former rule are finally = 0, any term, or the sum of any number of the terms, may be accurately determined; but if the difference do not vanish, the result is only an approximation; which, however, may be often very usefully applied in resolving various questions that may occur in this branch of the subject, and which will become continually nearer the truth as the differences diminish.

EXAMPLES.

1. Required the sum of n terms of the series, 1, 2, 3, 4, 5, 6, &c.

Here 1, 2, 3, 4, 5, 6, &c.
1, 1, 1, 1, 1, &c.
0, 0, 0, 0, &c.

Where 1 and 0 are the first terms of the differences

Let, therefore, $n = 1$, $d' = 1$, and $d'' = 0$.

Then will $na + n \cdot \frac{n-1}{2} d' = n + \frac{n^2-n}{2} = \frac{n^2+n}{2} =$ sum of n terms, as required.

2. Required the sum of n terms of the series $1^2, 2^2, 3^2, 4^2. 5^2$, &c., or 1, 4, 9, 16, 25, &c.

Here 1, 4, 9, 16, 25, &c.
3, 5, 7, 9, &c.
2, 2, 2, &c.
0, 0, &c.

Where 3 and 2 are the first terms of the differences.

Let, therefore, $a = 1$, $d' = 3$, and $d'' = 2$.

Then will $na + n \cdot \frac{n-1}{2} d' + n \cdot \frac{n-1}{2} \cdot \frac{n-2}{3} d'' = n + 3n \cdot \frac{n-1}{2} + 2n \cdot \frac{n-1}{2} \cdot \frac{n-2}{3} = n + \frac{3n^2-3n}{2} + \frac{n^3-3n^2+2n}{3}$

$= \frac{n \times (n+1) \times (2n+1)}{6} =$ sum of n terms as required.

3. Required the sum of n terms of the series $1^3, 2^3, 3^3, 4^3, 5^3$, &c., or 1, 8, 27, 64, 125, &c.

Here 1, 8, 27, 64, 125, &c.
7, 19, 37, 61, &c.
12, 18, 24, &c.
6, 6, &c.
0, &c.

Where the first terms of the differences are 7, 12, and 6.

Let, therefore, $a = 1$, $d' = 7$, $d'' = 12$, and $d''' = 6$.

Then will $na + n \cdot \frac{n-1}{2} d' + n \cdot \frac{n-1}{2} \cdot \frac{n-2}{3} d'' + n \cdot \frac{n-1}{2} \cdot \frac{n-2}{3} \cdot \frac{n-3}{4} d''' = n + 7n \cdot \frac{n-1}{2} + 12n \cdot \frac{n-1}{3} \cdot \frac{n-2}{3} + 6n \cdot \frac{n-1}{2} \cdot \frac{n-2}{3} \cdot \frac{n-3}{4} = n + \frac{7n^3-7n}{2} + 2n^3 - 6n^2 + 4n + \frac{n^4-6n^3+11n^2-6n}{4} = \frac{4n}{4} + \frac{14n^2-14n}{4} + \frac{8n^3-24n^2+16n}{4} +$

$\frac{n^4 - 6n^3 + 11n^2 - 6n}{4} = \frac{n^4 + 2n^3 + n^2}{4}$ = sum of n terms as required.

4. Required the sum of n terms of the series 2, 6, 12, 20, 30, &c. Ans. $\frac{n \times (n+1) \times (n+2)}{3}$.

5. Required the sum of n terms of the series 1, 3, 6, 10, 15, &c. Ans. $\frac{n}{1} \cdot \frac{n+1}{2} \cdot \frac{n+2}{3}$

6. Required the sum of n terms of the series 1, 4, 10, 20, 35, &c. Ans. $\frac{n}{1} \cdot \frac{n+1}{2} \cdot \frac{n+2}{3} \cdot \frac{n+3}{4}$.

7. Required the sum of n terms of the series 1^4, 2^4, 3^4, 4^4, &c., or 1, 16, 81, 256, &c. Ans. $\frac{n^5}{5} + \frac{n^4}{2} + \frac{n^3}{3} - \frac{n}{30}$.

8. Required the sum of n terms of the series 1^5, 2^5, 3^5, 4^5, 5^5, &c. Ans. $\frac{n^6}{6} + \frac{n^5}{2} + \frac{5n^4}{12} - \frac{n^2}{12}$.

PROBLEM 5.—The series a, b, c, d, e, &c., being given, whose terms are a unit's distance from each other, to find any intermediate term by interpolation.

RULE.—Let x be the distance of any term y, that is to be interpolated from the first term, and d', d'', d''', &c., the first terms of the differences.

Then will $a + xd' + x.\frac{x-1}{2}d'' + x.\frac{x-1}{2}.\frac{x-2}{3}d''' + x.\frac{x-1}{2}.\frac{x-2}{3}.\frac{x-3}{4}d^{iv}$, &c., $= y$.

EXAMPLES.

1. Given the logarithmic sines of 1° 0′, 1° 1′, 1° 2′, and 1° 3′, to find the log. sine of 1° 1′ 40″.

Here	1° 0′	1° 1	1° 2′	1° 3′
Sines	8.2418553	8.2490332	8.2560943	8.2630424
		71779	70611	69481
			−1168	−1130
				38

Whence the first terms of the differences are 71779, −1168, and 38.

Let, therefore, $x = 1° 1' 40'' - 1° 0' = 1' 40'' = 1\frac{2}{3}$ = distance of y, the term to be interpolated; and $d' = 71779$, $d'' = -1168$, and $d''' = 38$.

Then will $y = a + xd' + x.\frac{x-1}{2}d'' + x.\frac{x-1}{2}.\frac{x-2}{3}d''' = a +$ $\frac{5}{3}d' + \frac{5}{9}d'' - \frac{5}{81}d''' = 8.2418553 + .0119631 + 0000694 -$ $.0000002 = 8.2538232 =$ sine of $1^\circ\ 1'\ 40''$, as was required.

2. Given the series $\frac{1}{50}, \frac{1}{51}, \frac{1}{52}, \frac{1}{53}, \frac{1}{54}$, &c., to find the term which stands in the middle between the two terms $\frac{1}{52}$ and $\frac{1}{53}$.

Ans. $\frac{1}{52.5}$.

3. Given the natural tangents of $88^\circ\ 54'$, $88^\circ\ 55'$, $88^\circ\ 56'$, $88^\circ\ 57'$, $88^\circ\ 58'$, $88^\circ\ 59'$, to find the tangent of $88^\circ\ 58'18''$.

Ans. 55.711144.

PROBLEM 6.—Having given a series of equidistant terms, *a*, *b*, *c*, *d*, *e*, &c., whose first differences are small, to find any intermediate term by interpolation.

RULE.—Find the values of the unknown quantity in the equation which stands against the given number of terms, in the following table, and it will give the term required:—*

1. $a - b = 0.$
2. $a - 2b + c = 0.$
3. $a - 3b + 3c - d = 0.$
4. $a - 4b + 6c - 4d + e = 0.$
5. $a - 5b + 10c - 10d + 5e - f = 0.$
6. $a - 6b + 15c - 20d + 15e - 6f + g = 0.$

Or

$$a - nb + n.\frac{n-1}{2}c - n.\frac{n-1}{2}.\frac{n-2}{3}d + n.\frac{n-1}{2}.\frac{n-2}{3}.\frac{n-3}{4}e, \text{ \&c.}, = 0.$$

EXAMPLES.

1. Given the logarithms of 101, 102, 104, and 105, to find the logarithms of 103.

Here the number of terms is 4.

And against 4, in the table, we have $a - 4b + 6c - 4d + e = 0$; or $c = \frac{4 \times (b + d) - (a + e)}{6} =$ value of the unknown quantity, or term to be found.

* The more terms are given, in any series of this kind, the more accurately will the equation that is to be used approximate towards the true result, or answer required.

Where, taking the logs. of 01, 102, 104, and 105. | $a = 2.0043214$, $b = 2.0086002$, $d = 2.0170333$, $e = 2.0211893$

And consequently,

$$4 \times (b + d) = 16.1025340$$
$$a + e = 4.0255107$$
$$6)12.0770233$$
$$2.0128372 = \text{log. of } 103,$$

as required.

2. Given the cube roots of 45, 46, 47, 48, and 49, to find the cube root of 50. Ans. 3.684031.

3. Given the logarithms of 50, 51, 52, 54, 55, and 56, to find the logarithm of 53. Ans. 1.7242758695.

PROMISCUOUS EXAMPLES RELATING TO SERIES.

1. To find the sum (s) of n terms of the series 1, 2, 3, 4, 5, &c.

First, $1 + 2 + 3 + 4 + 5$, &c., . . . $n = s$.

And $n + (n - 1) + (n - 2) + (n - 3) + (n - 4)$, &c. . . . . $+ 1 = s$;

Therefore, by addition,

$(n + 1) + (n + 1) + (n + 1) + (n + 1) + (n + 1)$, &c. . . . . . $+ (n + 1) = 2s$.

And consequently, $n(n + 1) = 2s$; or $s = \frac{n^2 + n}{2} =$ sum required.

2. To find the sum (s) of n terms of the series 1, 3, 5, 7, 9, 11, &c.

First, $1 + 3 + 5 + 7 + 9$, &c. . . . . $(2n - 1) = s$.

And $(2n - 1) + (2n - 3) + (2n - 5) + \ldots + 1 = s$.

Therefore, by addition,

$2n + 2n + 2n + 2n + 2n +$, &c., ... $2n = 2s$.

And consequently, $2n \times n = 2s$;

$$\text{Or } s = \frac{2n^2}{2} = n^2 = \text{sum required.}$$

3. Required the sum (s) of n terms of the series $a + (a + d) + (a + 2d) + (a + 3d) + (a + 4d)$, &c.

First, $a + (a + d) + (a + 2d) + (a + 3d)$, &c., ... $+ \{a + (n - 1)d\} = s$.

And $a + (nd - d) + a + (nd - 2d) + a + (nd - 3d) + a + (nd - 4d)$, &c., ... $a = s$.

Therefore, by addition, $a + (nd - d) + 2a + (nd - d) + 2a + (nd - d)$, &c., ... $+ 2a + (nd - d) = 2s$.

And consequently, $(2a + nd - d) \times n = 2s$;

$$\text{Or } s = (2a + nd - d) \times \frac{n}{2} = \text{ sum required.}$$

Or the same may be done in a different manner, as follows:—

$a + (a + d) + (a + 2d) + (a + 3d) + (a + 4d)$, &c.

$$= \left| \begin{array}{l} (+ 1 + 1 + 1 + 1 + 1, \text{\&c.}) \times a \\ (+ 0 + 1 + 2 + 3 + 4, \text{\&c.}) \times a \end{array} \right| = s.$$

But n terms of $1 + 1 + 1 + 1 + 1$, &c., $= n$.

$$\text{And } n \text{ terms of } 0 + 1 + 2 + 3 + 4, \text{ \&c.}, = \frac{n \times (n - 1)}{2}$$

$$\text{Whence } s = na + \frac{n \times (n - 1)\, d}{2} = \{2a + d\,(n - 1)\} \times \frac{n}{2},$$

which is the same answer as before.

4. To find the sum (s) of n terms of the series $1, x, x^2, x^3, x^4$, &c.

First, $1 + x + x^2 + x^3 + x^4$, &c., ... $x^{n-1} = s$

And $x + x^2 + x^3 + x^4 + x^5$ &c., ... $x^n = sx$.

Whence, by subtraction, $x^n - 1 = sx - s$.

$$\text{Or } s = \frac{x^n - 1}{x - 1} = \text{ sum required.}$$

And when x is a proper fraction, the sum of the series, continued *ad infinitum*, may be found in the same manner.

Thus, putting $1 + x + x^2 + x^3 + x^4 + x^5$, &c., $= s$.

We shall have $x + x^2 + x^3 + x^4 + x^5$, &c., $= sx$,

And consequently, $- 1 = sx - s$; or $s - sx = 1$,

$$\text{Whence } s = \frac{1}{1 - x} = \text{ sum of an infinite number of terms,}$$

as was to be found.

5. Required the sum (s) of the circulating decimal .999999, &c., continued *ad infinitum*.

$$\text{First, } .999999, \text{ \&c.}, = \frac{9}{10} + \frac{9}{100} + \frac{9}{1000} + \frac{9}{10000}, \text{ \&c.}, =$$

$$9\left(\frac{1}{10} + \frac{1}{100} + \frac{1}{1000} + \frac{1}{10000} +, \text{ \&c.},\right) = s.$$

$$\text{Or, } \frac{1}{10} + \frac{1}{100} + \frac{1}{1000} + \frac{1}{10000} +, \text{ \&c.} = \frac{s}{9}.$$

$$\text{Therefore, } 1 + \frac{1}{10} + \frac{1}{100} + \frac{1}{1000} +, \text{ \&c.}, = \frac{10s}{9}.$$

$$\text{And consequently, } 1 = \frac{10s}{9} - \frac{s}{9} = \frac{9s}{9} = s;$$

Whence s = 1 = sum of the series.

6. Required the sum (s) of the series $a^2 + (a + d)^2 + (a + 2d)^2 + (a + 3d)^2 + (a + 4d)^2$, &c., continued to n terms.

Here,

First, $a^2 = a^2$

$$(a + d)^2 = a^2 + 2 \times 1ad + 1d^2$$
$$(a + 2d)^2 = a^2 + 2 \times 2ad + 4d^2$$
$$(a + 3d)^2 = a^2 + 2 \times 3ad + 9d^2$$
$$(a + 4d)^2 = a^2 + 2 \times 4ad + 16d^2,$$

&c. &c.

Whence

$$s = \left| \begin{array}{l} \text{Sum of } n \text{ terms of } (1 + 1 + 1 + 1 +, \text{ \&c.})\, a^2, \\ + \ldots \text{ ditto of } (0 + 1 + 2 + 3 + 4+, \text{ \&c.})\, 2ad^2, \\ + \ldots \text{ ditto of } (0 + 1 + 4 + 9 + 16 +, \text{ \&c.})\, d. \end{array} \right.$$

But n terms of $1 + 1 + 1 + 1 +$, &c., $= n$.

And of $0 + 1 + 2 + 3 + 4$, &c., $= \dfrac{n(n-1)}{1.2}$.

Also of $0 + 1 + 4 + 9 +$, &c., $= \dfrac{n(n-1)(2n-1)}{1.2.3}$.

Therefore $s = na^2 + n(n-1)\, ad + \dfrac{n(n-1)(2n-1}{1.2.3} d^2 =$ the whole sum of the series to n terms.

7. Required the sum (s) of the series $a^3 + (a + d)^3 + (a + 2d)^3 + (a + 3d)^3 + (a + 4d)^3$, &c., continued to n terms.

First, $a^3 = a^3$,

$$(a + d)^3 = a^3 + 3 \times 1a^2d + 3 \times 1ad^2 + 1d^3,$$
$$(a + 2d)^3 = a^3 + 3 \times 2a^2d + 3 \times 4ad^2 + 8d^3,$$
$$(a + 3d)^3 = a^3 + 3 \times 3a^2d + 3 \times 9ad^2 + 27d^3,$$
$$(a + 4d)^3 = a^3 + 3 \times 4a^2d + 3 \times 16ad^2 + 64d^3,$$
$$(x + 5d)^3 = a^3 + 3 \times 5a^2d + 3 \times 25ad^2 + 125d^3,$$

&c. &c.

Whence

$$s = \left| \begin{array}{l} \text{Sum of } n \text{ terms of } (1 + 1 + 1 + 1, \text{ \&c.})\, a^3, \\ + \ldots \text{ ditto of } (0 + 1 + 2 + 3 + 4, \text{ \&c.})\, 3a^2d, \\ + \ldots \text{ ditto of } (0 + 1 + 4 + 9 + 16, \text{ \&c.})\, 3ad^2, \\ + \ldots \text{ ditto of } (0 + 1 + 8 + 27 + 64, \text{ \&c.})\, d^3, \end{array} \right.$$

But n terms of $1 + 1 + 1 + 1 + 1$, &c., $= n$.

Ditto . . . of $0 + 1 + 2 + 3 + 4$, &c,. $\dfrac{n(n-1)}{1.2}$.

Ditto . . . of $0 + 1 + 4 + 9 + 16$, &c., $= \dfrac{n(n-1)(2n-1)}{1.2.3}$.

Ditto . . . of $0 + 1 + 8 + 27 + 64$, &c., $= \dfrac{n^4 - 2n^3 + n^2}{4}$.

Therefore, $s = na^3 + \frac{n(n-1)3a^2d}{2} + \frac{n(n-1)(2n-1)3ad^2}{6} + \frac{(n^4 - 2n^3 + n^2)d^3}{4} =$ sum of n terms, as was to be found.

8. Required the sum (s) of n terms of the series $1 + 3 + 7 + 15 + 31$, &c.

The terms of this series are evidently equal to 1, $(1 + 2)$, $(1 + 2 + 4)$, $(1 + 2 + 4 + 8)$, &c., or to the successive sums of the geometrical series 1, 2, 4, 8, 16, &c.

Let, therefore, $a = 1$ and $r = 2$, and we shall have
$a + ar + ar^2 + ar^3 + ar^4$, &c., $= 1 + 2 + 4 + 8 + 16$, &c.

But the successive sums of 1, 2, 3, 4, &c., terms of the series are,

1. $\frac{ar - a}{r - 1} = (r - 1) \times \frac{a}{r - 1}$.

2. $\frac{ar^2 - a}{r - 1} = (r^2 - 1) \times \frac{a}{r - 1}$.

3. $\frac{ar^3 - a}{r - 1} = (r^3 - 1) \times \frac{a}{r - 1}$.

4. $\frac{ar^4 - a}{r - 1} = (r^4 - 1) \times \frac{a}{r - 1}$.

&c. &c.

Therefore, $s = \frac{a}{r-1} \times \left| \begin{array}{l} n \text{ terms of } r + r^2 + r^3 + r^4, \text{ \&c.} \\ - n \text{ terms of } 1 + 1 + 1 + 1, \text{ \&c.} \end{array} \right.$

But $1 + 1 + 1 + 1 + 1 + 1 + 1$, &c., $= n$.

And $r + r^2 + r^3 + r^4 +$, &c., $= (r^n - 1) \times \frac{r}{r-1}$.

Whence $s = \frac{r(r^n - 1)}{r - 1} \times \frac{a}{r-1} n \times \frac{a}{r-1} = 2(2^n - 1) - n =$ whole sum required.

9. It is required to find the sum of n terms of the series $\frac{1}{1} + \frac{3}{2} + \frac{7}{4} + \frac{15}{8} + \frac{31}{16} + \frac{63}{32}$, &c.

Here the terms of this series are the successive sums of the geometrical progression $\frac{1}{1} + \frac{1}{2} + \frac{1}{4} + \frac{1}{8} + \frac{1}{16}$, &c.

Let, therefore, $a = 1$ and $r = 2$; then will

$\frac{1}{1} + \frac{1}{2} + \frac{1}{4} + \frac{1}{8}$ &c., $= a + \frac{a}{r} + \frac{a}{r^2} + \frac{a}{r^3} + \frac{a}{r^4} + \frac{a}{r^5} + \frac{a}{r^6}$, &c.

But the successive sums of 1, 2, 3, 4, &c., terms of the series are,

$$1. \frac{(r-1)\times a}{(r-1)\times 1} = (r-1)\times\frac{a}{r-1}.$$

$$2. \frac{(r^2-1)\times a}{(r-1)\times r} = (r-\frac{1}{r})\times\frac{a}{r-1}.$$

$$3. \frac{(r^3-1)\times a}{(r-1)\times r^2} = (r-\frac{1}{r^2})\times\frac{a}{r-1}.$$

$$4. \frac{(r^4-1)\times a}{(r-1)\times r^3} = (r-\frac{1}{r^3})\times\frac{a}{r-1}.$$

&c. &c.

Therefore,

$$s = \frac{a}{r-1}\times\left|\begin{array}{l} n \text{ terms of } r+r+r+r+r, \text{ \&c.} \\ -\,n \text{ terms of } \frac{1}{1}+\frac{1}{r}+\frac{1}{r^2}+\frac{1}{r^3}, \text{ \&c.}\end{array}\right.$$

These being the two series derived from the above expressions;

But $r+r+r+r+r+r$, &c., $= nr$.

$$\frac{1}{1}+\frac{1}{r}+\frac{1}{r^2}+\frac{1}{r^3}, \text{ \&c.,} = \frac{r^n-1}{(r-1)\,r^{n-1}}.$$

Whence

$$s = \frac{a}{r-1}\times\left(nr-\frac{r^n-1}{(r-1)\,r^{n-1}}\right) = \frac{(n-1)\,2^n+1}{2^{n-1}} = \text{sum}$$

required.

10. Required the sum (s) of the infinite series of the reciprocals of the triangular numbers $\frac{1}{1}+\frac{1}{3}+\frac{1}{6}+\frac{1}{10}+\frac{1}{15}$, &c.

Let $\frac{1}{1}+\frac{1}{3}+\frac{1}{6}+\frac{1}{10}$, &c., *ad infinitum* $=$ s.

Or, $\frac{1}{1.1}+\frac{1}{1.3}+\frac{1}{2.3}+\frac{1}{2.5}$, &c. . . . . $=$ s.

Then $\frac{1}{1.2}+\frac{1}{2.3}+\frac{1}{3.4}+\frac{1}{4.5}$, &c. . . . . $=\frac{s}{2}$.

That is, $\left(\frac{1}{1}-\frac{1}{2}\right)+\left(\frac{1}{2}-\frac{1}{3}\right)+\left(\frac{1}{3}-\frac{1}{4}\right)+\left(\frac{1}{4}-\frac{1}{5}\right)$, &c., $=\frac{s}{2}$.

$$\text{Or,}\quad \left|\begin{array}{l}\frac{1}{1}+\frac{1}{2}+\frac{1}{3}+\frac{1}{4}+\frac{1}{5}+\frac{1}{6}+\frac{1}{7}, \text{ \&c.} \\ \quad -\frac{1}{2}-\frac{1}{3}-\frac{1}{4}-\frac{1}{5}-\frac{1}{6}-\frac{1}{7}, \text{ \&c.}\end{array}\right| = \frac{s}{2}.$$

Whence $\frac{s}{2}=\frac{1}{1}$; or s $= 2 =$ sum required.

11. And if it be required to find the sum of n terms of the same series, $\frac{1}{1}+\frac{1}{3}+\frac{1}{6}+\frac{1}{10}+\frac{1}{15}$, &c.

Let $z=\frac{1}{1}+\frac{1}{2}+\frac{1}{3}+\frac{1}{4}+\frac{1}{5}$, &c., to $\frac{1}{n}$.

Then $z-\frac{1}{1}=\frac{1}{2}+\frac{1}{3}+\frac{1}{4}+\frac{1}{5}$, &c., to $\frac{1}{n}$.

And $z-\frac{1}{1}+\frac{1}{n+1}=\frac{1}{2}+\frac{1}{3}+\frac{1}{4}+\frac{1}{5}$, &c., to $\frac{1}{n+1}$.

Therefore $\frac{1}{1}-\frac{1}{n+1}=\frac{1}{2}+\frac{1}{6}+\frac{1}{12}+\frac{1}{20}$, &c. to $\frac{1}{n}-\frac{n}{n+1}$.

Or $\frac{n}{n+1}=\frac{1}{2}+\frac{1}{6}+\frac{1}{12}+\frac{1}{20}$, &c., to $\frac{1}{n(n+1)}$.

Whence $\frac{2n}{n+1}=\frac{1}{1}+\frac{1}{3}+\frac{1}{6}+\frac{1}{10}$, &c., to $\frac{2}{n(n+1)}$.

Or $\frac{1}{1}+\frac{1}{3}+\frac{1}{6}+\frac{1}{10}+\frac{1}{15}$, &c. to $\frac{2}{n(n+1)}=\frac{2n}{n+1}=$ sum of n terms of the series, as was required.

12. Required the sum of the infinite series $\frac{1}{1.2.3}+\frac{1}{2.3.4}+\frac{1}{3.4.5}+\frac{1}{4.5.6}$, &c.

Let $z=\frac{1}{1}+\frac{1}{2}+\frac{1}{3}+\frac{1}{4}+\frac{1}{5}$, &c., *ad infinitum.*

Then $z-\frac{1}{1}=\frac{1}{2}+\frac{1}{3}+\frac{1}{4}+\frac{1}{5}$, &c., by transposition.

And $1=\frac{1}{1.2}+\frac{1}{2.3}+\frac{1}{3.4}+\frac{1}{4.5}$, &c., by subtraction.

Or $1-\frac{1}{2}=\frac{1}{2.3}+\frac{1}{3.4}+\frac{1}{4.5}+\frac{1}{5.6}$, &c., by transposition.

Whence $\frac{1}{2}=\frac{4}{1.4.3}+\frac{6}{2.9.4}+\frac{8}{3.16.5}$, &c., by subtraction.

Or $\frac{1}{2}=\frac{2}{1.2.3}+\frac{2}{2.3.4}+\frac{2}{3.4.5}+\frac{2}{4.5.6}$, &c.

And $\frac{1}{2}\div 2=\frac{1}{1.2.3}+\frac{1}{2.3.4}+\frac{1}{3.4.5}+$, &c.

But $\frac{1}{2}\div 2=\frac{1}{4}$; therefore $\frac{1}{1.2.3}+\frac{1}{2.3.4}+\frac{1}{3.4.5}+\frac{1}{4.5.6}$,

19

ad infinitum, $=\frac{1}{4}$, which is the sum required.

13. And if it were required to find the sum of n terms of the same series $\frac{1}{1.2.3}+\frac{1}{2.3.4}+\frac{1}{3.4.5}+\frac{1}{4.5.6}$, &c.

$$\text{Let } z=\frac{1}{1.2}+\frac{1}{2.3}+\frac{1}{3.4}+\text{, \&c., to } \frac{1}{n(n+1)}.$$

$$\text{Then } z-\frac{1}{2}=\frac{1}{2.3}+\frac{1}{3.4}+\frac{1}{4.5}\text{, \&c., to } \frac{1}{n(n+1)}.$$

And $z-\frac{1}{2}+\frac{1}{(n+1)(n+2)}=\frac{1}{2.3}+\frac{1}{3.4}+\frac{1}{4.5}+\frac{1}{5.6}+\frac{1}{6.7}+\frac{1}{7.8}$, &c., continued to $\frac{1}{(n+1)(n+2)}$ terms.

Therefore $\frac{1}{2}-\frac{1}{(n+1)(n+n)}=\frac{2}{1.2.3}+\frac{2}{2.3.4}+\frac{2}{3.4.5}$, &c., to n terms, by subtraction.

Whence $\frac{1}{4}-\frac{1}{2(n+1)(n+2)}=\frac{1}{1.2.3}+\frac{1}{2.3.4}+\frac{1}{3.4.5}$, &c., to n terms, by division.

And consequently, $\frac{1}{1.2.3}+\frac{1}{2.3.4}+\frac{1}{3.4.5}$, &c., continued to n terms $=\frac{1}{4}-\frac{1}{2(n+1)(n+2)}=$ sum required.

14. Required the sum (s) of the series $\frac{1}{2}-\frac{1}{4}+\frac{1}{8}-\frac{1}{16}+\frac{1}{32}-$, &c., continued *ad infinitum*.

$$\text{Let } x=\frac{1}{2} \text{ and s} =\frac{x}{1+x}$$

$$\text{Then } \frac{z}{1+x}=x(1-x+x^2-x^3+x^4\text{, \&c.})$$

$$\text{And } z=(1+x)\times(x-x^2+x^3-x^4+x^5\text{, \&c.})$$

Whence, by multiplication,

$$\begin{array}{l} x-x^2+x^3-x^4+x^5\text{, \&c.} \\ 1+x \\ \hline x-x^2+x^3-x^4+x^5\text{, \&c.} \\ \quad +x^2-x^3+x^4-x^5\text{, \&c.} \\ \hline \end{array}$$

Whose sum is $=x+0+0+0+0$, &c.

Therefore, $z = x$, and $x - x^2 + x^3 - x^4 + x^5$, &c. $= \frac{x}{1+x}$.

Or $\frac{1}{2} - \frac{1}{4} + \frac{1}{8} - \frac{1}{16} + \frac{1}{32}$, &c., $= \frac{\frac{1}{2}}{1+\frac{1}{2}} = \frac{1}{3} =$ sum required.

15. Required the sum of the series $\frac{1}{2} + \frac{2}{4} + \frac{3}{8} + \frac{4}{16} +$, &c., continued *ad infinitum.*

Let $x = \frac{1}{2}$ and $s = \frac{z}{(1-x)^2}$.

Then, $\frac{z}{(1-x)^2} = x + 2x^2 + 3x^3 + 4x^4 + 5x^5$, &c.

And $z = (1-x)^2 \times (x + 2x^2 + 3x^3 + 4x^4 + 5x^5$, &c.)

Whence, by multiplication,

$$\begin{array}{l} x + 2x^2 + 3x^3 + 4x^4, \text{ \&c.} \\ 1 - 2x + x^2 \\ \hline x + 2x^2 + 3x^3 + 4x^4, \text{ \&c.} \\ \quad - 2x^2 - 4x^3 - 6x^4, \text{ \&c.} \\ \qquad\quad + x^3 + 2x^4, \text{ \&c.} \\ \hline \end{array}$$

Whose sum is $= x + 0 + 0 + 0 + 0$, &c.

Therefore $z = x$,

And $x + 2x^2 + 3x^3 + 4x^4 + 5x^5$, &c., $= \frac{x}{(1-x)^2}$.

Or $\frac{1}{2} + \frac{2}{4} + \frac{3}{8} + \frac{4}{16} + \frac{5}{32} + \frac{6}{64}$, &c., $= \frac{\frac{1}{2}}{(1-\frac{1}{2})^2} = 2 =$ sum of the infinite series required.

16. It is required to find the sum (s) of the series $\frac{1}{3} + \frac{4}{9} + \frac{9}{27} + \frac{16}{81} + \frac{25}{243}$, &c., continued *ad infinitum.*

Let $x = \frac{1}{3}$ and $\frac{z}{(1-x)^3} = s$.

Then $\frac{z}{(1-x)^3} = x + 4x^2 + 9x^3 + 16x^4 + 26x^5$, &c.

And $z = (1-x)^3 \times (x + 4x^2 + 9x^3 + 16x^4$, &c.) $= x + x^2$, as will be found by actual multiplication.

Therefore $x + x^2 = z$,

And $x + 4x^2 + 9x^3 + 16x^4$, &c., $+ \frac{x(1+x)}{(1-x)^3}$.

Or,

$$\frac{1}{3}+\frac{4}{9}+\frac{9}{17}+\frac{16}{81}, \text{\&c.}, = \frac{\frac{1}{3}(1+\frac{1}{3})}{(1-\frac{1}{3})^3} = \frac{3}{2} = 1\tfrac{1}{2} = \text{sum required.}$$

17. Required the sum (s) of the series $\frac{a}{m}+\frac{a+d}{mr}+\frac{a+2d}{mr^2}+\frac{a+3d}{mr^3}$, &c., continued *ad infinitum.*

$$\text{Let } x = \frac{1}{r}, \text{ and } s = \frac{z}{m(1-x)^2}.$$

$$\text{Then, } \frac{z}{m(1-x)^2} = \frac{a}{m}+\frac{a+d}{mr}+\frac{a+2d}{mr^2}+\frac{a+3d}{mr^3}, \text{\&c.}$$

$$\text{Or, } \frac{z}{(1-x)^2} = a+\frac{a+d}{r}+\frac{a+2d}{r^2}+\frac{a+3d}{r^3}, \text{\&c.}$$

$$\text{That is } \frac{z}{(1-x)^2} =$$

$$a+(a+d)x+(a+2d)x^2+(a+3d)x^3+(a+4d)x^4, \text{\&c.}$$

$$\text{And } z=(1-x)^2\times\{a+(a+d)x+(a+2d)x^2+(a+3d)x^3, \text{\&c.},\} = (1-x)a+dx,$$

as will appear by actually multiplying by $(1-x)^2$.

Therefore, $z=(1-x)a+dx$; and consequently, $\frac{a}{m}+\frac{a+d}{mr}+\frac{a+2d}{mr^2}$, &c., $=\frac{r}{m}\left\{\frac{a(r-1)+d}{(r-1)^2}\right\}$ = sum of the infinite series required.

EXAMPLES.

1. Required the sum of 100 terms of the series 2, 5, 8, 11, 14, &c. Ans. 15050.

2. Required the sum of 50 terms of the series $1+2^2+3^2+4^2+5^2$, &c. Ans. 42925.

3. It is required to find the sum of the series $1+3x+6x^2+10x^3+15x^4$, continued *ad infinitum*, &c., when x is less than 1. Ans. $\frac{1}{(1-x)^3}$.

4. It is required to find the sum of the series $1+4x+10x^2+20x^3+35x^4$, &c., continued *ad infinitum*, when x is less than 1. Ans. $\frac{1}{(1-x)^4}$

5. It is required to find the sum of the infinite series, $\frac{1}{1.3}+\frac{1}{3.5}+\frac{1}{5.7}+\frac{1}{7.9}$, &c. Ans. $\frac{5}{10}$, or $\frac{1}{2}$.

6. Required the sum of 40 terms of the series $(1 \times 2) + (3 \times 4) + (5 \times 6) + (7 \times 8)$, &c. Ans. 86884.

7. Required the sum of n terms of the series $\frac{2x-1}{2x} + \frac{2x-3}{2x} + \frac{2x-5}{2x} + \frac{2x-7}{2x}$, &c. Ans. $n\left(\frac{2x-n}{2x}\right)$.

8. Required the sum of the infinite series $\frac{1}{1.2.3.4} + \frac{1}{2.3.4.5} + \frac{1}{3.4.5.6} + \frac{1}{4.5.6.7}$, &c. Ans. $\frac{1}{18}$.

9. Required the sum of the series $\frac{1}{1} + \frac{1}{4} + \frac{1}{10} + \frac{1}{20} + \frac{1}{35}$, &c., continued *ad infinitum*. Ans. $\frac{3}{2}$, or $1\frac{1}{2}$.

10. It is required to find the sum of n terms of the series $1 + 8x + 27x^2 + 64x^3 + 125x^4$, &c., continued *ad infinitum*. Ans. $\frac{1 + 4x + x^2}{(1-x)^4}$.

11. Required the sum of n terms of the series $\frac{1}{r} + \frac{2}{r^2} + \frac{3}{r^3} + \frac{4}{r^4} + \frac{5}{r^5} + \frac{6}{r^6}$, &c. Ans. $\frac{1}{(r-1)^2} - \frac{1}{r^n}\left\{\frac{nr + r - 1}{(r-1)^2}\right\}$.

12. Required the sum of the series $\frac{1}{2.6} + \frac{1}{4.8} + \frac{1}{6.10} + \frac{1}{8.12}$, &c., . . . $+ \frac{1}{2n(1+2n)}$.*

Ans. $\Sigma = \frac{3}{16}$, $s = \frac{5n + 3n^2}{32 + 48n - 16n^2}$.

13. Required the sum of the series $\frac{1}{3.8} + \frac{1}{6.12} + \frac{1}{9.16} + \frac{1}{12.20}$, &c., . . . $\frac{1}{3n(4+4n)}$.

Ans. $\Sigma = \frac{1}{12}$, $s = \frac{n}{12 + 12n}$.

---

* The symbol Σ, made use of in these and some of the following series, denotes the sum of an infinite number of terms, and S the sum of n terms.

14. Required the sum of the series $\frac{6}{2.7} + \frac{6}{7.12} + \frac{6}{12.17} + \frac{6}{17.22}$, &c., . . . $+ \frac{6}{(5n - 3) . (5n + 2)}$.

Ans. $\Sigma = \frac{3}{5}$, $s = \frac{3n}{2 + 5n}$

15. Required the sum of the series $\frac{1}{3.6} - \frac{1}{6.8} + \frac{1}{9.10} - \frac{1}{12.12} +$, &c., . . . $\pm \frac{1}{3n (4 + 2n)}$.

Ans. $\Sigma = \frac{1}{24}$, $s = \frac{n}{2 (3 + 6n)} - \frac{n}{4 (6 + 6n)}$.

16. Required the sum of the series $\frac{2}{3.5} - \frac{3}{5.7} + \frac{4}{7.9} - \frac{5}{9.11} +$, &c., . . . $\pm \frac{1 + n}{(1 + 2n) . (3 + 2n)}$.

Ans. $\Sigma = \frac{1}{12}$, $s = \frac{1}{12} - \frac{1}{4 (3 + 4n)}$.

17. Required the sum of the series $\frac{5}{1.2.3} + \frac{6}{2.3.4} + \frac{7}{3.4.5} + \frac{8}{4.5.6}$, &c., . . . $+ \frac{4 + n}{n (1 + n) . (2 + n)}$.

Ans. $\Sigma = \frac{3}{2}$, $s = \frac{3}{2} - \frac{2}{1 + n} + \frac{1}{2 + n}$*.

OF LOGARITHMS.

Logarithms are a set of numbers that have been computed and formed into tables, for the purpose of facilitating many difficult arithmetical calculations; being so contrived, that the addition and subtraction of them answers to the multiplication

---

* The series here treated of are such as are usually called algebraical, which, of course, embrace only a small part of the whole doctrine. Those, therefore, who may wish for farther information on this abstruse but highly curious subject, are referred to the *Miscellanea Analytica* of of Demoivre, Sterling's *Method Differ.*, James Bernouilli's *de Seri. Infin.*, Simpson's *Math. Dissert.*, Waring's *Medii Analyt.*, Clark's translation of *Lorgna's Series*, the various works of Euler, and Lacroix *Traite du Calcul. Diff. et Int.*, where they will find nearly all the materials that have been hitherto collected respecting this branch of analysis.

and division of natural numbers with which they are made to correspond.*

Or, when taken in a similar but more general sense, logarithms may be considered as the exponents of the powers to which a given or invariable number must be raised, in order to produce all the common, or natural numbers. Thus, if

$$a^x = y,\ a^{x'} = y',\ a^{x''} = y'',\ \&c.$$

then will the indices x, x', x'', &c., of the several powers of a, be the logarithms of the numbers y, y', y'', &c., in the scale, or system, of which a is the base.

So that from either of these formulæ it appears, that the logarithm of any number, taken separately, is the index of that power of some other number, which, when involved in the usual way, is equal to the given number.

And since the base a, in the above expressions, can be assumed of any value, greater or less than 1, it is plain that there may be an endless variety of systems of logarithms, answering to the same natural number.

It is likewise farther evident, from the first of these equations, that when $y = 1$, x will be $= 0$, whatever may be the value of a; and consequently, the logarithm of 1 is always 0, in every system of logarithms.

And if $x = 1$, it is manifest from the same equation, that the base a will be $= y$; which base is therefore the number whose proper logarithm, in the system to which it belongs, is 1.

---

* This mode of computation, which is one of the happiest and most useful discoveries of modern times, is due to Lord Napier, Baron of Merchiston, in Scotland, who first published a table of these numbers in the year 1614, under the title of *Canon Mirificum Logarithmorum;* which performance was eagerly received by the learned throughout Europe, whose efforts were immediately directed to the improvement and extensions of the new calculus that had so unexpectedly presented itself.

Mr. Henry Briggs, in particular, who was, at that time, professor of geometry in Gresham College, greatly contributed to the advancement of this doctrine, not only by the very advantageous alteration which he first introduced into the system of these numbers, by making 1 the logarithm of 10, instead of 2.3025852, has had been done by Napier; but also by the publication, in 1624 and 1633, of his two great works, the *Arithmetica Logarithmica* and the *Trigonometrica Britanica*, both of which were formed upon the principle abovementioned; as are, likewise, all our common logarithmic tables at present in use.

See, for farther details on this part of the subject, the Introduction to my *Treatise of Plane and Spherical Trigonometry*, 8vo. 2d edit., 1813; and for the construction and use of the tables, consult those of Sherwin, Hutton, Taylor, Callet, and Borda, where every necessary information of this kind may be readily obtained.

Also, because $a^x = y$, and $a^{x'} = y'$, it follows from the multiplication of powers, that $a^x \times a^{x'}$, or $a^{x + x'} = yy'$; and consequently, by the definition of logarithms, given above, $x + x' = \log. yy'$, or,

$$\log. yy' = \log. y + \log. y'.$$

And, for a like reason, if any number of the equations $a^x = y$, $x^{x'} = y'$, $a^{x''} = y''$, &c., be multiplied together, we shall have $a^{x + x' + x''}$, &c., $= yy'y''$, &c.; and consequently, $x + x' + x''$, &c., $= \log. yy'y,''$ &c.; or,

$$\log. yy'y'', \&c.; = \log. y + \log. y' + \log. y'', \&c.$$

From which it is evident, that the logarithm of the product of any number of factors is equal to the sum of the logarithms of those factors.

Hence, if all the factors of a given number, in any case of this kind, be supposed equal to each other, and the sum of them be denoted by m, the preceding property will then become

$$\log. y^m = m \log. y.$$

From which it appears, that the logarithm of the mth power of any number is equal to m times the logarithm of that number.

In like manner, if the equation $a^x = y$ be divided by $a^{x'} = y'$, we shall have, from the nature of powers, as before, $\frac{a^x}{a^{x'}}$, or $a^{x-x'} = \frac{y}{y'}$; and, by the definition of logarithms laid down in the first part of this article, $x - x' = \log. \frac{y}{y'}$, or

$$\log. \frac{y}{y'} = \log. y - \log. y'.$$

Hence, the logarithm of a fraction, or of the quotient arising from dividing one number by another, is equal to the logarithm of the numerator *minus* the logarithm of the denominator.

And if each member of the common equation $a^x = y$ be raised to the fractional power denoted by $\frac{m}{n}$, we shall have, in that case, $a^{\frac{m}{n}x} = y^{\frac{m}{n}}$;

And consequently, by taking the logarithms, as before, $\frac{m}{n} x = \log. y^{\frac{m}{n}}$, or $\log. y^{\frac{m}{n}} = \frac{m}{n} \log. y$.

Where it appears that the logarithm of a mixed root, or power, of any number, is found by multiplying the logarithm of the given number by the numerator of the index of that power, and dividing the result by the denominator.

And if the numerator m, of the fractional index, be in this case taken equal to 1, the above formula will then become

$$\log.\, y^{\frac{1}{n}} = \frac{1}{n} \log,\, y.$$

From which it follows, that the logarithm of the nth root of any number is equal to the nth part of the logarithm of that number.

Hence, besides the use of logarithms, in abridging the operations of multiplication and division, they are equally applicable to the raising of powers and extracting of roots; which are performed by simply multiplying the given logarithm by the index of the power, or dividing it by the number denoting the root.

But although the properties here mentioned are common to every system of logarithms, it was necessary, for practical purposes, to select some one of them from the rest, and to adapt the logarithms of all the natural numbers to that particular scale.

And, as 10 is the base of our present system of arithmetic, the same number has accordingly been chosen for the base of the logarithmic system, now generally used.

So that, according to this scale, which is that of the common logarithmic tables, the numbers

$$\ldots 10^{-4},\ 10^{-3},\ 10^{-2},\ 10^{-1},\ 10^{0},\ 10^{1},\ 10^{2},\ 10^{3},\ 10^{4},\ \&c.$$

Or,

$$\ldots \frac{1}{10000},\ \frac{1}{1000},\ \frac{1}{100},\ \frac{1}{10},\ 1,\ 10, 100,\ 1000,\ 10000,\ \&c.,$$

have for their logarithms

$$\ldots\ldots -4,\ -3,\ -2,\ -1,\ 0,\ 1,\ 2,\ 3,\ 4,\ \&c.$$

Which are evidently a set of numbers in arithmetical progression, answering to another set in geometrical progression; as is the case in every system of logarithms.

And therefore, since the common or tabular logarithm of any number (n) is the index of that power of 10, which, when involved, is equal to the given number, it is plain, from the following equation,

$$10^{x} = n, \text{ or } 10^{-x} = \frac{1}{n}.$$

that the logarithms of all the intermediate numbers, in the above series, may be assigned by approximation, and made to occupy their proper places in the general scale.

It is also evident, that the logarithms of 1, 10, 100, 1000,

&c., being 0, 1, 2, 3, &c., respectively, the logarithm of any number, falling between 0 and 1, will be 0 and some decimal parts; that of a number between 10 and 100, 1 and some decimal parts; of a number between 100 and 1000, 2 and some decimal parts; and so on, for other numbers of this kind.

And, for a similar reason, the logarithms of $\frac{1}{10}, \frac{1}{100}, \frac{1}{1000}$, &c., or of their equals .1, .01, .001, &c., in the descending part of the scale, being -1, -2, -3, &c., the logarithm of any number, falling between 0 and 1, will be -1, and some positive decimal parts; that of a number between .1 and .01, -2, and some positive decimal parts; of a number between .01 and .001, -3, and some positive decimal parts; &c.

Hence, likewise, as the multiplying or dividing of any number by 10, 100, 1000, &c., is performed by barely increasing or diminishing the integral part of its logarithm by 1, 2, 3, &c., it is obvious that all numbers which consist of the same figures, whether they be integral, fractional, or mixed, will have, for the decimal part of their logarithms, the same positive quantity.

So that, in this system, the integral part of any logarithm, which is usually called its index or characteristic, is always less by 1 than the number of integers which the natural number consists of; and for decimals, it is the number which denotes the distance of the first significant figure from the place of units.

Thus, according to the logarithmic tables in common use, we have

Numbers.	*Logarithms.*
1.36820	0.1361496
20.0500	1.3021144
335.260	2.5253817
.46521	$\bar{1}$.6676490
.06154	$\bar{2}$.7891575
&c.	&c.

Where the sign $-$ is put over the index, instead of before it, then that part of the logarithm is negative, in order to distinguish it from the decimal part, which is always to be considered as $+$, or affirmative.

Also, agreeably to what has been before observed, the logarithm of 38540 being 4.5859117, the logarithms of any other numbers, consisting of the same figures, will be as follows:—

Numbers.	*Logarithms.*
3854	3.5859117
385.4	2.5859117
38.54	1.5859117
3.854	0.5859117
.3854	$\bar{1}$.5859117
.03854	$\bar{2}$.5859117
.003854	$\bar{3}$.5859117

Which logarithms, in this case, as well as in all others of a similar kind, whether the number contains ciphers or not, differ only in their indices, the decimal, or positive part, being the same in them all.*

And, as the indices, or integral parts, of the logarithms of any numbers whatever, in this system, can always be thus readily found from the simple consideration of the rule above-mentioned, they are generally omitted in the tables, being left to be supplied by the operator, as occasion requires.

It may here, also, be farther added, that when the logarithm of a given number in any particular system, is known, it will be easy to find the logarithm of the same number in any other system, by means of the following equations:—

$$a^x = n, \text{ and } e^{x'} = n, \text{ or log. } n = x, \text{ and } l.\, n = x'.$$

Where log. denotes the logarithm of n, in the system of which a is the base, and $l.$ its logarithm in the system of which e is the base.

For, since $a^x = e^{x'}$, or $a^{\frac{x}{x'}} = e$, and $e^{\frac{x'}{x}} = a$, we shall have for the base a, $\frac{x}{x'} = \text{log. } e$, or $x = x' \text{ log. } e$;

and for the base e, $\frac{x'}{x} = l.\, a$, or $x' = x\, l.\, a$.

---

* The great advantages attending the common, or Briggean system of logarithms, above all others, arise chiefly from the readiness with which we can always find the characteristic or integral part of any logarithm from the bare inspection of the natural number to which it belongs; and the circumstance, that multiplying or dividing any number by 10, 100, 1000, &c., only influences the characteristic of its logarithm, without affecting the decimal part. Thus, for instance, if i be made to denote the index or integral part of the logarithm of any number N, and d its decimal part, we shall have log. $\text{N} = i + d$; log. $10^m \times \text{N} = (i + m) + d$; log. $\frac{\text{N}}{10^m} = (i - m) + d$; where it is plain that the decimal part of the logarithm, in each of these cases, remains the same.

Whence, by substitution from the former equations

$$\log. n = l. n \times \log. e; \text{ or } \log. n = l. n \times \frac{1}{l.a}.$$

Where the multiplier, log. e, or its equal $\frac{1}{l.a}$, expresses the constant relation which the logarithms of n have to each other in the systems to which they belong.

But the only system of these numbers deserving of notice, except that above described, is the one that furnishes what have been usually called hyperbolic or Naperian logarithms, the base e of which is 2.718281828459 . . .

Hence, in comparing these with the common or tabular logarithms, we shall have, by putting a in the latter of the above formulæ $= 10$, the expression

$$\log. n = l. n \times \frac{1}{l.10}, \text{ or } l. n = \log. n \times l. 10.$$

Where log. in this case denotes the common tabular logarithm of the number n, and l its hyperbolic logarithm; the constant factor, or multiplies, $\frac{1}{l.10}$, which is $\frac{1}{2,3025850929}$, or its equal .4342944819, being what is usually called the *modulus* of the common system of logarithms.*

PROBLEM 1.—To compute the logarithm of any of the natural numbers 1, 2, 3, 4, 5, &c.

RULE 1.—1. Take the geometrical series 1, 10, 100, 1000, 10000, &c., and apply to it the arithmetical series, 0, 1, 2, 3, 4, &c., as logarithms.

2. Find a geometric mean between 1 and 10, 10 and 100, or any other two adjacent terms of the series, betwixt which the number proposed lies.

3. Also, between the mean, thus found, and the nearest extreme, find another geometrical mean in the same manner;

---

* It may here be remarked, that although the common logarithms have superseded the use of hyperbolic or Naperian logarithms, in all the ordinary operations to which these numbers are generally applied, yet the latter are not without some advantages peculiar to themselves; being of frequent occurrence in the application of the Fluxionary Calculus, to many analytical and physical problems, where they are required for the finding of certain fluents, which could not be so readily determined without their assistance; on which account great pains have been taken to calculate tables of hyperbolic logarithms, to a considerable extent, chiefly for this purpose. Mr. Barlow, in a *Collection of Mathematical Tables* lately published, has given them for the first 10000 numbers.

and so on, till you are arrived within the proposed limit of the number whose logarithm is sought.

4. Find, likewise, as many arithmetical means between the corresponding terms of the other series 0, 1, 2, 3, 4, &c., in the same order as you found the geometrical ones, and the last of these will be the logarithm answering to the number required.

EXAMPLES.

Let it be required to find the logarithm of 9.

Here the proposed number lies between 1 and 10.

First, then, the log. of 10 is 1, and the log. of 1 is 0.

There $\sqrt{(10 \times 1)} = \sqrt{10} = 3.1622777$ is the geometrical mean;

And $\frac{1}{2}(1 + 0) = \frac{1}{2} = .5$ is the arithmetical mean;

Hence the log. of 3.1622777 is .5.

Secondly, the log. of 10 is 1, and the log. of 3.1622777 is .5.

Therefore $\sqrt{(10 \times 3.1622777)} = 5.6234132$ is the geometrical mean;

And $\frac{1}{2}(1 + .5) = .75$ is the arithmetical mean;

Hence the log. of 5.6234132 is .75.

Thirdly, the log. of 10 is 1, and the log. of 5.6234132 is 75;

Therefore $\sqrt{(10 \times 5.6234132)} = 7.4989422$ is the geometrical mean;

And $\frac{1}{2}(1 + .75) = .875$ is the arithmetical mean;

Hence the log. of 7.4989422 is .875.

Fourthly, the log. of 10 is 1, and the log. of 7.498422 is 875;

Therefore $\sqrt{(10 \times 7.4989422)} = 8.6596431$ is the geometrical mean;

And $\frac{1}{2}(1 + .875) = .9375$ is the arithmetical mean;

Hence the log. of 8.6596431 is .9375.

Fifthly, the log. of 10 is 1, and the log. of 8.6596431 is .9375.

Therefore $\sqrt{(10 \times 8.6596431)} = 9.3057204$ is the geometrical mean.

And $\frac{1}{2}(1 + .9375) = .96875$ is the arithmetical mean;

Hence the log. of 9.3057204 is .96875.

Sixthly, the log. of 8.6596431 is .9375, and the log. of 9.3057204 is .96875;

Therefore $\sqrt{(8.6596431 \times 9.3057204)} = 8.9768713$ is the geometrical mean.

And $\frac{1}{2}(.9375 + .96875) = .953125$ is the arithmetical mean;

Hence the log. of 8.9768713 is .953125.

And, by proceeding in this manner, it will be found, after 25 extractions, that the logarithm of 8.9999998 is .9542425, which may be taken for the logarithm of 9, as it differs from it so little that it may be considered as sufficiently exact for all practical purposes.

And in this manner were the logarithms of all the prime numbers at first computed.

Rule 2. — When the logarithm of any number (n) is known, the logarithm of the next greater number may be readily found from the following series, by calculating a sufficient number of its terms, and then adding the given logarithm to their sum.

$$\text{Log.}\ (n+1) = \text{log.}\ n + \text{M}' \left\{ \frac{1}{2n+1} + \frac{1}{3(2n+1)^3} + \frac{1}{5(2n+1)^5} + \frac{1}{7(2n+1)^7} + \frac{1}{9(2n+1)^9} + \frac{1}{11(2n+1)^{11}}, \&\text{c.} \right\}$$

Or,

$$\text{Log.}\ (n+1) = \text{log.}\ n + \left\{ \frac{\text{M}'}{2n+1} + \frac{\text{A}}{3(2n+1)^2} + \frac{3\text{B}}{5(2n+1)^2} + \frac{5\text{C}}{7(2n+1)^2} + \frac{7\text{D}}{9(2n+1)^2} + \frac{9\text{E}}{11\,(2n+1)^2}, \&\text{c.} \right\}$$

Where A, B, C, &c., represent the terms immediately preceding those in which they are first used, and $\text{M}' =$ twice the modulus $= .8685889638 \ldots$ *

EXAMPLES.

1. Let it be required to find the common logarithm of the number 2.

Here, because $n+1=2$, and consequently $n=1$ and $2n+1=3$, we shall have

$$\frac{\text{M}'}{2n+1} = \frac{8685889638}{3} = .289529654\ (\text{A})$$

$$\frac{\text{A}}{3(2n+1)^2} = \frac{.289529654}{3.3^2} = .010723321\ (\text{B})$$

* It may here be remarked, that the difference between the logarithms of any two consecutive numbers is so much the less as the numbers are greater; and consequently, the series which comprises the latter part of the above expression will in that case converge so much the faster. Thus, log. n and log. $(n+1)$, or its equal, log. $n+$log. $(1+\frac{1}{n})$, will obviously differ but little from each other when n is a larger number.

$$\frac{3\text{B}}{5(2n+1)^2} = \frac{3 \times .010723321}{5.3^2} = .000714888 \text{ (C)}$$

$$\frac{5\text{C}}{7(2n+1)^2} = \frac{5 \times .000714888}{7.3^2} = .000056737 \text{ (D)}$$

$$\frac{7\text{D}}{9(2n+1)^2} = \frac{7 \times .000056737}{9.3^2} = .000004903 \text{ (E)}$$

$$\frac{9\text{E}}{11(2n+1)^2} = \frac{9 \times .000004903}{11.3^2} = .000000446 \text{ (F)}$$

$$\frac{11\text{F}}{13(2n+1)^2} = \frac{11 \times .000000446}{13.3^2} = .000000042 \text{ (G)}$$

$$\frac{13\text{G}}{15(2n+1)^2} = \frac{13 \times .000000042}{15.3^2} = .000000004 \text{ (H)}$$

Sum of 8 terms . .	.301029995
Add log. of 1 . .	.000000000
Log. of 2 . . .	.301029995

Which logarithm is true to the last figure inclusively.

2. Let it be required to compute the logarithm of the number 3.

Here, since $n+1=3$, and consequently, $n=2$, and $2n+1=5$, we shall have

$$\frac{\text{M}'}{2n+1} = \frac{.868588964}{5} \;..\; = .173717793 \text{ (A)}$$

$$\frac{\text{A}}{3(2n+1)^2} = \frac{.173717793}{3.5^2} \;....\; = .002316237 \text{ (B)}$$

$$\frac{3\text{B}}{5(2n+1)^2} = \frac{3 \times .002316237}{5.5^2} = .000055590 \text{ (C)}$$

$$\frac{5\text{C}}{7(2n+1)^2} = \frac{5 \times .000055590}{7.5^2} = .000001588 \text{ (D)}$$

$$\frac{7\text{D}}{9(2n+1)^2} = \frac{7 \times .000001588}{9.5^2} = .000000050 \text{ (E)}$$

$$\frac{9\text{E}}{11(2n+1)^2} = \frac{9 \times .000000050}{11.5^2} = .000000002 \text{ (F)}$$

Sum of 6 terms . .	.176091260
Log. of 2 . . .	.301029995
Log. of 3 . . .	.477121255

Which logarithm is also correct to the nearest unit in the last figure.

And in same way we may proceed to find the logarithm of any prime number.

Also, because the sum of the logarithms of any two numbers gives the logarithm of their product, and the difference of the logarithms the logarithm of their quotient, &c.; we may readily compute, from the above two logarithms, and the logarithm of 10, which is 1, a great number of other logarithms, as in the following examples:—

3. Because $2 \times 2 = 4$, therefore log. 2	.301029995
Mult. by 2	2
gives log. 4	.602059990

4. Because $2 \times 3 = 6$, therefore to log. 2	.301029995
add log. 3	.477121255
gives log. 6	.778151250

5. Because $2^3 = 8$, therefore log. 2	.301029995
Mult. by 3	3
gives log. 8	.903089985

6. Because $3^2 = 9$, therefore log. 3	.477121255
Mult. by 2	2
gives log. 9	.954242510

7. Because $\frac{10}{2} = 5$, therefore from log. 10	1.000000000
take log. 2	.301029995
gives log. 5	.698970005

8. Because $3 \times 4 = 12$, therefore to log. 3	.477121255
add log. 4	.602059991
gives log. 12	1.079181246

And thus, by computing, according to the general formula,

the logarithms of the next succeeding prime numbers 7, 11, 13, 17, 19, 23, &c., we can find, by means of the simple rules before laid down for multiplication, division, and the raising of powers, as many other logarithms as we please, or may speedily examine any logarithm in the table.

MULTPILICATION BY LOGARITHMS.

TAKE out the logarithms of the factors from the table, and add them together; then the natural number, answering to the sum, will be the product required.

Observing, in the addition, that what is to be carried from the decimal part of the logarithms is also affirmative, and must, therefore, be added to the indices, or integral parts, after the manner of positive and negative quantities in algebra.

Which method will be found much more convenient, to those who possess a slight knowledge of this science, than that of using the arithmetical complements.

EXAMPLES.

1. Multiply 37.153 by 4.086, by logarithms.

Nos.	*Logs.*
37.153	1.5699939
4.086	0.6112984
Prod. 151.8071 .	2.1812923

2. Multiply 112.246 by 13.958, by logarithms.

Nos.	*Logs.*
112.246	2.0491709
13.958	1,1448232
Prod. 1563.128 .	3.1939941

3. Multiply 46.7512 by .3275, by logarithms.

Nos.	*Logs.*
46.7512	1.6697928
.3275	$\bar{1}$.5152113
Prod. 15.31102 .	1.1850041

Here, the + 1, that is to be carried from the decimals, cancel the − 1, and consequently there remains 1 in the upper line to be set down.

4. Multiply, .37816 by .04782, by logarithms.

Nos.		*Logs.*
.37816	.	$\bar{1}$.5776756
.04782	. .	$\bar{2}$.6796096
Prod. .0180836	.	$\bar{2}$.2572852

Here the + 1, that is to be carried from the decimals, destroys the — 1, in the upper line, as before, and there remains the — 2 to be set down.

5. Multiply 3.768, 2.053, and .007693, together.

Nos.		*Logs.*
7.768	. . . .	0.5761109
2.053	. . . .	0.3123889
.007693	. . .	$\bar{3}$.8860997
Prod. .059511	.	$\bar{2}$.7745995

Here the + 1, that is to be carried from the decimals, when added to — 3, makes — 2 to be set down.

6 Multiply 3.586, 2.1046, 8372, and 0294, together.

Nos.		*Logs.*
3.586	. . . . .	0.554610
2.1046	. . . .	0.323170
.8372	. . . . .	$\bar{1}$.922829
.0294	. . . .	$\bar{2}$.468347
Prod. .1857618	.	$\bar{1}$.268956

Here the + 2, that is to be carried, cancels the — 2, and there remains the — 1 to be set down.

7. Multiply 23.14 by 5.062 by logarithms.

Ans. 117.1347.

8. Multiply 4.0763 by 9.8432, by logarithms.

Ans. 40.12383.

9. Multiply 498.256 by 41.2467, by logarithms.

Ans. 20551.41.

10. Multiply 4.026747 by .012345, by logarithms.

Ans. 0497102.

11. Multiply 3.12567, .02868, and .12379, together, by logarithms. Ans. .01109705.

12. Multiply 2876.9, .10674 .098762, and .0031598, by logarithms. Ans. .0958299.

DIVISION BY LOGARITHMS.

From the logarithm of the dividend, as found in the tables, subtract the logarithm of the divisor, and the natural number answering to the remainder, will be the quotient required.

Observing, if the subtraction cannot be made in the usual way, to add, as in the former rule, the 1 that is to be carried from the decimal part, when it occurs, to the index of the logarithm of the divisor, and then this result, with its sign changed, to the remaining index, for the index of the logarithm of the quotient.

EXAMPLES.

1. Divide 4768.2 by 36.954, by logarithms.

Nos.	*Logs.*
4768.2	3.6783545
36.954	1.5676615
Quot. 129.032 . .	2.1106930

2. Divide 21.754 by 2.4678, by logarithms.

Nos.	*Logs.*
21.754	1.3375391
2.4678	0.3923100
Quot. 8.1518 . .	0.9452291

3. Divide 4.6257 by .17608, by logarithms.

Nos.	*Logs.*
4.6257	0.6651725
.17608	$\bar{1}$.2457100
Quot. 26.2741 . .	1.4194625

Here − 1, in the lower index, is changed into + 1, which is then taken for the index of the result.

4. Divide .27684 by 5.1576, by logarithms.

Nos.	*Logs.*
.27684	$\bar{1}$.4422288
5.1576	0.7124477
Quot. .0536761	$\bar{1}$.7297811

Here the 1 that is to be carried from the decimals, is taken as — 1, and then added to — 1, in the upper index, which gives — 2 for the index of the result.

5. Divide 6.9875 by .075789, by logarithms.

Nos.	*Logs.*
6.9875	0.8443218
.075789	$\bar{2}$.8796062
Quot. 92.1967 .	1.9647156

Here the 1, that is to be carried from the decimals, is added to — 2, which makes — 1, and this put down, with its sign changed, is + 1.

6. Divide .19876 by .0012345, by logarithms.

Nos.	*Logs.*
.19786	$\bar{1}$.2983290
.001.2345 . . .	$\bar{3}$.0914911
Quot. 161.0051 .	2.2068379

Here — 3, in the lower index, is changed into + 3, and this added to — 1, the other index, gives + 3 — 1 or 2.

7. Divide 125 by 1728, by logarithms. Ans. 0723379.

8. Divide 1728.95 by 1.10678, by logarithms. Ans. 1562.144.

9. Divide 10.23674 by 4.96523, by logarithms. Ans. 2.061685.

10. Divide 19956.7 by .048235, by logarithms Ans. .413739.

11. Divide .067859 by 1234.59, by logarithms. Ans. .0000549648.

THE RULE OF THREE, OR PROPORTION, BY LOGARITHMS.

For any single proportion, add the logarithms of the second and third terms together, and subtract the logarithm of the first from their sum, according to the foregoing rules; then the natural number answering to the result will be the fourth term required.

But if the proportion be compound, add together the logarithms of all the terms that are to be multiplied, and from the result take the sum of the logarithms of the other terms, and the remainder will be the logarithm of the term sought.

Or, the same may be performed most conveniently thus :—

Find the complement of the logarithm of the first term of the proportion, or what it wants of 10, by beginning at the lefthand, and taking each of its figures from 9, except the last significant figure on the right, which must be taken from 10; then add this result and the logarithms of the other two terms together, and the sum, abating 10 in the index, will be the logarithm of the fourth term, as before.

And if two or more logarithms are to be subtracted, as in the latter part of the above rule, add their complements and the logarithms of the terms to be multiplied together, and the result, abating as many 10s in the index as there are logarithms to be subtracted, will be the logarithm of the term required; observing when the index of the logarithm, whose complement is to be taken, is negative, to add it, as if it were affirmative, to 9; and then take the rest of the figures from 9, as before.

EXAMPLES.

1. Find a fourth proportion to 37.125, 14.768, and 135,279, by logarithms.

Log. of 37.125	1.5696665
Complement	8.4303335
Log. of 14.768	1.1693217
Log. of 135.279	2.1312304
Ans. 53.81099 . . .	1.7308856

2. Find a fourth proportional to .05764, .7186, and .34721, by logarithms.

Log. of .05764	$\bar{2}$.7607240
Complement	11.2392760
Log. of .7186	$\bar{1}$.8564872
Log. of .34721	$\bar{1}$.5405992
Ans. 4.328681 . . .	0.6363554

3. Find a third proportional to 12.796, and 3.24718, by logarithms.

Log. of 12.796	1.1070742
Complement	8.8929258
Log. of 3.24718	0.5115064

Log. of 3.24718	0.5115064
Ans. .8240216 . . .	$\bar{1}$.9159386

4. Find the interest of 279*l.* 5*s.* for 274 days, at 4½ per cent. per annum, by logarithms.

Comp. log. of 100 . . .	8.0000000
Comp. log. of 365 . . .	7.4377071
Log. of 279.25	2.4459932
Log. of 274	2.4377506
Log. of 4.5	0.6532125
Ans. 9.433296 . . .	0.9746634

5. Find a fourth proportional to 12.678, 14.065, and 100.979, by logarithms. Ans. 112.0263.

6. Find a fourth proportional to 1.9864, .4678, and 50.4567, by logarithms. Ans. 11.88262.

7. Find a fourth proportional to .09658, .24958, and .005967, by logarithms. Ans. .02317234.

8. Find a third proportional to .498621, and 2.9587, and a third proportional to 12.796, and 3.24718, by logarithms.
Ans. 17.55623, and .8240216.

INVOLUTION, OR THE RAISING OF POWERS BY LOGARITHMS.

Take out the logarithm of the given number from the tables, and multiply it by the index of the proposed power; then the natural number answering to the result, will be the power required.

Observing, if the index of the logarithm be negative, that this part of the product will be negative; but as what is to be carried from the decimal part will be affirmative, the index of the result must be taken accordingly.

EXAMPLES.

1. Find the square of 2.7568, by logarithms.

Log. of 2.7568	0.4402477
Square 7.599946 . . .	0.8804954

2. Find the cube of 7.0851, by logarithms.

Log. of 7.0851 , . . . 0.8503399
3

Cube 355.6475 . . . . 2.5510197

3 Find the fifth power of .87451, by logarithms.

Log. of .87451 . . . . $\bar{1}$.9417648
5

Fifth power .5114695 . 1.7088240

Where 5 times the negative index — 1, being — 5, and + 4 to carry, the index of the power is $\bar{1}$.

4. Find the 365th power of 1.0045, by logarithms.

Log. 1.0045* 0.0019499
365

97495
116994
58497

Power 5.148888 . . 07117135

5. Required the square of 6.05987, by logarithms.
Ans. 36.72603.

6. Required the cube of .176546, by logarithms.
Ans. 005502674.

7. Required the 4th power of .076543, by logarithms.
Ans. 0000343259.

8. Required the 5th power of 2.97643, by logarithms.
Ans. 233.6031.

9. Required the 6th power of 21.0576, by logarithms.
Ans. 87187340.

10. Required the 7th power of 1.09684, by logarithms.
Ans. 1.909864.

* The answer, 5.148888, though found strictly according to the general rule, is not correct in the last four figures, 8888; nor can the answers to such questions, relating to very high powers, be generally found true to 6 places of figures by the tables of Log. commonly used; if any power above the hundred thousandth were required, not one figure of the answer here given could be depended on. The Log. of 1.0045 is 00194994108 true to eleven places, which, multiplied by 365, gives .7117285 true to 7 places, and the corresponding number true to 7 places is 5.149067. See Doctor Adrain's edition of Hut. Math. Vol. 1. p. 169.

EVOLUTION, OR THE EXTRACTION OF ROOTS, BY LOGARITHMS.

Take out the logarithm of the given number from the table, and divide it by 2 for the square root, 3 for the cube root, &c., and the natural number, answering to the result, will be the root required.

But if it be a compound root, or one that consists of both a root and a power, multiply the logarithm of the given number by the numerator of the index, and divide the product by the denominator, for the logarithm of the root sought.

Observing, in either case, when the index of the logarithm is negative, and cannot be divided without a remainder, to increase it by such a number as will render it exactly divisible; and then carry the units borrowed, as so many tens, to the first figure of the decimal part, and divide the whole accordingly.

EXAMPLES.

1. Find the square root of 27.465, by logarithms.

Log. of 27.465 . . . 2)1.4387796

Root 5.2407 7193898

2. Find the cube root of 35.6415, by logarithms.

Log. of 35,6415 . . . 3)1.5519560

Root 3.29093 5173186

3. Find the 5th root of 7.0825, by logarithms.

Log. of 7.0825 . . . . 5)0.8501866

Root 1.479235 1700373

4. Find the 365th root of 1.045, by logarithms.

Log. of 1.045 . . . 365)0.0019116

Root 1.000121 . . . . .0.0000524

5. Find the value of $(.001234)^{\frac{2}{3}}$, by logarithms.

Log. of .001234 . . . . $\bar{3}.0913152$
$\quad 2$

$3)\bar{6}.1826304$

Ans. .00115047 . . . $\bar{2}.0608768$.

Here, the divisor 3 being contained exactly twice in the negative index -6, the index of the quotient, to be put down, will be -2.

6. Find the value of $(.024554)^{\frac{2}{3}}$, by logarithms.

Log. of .024554 . . . . $\bar{2}.3901223$

$2)\bar{5}.1703669$

Ans. .00384754 . . . $\bar{3}.5851834$.

Here 2 not being contained exactly in -5, 1 is added to it, which gives -3 for the quotient; and the 1 that is borrowed, being carried to the next figure, makes 11, which, divided by 2, gives .58, &c.

7. Required the square root of 365.5674, by logarithms.
Ans. 19.11981.

8. Required the cube root of 2.987635, by logarithms.
Ans. 1.440265.

9. Required the 4th root of .967845, by logarithms.
Ans. .9918624.

10. Required the 7th root of .098674, by logarithms.
Ans. .7183146.

11. Required the value of $\left(\frac{21}{373}\right)^{\frac{2}{3}}$, by logarithms.
Ans. .146895.

12. Required the value of $\left(\frac{112}{1727}\right)^{\frac{3}{5}}$, by logarithms.
Ans. .1937115.

MISCELLANEOUS EXAMPLES IN LOGARITHMS.

1. Required the square root of $\frac{2}{123}$, by logarithms.
Ans. .1275153.

2. Required the cube root of $\frac{1}{3.14159}$, by logarithms.
Ans. .6827842.

3. Required the .07 power of .00563, by logarithms.

Ans. .6958821.

4. Required the value of $\frac{(\frac{2}{3})^{\frac{1}{2}} \times (\frac{3}{4})^{\frac{1}{3}}}{17\frac{1}{3}}$, by logarithms.

Ans. .04279825.

5. Required the value of $\frac{1}{7} \sqrt{\frac{5}{8}} \times .012 \sqrt[3]{\frac{7}{11}}$, by logarithms.

Ans. .001165713.

6. Required the value of $\frac{\frac{1}{9} \sqrt{\frac{11}{21}} \times .03 \sqrt[3]{15\frac{1}{5}}}{7\frac{1}{3} \sqrt[3]{12\frac{1}{5}} \times .19 \sqrt[4]{17\frac{1}{8}}}$, by logarithms.

Ans. .0009158638.

7. Required the value of $\frac{127}{4}\left(\frac{\frac{5}{6}\sqrt{19} + \frac{4}{7}\sqrt[3]{35\frac{1}{3}}}{14\frac{7}{19} - \frac{1}{11}\sqrt[5]{28\frac{2}{3}}}\right)$, by logarithms.

Ans. 49.38712.

MISCELLANEOUS QUESTIONS.

1. A person being asked what o'clock it was, replied that it was between eight and nine, and that the hour and minute hands were exactly together; what was the time?

Ans. 8 h. 43 min. $38\frac{2}{11}$ sec.

2. A certain number, consisting of two places of figures, is equal to the difference of the squares of its digits, and if 36 be added to it, the digits will be inverted; what is the number? Ans. 48.

3. What two numbers are those, whose difference, sum, and product, are to each other as the numbers 2, 3, and 5, respectively? Ans. 2 and 10.

4. A person, in a party at cards, bet three shillings to two upon every deal, and after twenty deals found he had gained five shillings; how many deals did he win? Ans. 13.

5. A person wishing to enclose a piece of ground with palisades, found, if he set them a foot asunder, that he should have too few by 150, but if he set them a yard asunder he should have too many by 70; how many had he? Ans. 180.

6. A cistern will be filled by two cocks, A and B, running together, in twelve hours, and by the cock A alone in twenty hours; in what time will it be filled by the cock B alone?

Ans. 30 hours.

7. If three agents, A, B, C, can produce the effects, a, b, c, in the times e, f, g, respectively; in what time would they jointly produce the effect d.

Ans. $d \div \left(\frac{a}{e} + \frac{d}{f} + \frac{c}{g}\right)$.

8. What number is that, which being severally added to 3, 19, and 51, shall make the results in geometrical progression? Ans. 13.

9. It is required to find two geometrical mean proportionals between 3 and 24, and four geometrical means between 3 and 96. Ans. 6 and 12; and 6, 12, 24, and 48.

10. It is required to find six numbers in geometrical progression such, that their sum shall be 315, and the sum of the two extremes 165. Ans. 5, 10, 20, 40, 80, and 160.

11. The sum of two numbers is a, and the sum of their reciprocals is b, required the numbers.

$$\text{Ans } \frac{a}{2} \pm \frac{1}{2} \sqrt{\left\{\frac{a}{q}(ab-4)\right\}}.$$

12. After a certain number of men had been employed on a piece of work for 24 days, and had half finished it, 16 men more were set on, by which the remaining half was completed in 16 days; how many men were employed at first; and what was the whole expense, at 1*s*. 6*d*. a day per man? Ans. 32 the number of men; and the whole expense 115*l*. 4*s*.

13. It is required to find two numbers such, that if the squares of the first be added to the second, the sum shall be 62, and if the square of the second be added to the first, it shall be 176. Ans. 7 and 13.

14. The fore wheel of a carriage makes six revolutions more than the hind wheel, in going 120 yards; but if the circumference of each wheel was increased by three feet, it would make only four revolutions more than the hind wheel in the same space; what is the circumference of each wheel? Ans 12 and 15 feet.

15. It is required to divide a given number a into two such parts, x and y, that the sum of mx and ny shall be equal to some other given number b.

$$\text{Ans. } x = \frac{b - an}{m - n} \text{ and } y = \frac{am - b}{m - n}.$$

16. Out of a pipe of wine, containing 84 gallons, 10 gallons were drawn off, and the vessel replenished with 10 gallons of water; after which 10 gallons of the mixture were again drawn off, and then 10 gallons more of water poured in; and so on for a third and fourth time; which being done, it is required to find how much pure wine remained in the vessel, supposing the two fluids to have been thoroughly mixed each time? Ans. $48\frac{4}{5}$ gallons.

17. A sum of money is to be divided equally among a

certain number of persons; now, if there had been 3 claimants less, each would have had 150*l.* more, and if there had been 6 more, each would have had 120*l.* less; required the number of persons and the sum divided. Ans. 9 persons; sum 2700*l.*

18. From each of 16 pieces of gold, a person filed the worth of half a crown, and then offered them in payment for their original value, but the fraud being detected, and the pieces weighed, they were found to be worth in the whole, no more than eight guineas; what was the original value of each piece? Ans. 13*s.*

19. A composition of tin and copper, containing 100 cubic inches, was found to weigh 505 ounces; how many ounces of each did it contain, supposing the weight of a cubic inch of copper to be $5\frac{1}{4}$ ounces, and that of a cubic inch of tin $4\frac{1}{4}$ ounces? Ans. 420 oz. of copper, and 85 oz. of tin.

20. A privateer, running at the rate of 10 miles an hour, discovers a vessel 18 miles ahead of her, making way at the rate of 8 miles an hour; how many miles will the latter run before she is overtaken? Ans. 72 miles.

21. In how many different ways is it possible to pay 100*l.* with seven shilling pieces, and dollars of 4*s.* 6*d.* each?

Ans. 31 different ways.

22. Given the sum of two numbers $= 2$, and the sum of their ninth powers $= 32$, to find the numbers by a quadratic equation. Ans. $1 \pm \frac{1}{3}\sqrt{(6\sqrt{34} - 33)}$.

23. Given $y^3 - xy = 666$, and $x^3 + xy = 406$, to find x and y. Ans. $x = 7$, and $y = 9$.

24. The arithmetical mean of two numbers exceeds the geometrical mean by 13, and the geometrical mean exceeds the harmonical mean by 12; what are the numbers?

Ans. 234 and 104.

25. Given $x^3y + y^3x = 3$, and $x^6 y^2 + y^6x^2 = 7$, to find the values of x and y. Ans. $x = \frac{1}{2}(\sqrt{5} + 1)$, $y = \frac{1}{2}(\sqrt{5} - 1)$.

26. Given $x + y + z = 23$, $xy + xz + yz = 167$, and $xyz = 385$, to find x, y, and z. Ans. $x = 5$, $y = 7$, $z = 11$.

27. To find four numbers, x, y, z, and w, having the product of every three of them given; viz. $xyz = 231$, $xyw = 420$, $yzw = 1540$, and $xzw = 660$.

Ans. $x = 3$, $y = 17$, $z = 11$, and $w = 20$.

28. Given $x + yz - 384$, $y + xz = 237$, and $z + xy = 192$, to find the values of x, y, and z.

Ans. $x = 10$, $y = 7$, and $z = 22$.

29. Given $x^2 + xy = 108$, $y^2 + yz = 69$, and $x^2 + xz = 580$, to find the values of x, y, and z.

Ans. $x = 9$, $y = 3$, and $z = 20$

30. Given $x^2 + xy + y^2 = 5$, and $x^4 + x^2y^2 + y^4 = 11$, to find the values of x and y by a quadratic.

Ans. $x = \frac{2}{5}\sqrt{10} + \frac{1}{5}\sqrt{5}$, $y = \frac{2}{5}\sqrt{10} - \frac{1}{5}\sqrt{5}$.

31. Given the equation $x^4 - 6x^3 + 13x^2 - 12x = 5$, to find the value of x by a quadratic. Ans. $\frac{3}{2} \pm \frac{1}{2}\sqrt{13}$.

32. It is required to find by what part of the population a people must increase annually, so that they may be double at the end of every century. Ans. By 144th part *nearly*.

33. Required the least number of weights, and the weight of each, that will weigh any number of pounds from one pound to a hundred weight. Ans. 1, 3, 9, 27, 81.

34. It is required to find four whole numbers such, that the squares of the greatest may be equal to the sum of the squares of the other three. Ans. 3, 4, 12, and 13.

35. It is required to find the least number, which, being divided by 6, 5, 4, 3, and 2, shall leave the remainders, 5, 4, 3, 2, and 1, respectively. Ans. 59.

36. Given the cycle of the sun 18, the golden number 8, and the Roman indiction 10, to find the year. Ans. 1717.

37. Given $256x - 87y = 1$, to find the least possible values of x and y in whole numbers. Ans. $x = 52$, and $y = 153$.

38. It is required to find two different isosceles triangles such, that their perimeters and areas shall be both expressed by the same numbers.

Ans. Sides of the one 29, 29, 40; and of the other 37, 37, 24.

39. It is required to find the sides of three right-angled triangles, in whole numbers, such, that their areas shall be all equal to each other.

Ans. 58, 40, 42; 74, 24, 70; 113, 15, 112.

40. Given $x^{\frac{1}{x}} = 1.2655$, to find a near approximate value of x. Ans. 3.82013.

41. Given $xy = 5000$, and $y^x = 3000$, to find the values of x and y. Ans. $x = 4.691445$, and $y = 5.510132$.

42. Given $x^x + y^y = 285$, and $y^x - x^y = 14$, to find the values of x and y. Ans. $x = 4.016698$, and $y = 2.825716$.

43. To find two whole numbers such, that if unity be added to each of them, and also to their halves, the sums, in both cases, shall be squares. Ans. 48 and 1680.

44. Required the two least nonquadrate numbers x and y such, that $x^2 + y^2$ and $x^3 + y^3$ shall be both squares.

Ans. $x = 364$, and $y = 273$.

45. It is required to find two whole numbers such, that

their sum shall be a cube, and their product and quotient squares. Ans. 25 and 100, or 100 and 900, &c.

46. It is required to find three biquadratic numbers such, that their sum shall be a square. Ans. 12^4, 15^4, and 20^4.

47. It is required to find three numbers in continued geometrical progression such, that their three differences shall be all squares. Ans. 567, 1008, and 1792.

48. It is required to find three whole numbers such, that the sum of difference of any two of them shall be square numbers. Ans. 434657, 420968, and 150568.

49. It is required to find two whole numbers such, that their sum shall be a square, and the sum of their squares a biquadrate. Ans. 4565486027761, and 1061652293520.

50. It is required to find four whole numbers such, that the difference of every two of them shall be a square number.

Ans. 1873432, 2288168, 2399057, and 6560657.

51. It is required to find the sum of the series $\frac{1}{3}+\frac{2}{9}+\frac{3}{27}+\frac{4}{81}$, &c., continued to infinity. Ans. $\frac{3}{4}$.

52. It is requlred to find the sum of the infinite series $\frac{3}{4}-\frac{9}{16}+\frac{27}{64}-\frac{81}{256}+\frac{243}{1024}$, &c. Ans. $\frac{3}{7}$.

53. Required the sum of the series $5+6+7+8+9+$, &c., continued to n terms Ans. $\frac{n}{2}(n+9)$.

54. It is required to find how many figures it would take to express the 25th term of the series $2^1+2^2+2^4+2^8+2^{16}$, &c. Ans. 5050446 figures.

55. It is required to find the sum of 100 terms of the series $(1\times 2)+(3\times 4)+(5\times 6)+(7\times 8)+(9\times 10)$, &c.

Ans. 1343300.

56. Required the sum of $1^2+2^2+3^2+4^2+5^2$, &c., ... $+ 50$, which gives the number of shot in a square pile, the side of which is 50. Ans. 42925.

57. Required the sum of 25 terms of the series $35+36\times 2+37\times 3+38\times 4+39\times 5$, &c., which gives the number of shot in a complete oblong pile, consisting of 25 tiers, the number of shot in the uppermost row being 35.

Ans. 16575.

APPENDIX.

OF THE APPLICATION OF ALGEBRA TO GEOMETRY.

In the preceding part of the present performance, I have considered Algebra as an independent science, and confined myself chiefly to the treating on such of its most useful rules and operations as could be brought within a moderate compass; but as the numerous applications, of which it is susceptible, ought not to be wholly overlooked, I shall here show, in compliance with the wishes of many respectable teachers, its use in the resolution of geometrical problems, referring the reader to my larger work on this subject for what relates more immediately to the general doctrine of curves.*

For this purpose it may be observed, that when any proposition of the kind here mentioned is required to be resolved algebraically, it will be necessary, in the first place, to draw a figure that shall represent the several parts, or conditions, of the problem under consideration, and regard it as the true one.

Then, having properly considered the nature of the question, the figure so formed, must, if necessary, be still farther prepared for solution, by producing, or drawing, such lines in it as may appear, by their connexion or relations to each other, to be most conducive to the end proposed.

This being done, let the unknown line, or lines, which it is judged will be the easiest to find, together with those that are known, be denoted by the common algebraical symbols, or letters; then, by means of the proper geometrical theorems, make out as many independent equations as there are unknown

---

* The learner, before he can obtain a competent knowledge of the method of application abovementioned, must first make himself master of the principal propositions of Euclid, or of those contained in my *Elements of Geometry;* in which work he will find all the essential principles of the science comprised within a much shorter compass than in the former.

And in such cases where it may be requisite to extend this mode of application to trigonometry, mechanics, or any other branch of mathematics, a previous knowledge of the nature and principles of these subjects will be equally necessary.

quantities employed; and the resolution of these, in the usual manner, will give the solution of the problem.

But as no general rules can be laid down for drawing the lines here mentioned, and selecting the properest quantities to substitute for, so as to bring out the most simple conclusions, the best means of obtaining experience in these matters will be to try the solution of the same problem in different ways; and then to apply that which succeeds the best to other cases of the same kind, when they afterwards occur.

The following directions, however, which are extracted, with some alterations, from Newton's *Universal Arithmetic* and Simpson's *Algebra* and *Select Exercises*, will often be found of considerable use to the learner, by showing him how to proceed in many cases of this kind, where he would otherwise be left to his own judgment:—

1st. In preparing the figure in the manner abovementioned, by producing or drawing certain lines, let them be either parallel or perpendicular to some other lines in it, or be so drawn as to form similar triangles; and, if an angle be given, let the perpendicular be drawn opposite to it, and so as to fall, if possible, from one end of a given line.

2d. In selecting the proper quantities to substitute for, let those be chosen, whether required or not, that are nearest to the known or given parts of the figure, and by means of which the next adjacent parts may be obtained by addition or subtraction only, without using surds.

3d. When in any problem there are two lines, or quantities, alike related to other parts of the figure, or problem, the best way is not to make use of either of them separately, but to substitute for their sum, difference, or rectangle, or the sum of their alternate quotients; or for some other line or line in the figure, to which they have both the same relation.

4th. When the area, or the perimeter, of a figure is given, or such parts of it as have only a remote relation to the parts that are to be found, it will sometimes be of use to assume another figure similar to the proposed one, that shall have one of its sides equal to unity, or to some other known quantity; as the other parts of the figure, in such cases, may then be determined by the known proportions of their like sides, or parts, and thence the resulting equation required.

These being the most general observations that have hitherto been collected upon this subject, I shall now proceed to eludicate them by proper examples; leaving such farther remarks as may arise out of the mode of proceeding here used, to be applied by the learner, as occasion requires, to the solutions of the miscellaneous problems given at the end of the present article.

PROBLEM 1.—The base, and the sum of the hypothenuse and perpendicular of a right-angled triangle being given, it is required to determine the triangle.

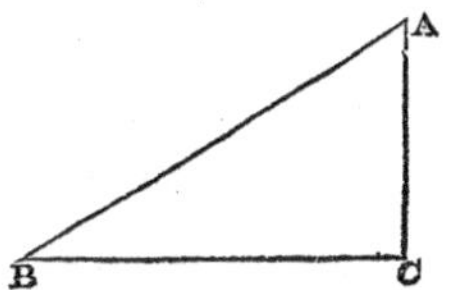

Let ABC, right-angled at C, be the proposed triangle; and put BC $= b$, and AC $= x$.

Then, if the sum of AB and AC be represented by s, the hypothenuse AB will be expressed by $s - x$.

But, by the well-known property of right-angled triangles (Euc. I, 47,)

$$AC^2 + BC^2 = AB^2, \text{ or}$$
$$x^2 + b^2 = s^2 - 2sx + x^2.$$

Whence, omitting x^2, which is common to both sides of the equation, and transposing the other terms, we shall have

$$2sx = s^2 - b^2, \text{ or}$$
$$x = \frac{s^2 - b^2}{2s}, \quad . \; . \; . \; . \; *$$

which is the value of the perpendicular AC; where s and b may be any numbers whatever, provided s be greater than b.

In like manner, if the base and the difference between the hypothenuse and perpendicular be given, we shall have, by putting x for the perpendicular and $d + x$ for the hypothenuse,

$$x^2 + 2dx + d^2 = b^2 + x^2, \text{ or}$$
$$x = \frac{b^2 - d^2}{2d}.$$

Where the base (b) and the given difference (d) may be any numbers as before, provided b be greater than d.

PROBLEM 2.—The difference between the diagonal of a square and one of its sides being given, to determine the square.

---

* The edition of *Euclid* referred to in this and all the following problems, is that of *Dr. Simson*, London, 1801; which may also be used in the geometrical construction of these problems, should the student be inclined to exercise his talents upon this elegant but more difficult branch of the subject. The New York edition of Brewster's Legendre, or Playfair's Euclid, will answer the same purpose.

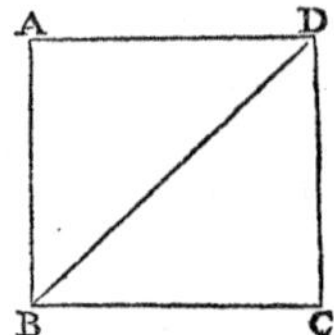

Let AC be the proposed square, and put the side BC, or CD, $= x$.

Then, if the difference of BD and BC be put $= d$, the hypothenuse BD will be $= x + d$.

But since, as in the former problem. $BC^2 + CD^2$, or $2BC^2 = BD^2$, we shall have

$$2x^2 = x^2 + 2dx + d^2, \text{ or}$$
$$x^2 - 2dx = d^2;$$

Which equation, being resolved according to the rule laid down for quadratics in the preceding part of the work, gives

$$x = d + d\sqrt{2},$$

Which is the value of the side BC, as was required.

PROBLEM 3.—The diagonal of a rectangle ABCD, and the perimeter, or sum of all its four sides, being given, to find the sides.

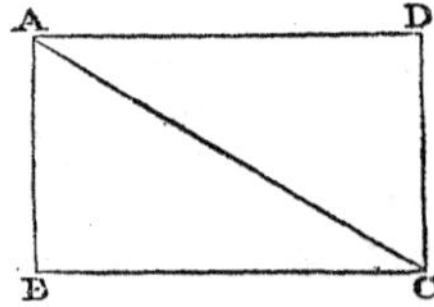

Let the diagonal $AC = d$, half the perimeter $AB + BC = a$ and the base $BC = x$; then will the altitude $AB = a - x$.

And since, as in the former problem, $AB^2 + BC^2 = AC^2$, we shall have

$$a^2 - 2ax + x^2 + x^2 = d^2, \text{ or}$$
$$x^2 - ax = \frac{d^2 - a^2}{2}.$$

Which last equation, being resolved, as in the former instance, gives

$$x = \tfrac{1}{2}a \pm \tfrac{1}{2}\sqrt{(2d^2 - a^2)}.$$

Where a must be taken greater than d and less than $d\sqrt{2}$.

PROBLEM 4.—The base and perpendicular of any plane triangle ABC being given to find the side of its inscribed square.

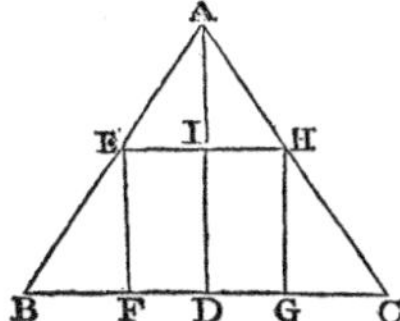

Let EG be the inscribed square; and put BC $= b$, AD $= p$, and the side of the square EH or EF $= x$.

Then, because the triangles ABC, AEH, are similar, we shall have

$$\text{AD} : \text{BC} :: \text{AI} : \text{EH}, \text{ or}$$
$$p : b :: (p - x) : x.$$

Whence, taking the products of the means and extremes, there will arise

$$px = bp - bx,$$

Which, by transposition and division, gives

$$x = \frac{bp}{b + p}.$$

Where b and p may be any numbers whatever, either whole or fractional.

PROBLEM 5.—Having the lengths of three perpendiculars, EF, EG, EH, drawn from a certain point E, within an equilateral triangle ABC, to its three sides, to determine the sides.

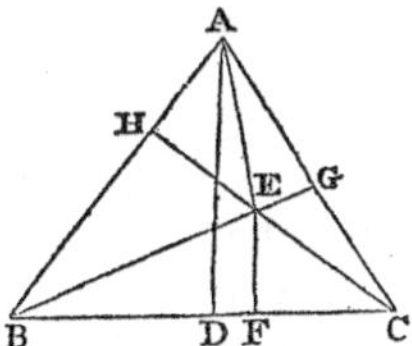

Draw the perpendicular AD, and having joined EA, EB, and EC, put EF $= a$, EG $= b$, EH $= c$, and BD (which is $\frac{1}{2}$ BC) $= x$.

Then, since AB, BC, or CA, are each $= 2x$, we shall have, by Euc. I, 47,

$$\text{AD} = \sqrt{(\text{AB}^2 - \text{BD}^2)} = \sqrt{(4x^2 - x^2)} = \sqrt{3x^2} = x\sqrt{3}.$$

And because the area of any plane triangle is equal to half the rectangle of its base and perpendicular, it follows that

$$\Delta\,\text{ABC} = \tfrac{1}{2}\,\text{BC} \times \text{AD} = x \times x\sqrt{3} = x^2\sqrt{3},$$
$$\Delta\,\text{BEC} = \tfrac{1}{2}\,\text{BC} \times \text{EF} = x \times a \qquad = ax,$$
$$\Delta\,\text{AEC} = \tfrac{1}{2}\,\text{AC} \times \text{EG} = x \times b \qquad = bx,$$
$$\Delta\,\text{AEB} = \tfrac{1}{2}\,\text{AB} \times \text{EH} = x \times c \qquad = cx.$$

But the last three triangles BEC, AEC, AEB, are together equal to the whole triangle ABC; whence

$$x^2\sqrt{3} = ax + bx + cx,$$

And consequently if each side of this equation be divided by x, we shall have

$$x\sqrt{3} = a + b + c, \text{ or}$$
$$x = \frac{a + b + c}{\sqrt{3}}.$$

Which is, therefore, half the length of either of the three equal sides of the triangle.

COR.—Since, from what is above shown, AD is $= x\sqrt{3}$, it follows, that the sum of all the perpendiculars, drawn from any point in an equilateral triangle to each of its sides, is equal to the whole perpendicular of the triangle.

PROBLEM 6.—Through a given point P, in a given circle ACBD, to draw a chord CD, of a given length.

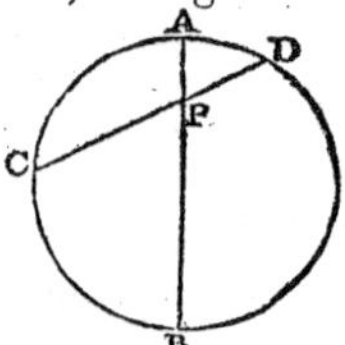

Draw the diameter APB; and put CD $= a$, AP $= b$, PB $= c$, and CP $= x$; then will PD $= a - x$.

But, by the property of the circle (Euc. III. 35,) CP $\times$ PD $=$ AP $\times$ PB; whence

$$x(a - x) = bc, \text{ or}$$
$$x^2 - ax = -bc.$$

Which equation, being resolved in the usual way, gives

$$x = \tfrac{1}{2}a \pm \sqrt{(\tfrac{1}{4}a^2 - bc)};$$

Where x has two values, both of which are positive.

PROBLEM 7.—Through a given point P, without a giver circle ABDC, to draw a right line so that the part CD, intercepted by the circumference, shall be of a given length.

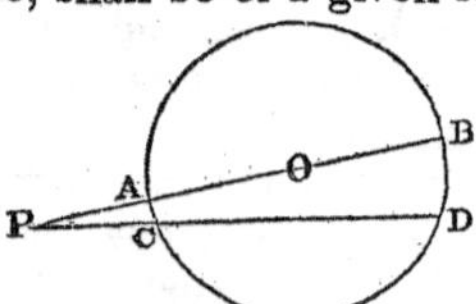

Draw PAB through the centre O; and put CD $= a$, PA $= b$, PB $= c$, and PC $= x$; then will PD $= x + a$.

But, by the property of the circle, (Euc. III, 36, cor.,) PC $\times$ PD $=$ PA $\times$ PB; whence

$$x(x + a) = bc, \text{ or}$$
$$x^2 + ax = bc.$$

Which equation being resolved, as in the former problem, gives

$$x = -\tfrac{1}{2}a \pm \sqrt{(\tfrac{1}{4}a^2 + bc)};$$

Where one value of x is positive and the other negative.*

PROBLEM 8.—The base of BC, of any plane triangle ABC, the sum of the sides AB, AC, and the line AD, drawn from the vertex to the middle of the base, being given, to determine the triangle.

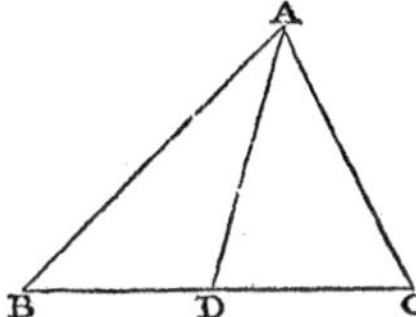

Put BD or DC $= a$, AD $= b$, AB + AC $= s$, and AB $= x$; then will AC $= s - x$.

But, by my Geometry, B. II, Prop. 19, $AB^2 + AC^2 = 2BD^2 + 2AD^2$; whence

$$x^2 + (s - x)^2 = 2a^2 + 2b^2, \text{ or}$$
$$x^2 - sx = a^2 + b^2 - \tfrac{1}{2}s^2.$$

Which last equation, being resolved as in the former instances, gives

$$x = \tfrac{1}{2}s \pm s\sqrt{(a^2 + b^2 - \tfrac{1}{4}s^2)},$$

for the values of the two sides AB and AC of the triangle; taking the sign + for one of them, and − for the other, and observing that $a^2 + b^2$ must be greater than $\tfrac{1}{4}s^2$.

PROBLEM 9.—The two sides AB, AC, and the line AD, bisecting the vertical angle of any plane triangle ABC, being given, to find the base BC.

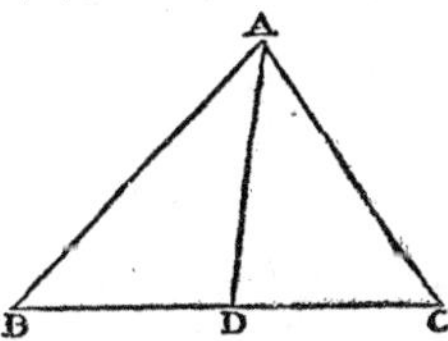

* The two last problems, with a few slight alterations, may be readily employed for finding the roots of quadratic equations by construction; but this, as well as the methods frequently given for constructing cubic and some of the higher order of equations, is a matter of little importance in the present state of mathematical science; analysis, in these cases, being generally thought a more commodious instrument than geometry.

Put AB $= a$, AC $= b$, AD $= c$, and BC $= x$; then, by Euc. VI., 3, we shall have

$$\text{AB } (a) : \text{AC } (b) :: \text{BD} : \text{DC}.$$

And consequently, by the compositions of ratios, (Euc. v, 18,)

$$a + b : a :: x : \text{BD} = \frac{ax}{a+b}, \text{ and}$$

$$a + b : b :: x : \text{DC} = \frac{bx}{a+b}.$$

But, by Euc. VI., 13, BD $\times$ DC $+$ AD$^2 =$ AB $\times$ AC; wherefore, also,

$$\frac{abx^2}{(a+b)^2} + c^2 = ab, \text{ or}$$

$$abx^2 = (a+b)^2 \times (ab - c^2).$$

From which last equation, we have

$$x = (a+b)\sqrt{\frac{ab - c^2}{ab}},$$

Which is the value of the base BC, as required.

PROBLEM 10.—Having given the lengths of two lines AD, BE, drawn from the acute angles of a right-angled triangle ABC, to the middle of the opposite sides, it is required to determine the triangle.

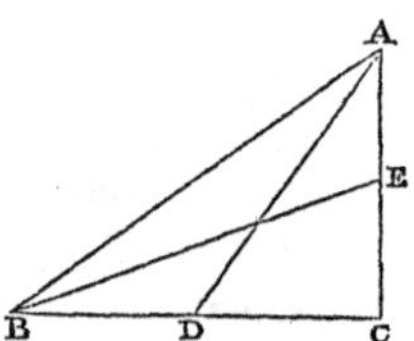

Put AD $= a$, BE $= b$, CD or $\frac{1}{2}$CB $= x$, and CE or $\frac{1}{2}$ CA $= y$; then, since (Euc. I, 47) CD$^2 +$ CA^2, $=$ AD^2, and CE$^2 -$ CB$^2 =$ BE^2, we shall have

$$x^2 + 4y^2 = x^2,$$

$$y^2 + 4x^2 = b^2.$$

Whence, taking the second of these equations from four times the first, there will arise

$$15y^2 = 4a^2 - b^2, \text{ or}$$

$$y = \sqrt{\frac{4a^2 - b^2}{15}}.$$

And, in like manner, taking the first of the same equations from four times the second, there will arise

$$15x^2 = 4b^2 - a^2, \text{ or}$$

$$x = \sqrt{\frac{4b^2 - a^2}{15}}.$$

Which values of x and y are half the lengths of the base and perpendicular of the triangle; observing that b must be less than $2a$ and greater than $\frac{1}{2}a$.

PROBLEM 11.—Having given the ratio of the two sides of a plane triangle ABC, and the segments of the base, made by a perpendicular falling from the vertical angle, to determine the triangle.

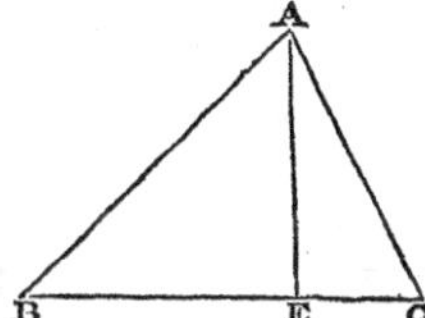

Put BD $= a$, DC $= b$, AB $= x$, AC $= y$, and the ratio of the sides as m to n.

Then, since by the question, AB : AC : : $m : n$, and by B. II, Prop. 16, of my *Elements of Geometry*, $AB^2 - AC^2 = BD^2 - DC^2$, we shall have

$$x : y :: m : n, \text{ and}$$
$$x^2 - y^2 = a^2 - b^2.$$

But, by the first of these expressions, $nx = my$, or $y = \frac{nx}{m}$; whence, if this be substituted for y in the second, there will arise

$$x^2 - \frac{n^2}{m^2}x^2 = a^2 - b^2, \text{ or}$$
$$(m^2 - n^2)\,x^2 = m^2\,(a^2 - b^2).$$

And consequently, by division and extracting the square root, we shall have

$$x = m\,\sqrt{\frac{a^2 - b^2}{m^2 - n^2}}, \text{ and}$$
$$y = n\,\sqrt{\frac{a^2 - b^2}{m^2 - n^2}};$$

which are the values of the two sides AB, AC, of the triangle, as was required.

PROBLEM 12.—Given the hypothenuse of a right-angled triangle ABC, and the sides of its inscribed square DC, to find the other two sides of the triangle.

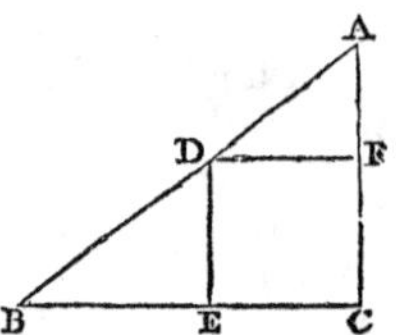

Put $AB = h$, DE or $DF = s$, $AC = x$, and $CB = y$; then, by similar triangles, we shall have

$$AC\,(x) : CB\,(y) :: AF\,(x - s) : FD\,(s).$$

And consequently, by multiplying the means and extremes,

$$xy - sy = sx, \text{ or}$$
$$xy = s\,(x + y), \quad \ldots \quad (1).$$

But since, by Euc I, 47, $AC^2 + CB^2 = AB^2$, we shall likewise have $\quad x^2 + y^2 = h^2 \quad \ldots \ldots \quad (2).$

Whence, if twice equation (1) be added to equation (2), there will arise

$$x^2 + 2xy + y^2 = h^2 + 2s\,(x + y), \text{ or}$$
$$(x + y)^2 - 2s\,(x + y) = h^2;$$

Which equation, being resolved after the manner of a quadratic, gives

$$x + y = s \pm \sqrt{(h^2 + s^2)}, \text{ or}$$
$$y = s - x \pm \sqrt{(h^2 + s^2)}.$$

Hence, if this value be substituted for y in equation (1), there will arise

$$x\,\{s - x \pm \sqrt{(h^2 + s^2)}\} = s\,\{s \pm \sqrt{(h^2 + s^2)}\}, \text{ or}$$
$$x^2 - \{s \pm \sqrt{(h^2 + s^2}\}\,x = -\,s\,\{s \pm \sqrt{(h^2 + s^2)}\}.$$

And consequently, by resolving this last equation, we shall have

$$x = \tfrac{1}{2}\{s \pm \sqrt{(h^2 + s^2)}\} \pm \sqrt{\{\tfrac{1}{4}h^2 - \tfrac{1}{2}s^2 \mp \frac{s}{2}\sqrt{(h^2 + s^2)}\}};$$

and

$$y = \tfrac{1}{2}\{s \pm \sqrt{(h^2 + s^2)}\} \mp \sqrt{\{\tfrac{1}{4}h^2 - \tfrac{1}{2}s^2 \mp \sqrt{\frac{s}{2}}\sqrt{(h^2 + s^2}\}};$$

Which are the values of the perpendicular AC and base BC, as was required.

PROBLEM 13.—Having given the perimeter of a right-angled triangle ABC, and the perpendicular CD, falling from the right angle on the hypothenuse, to determine the triangle.

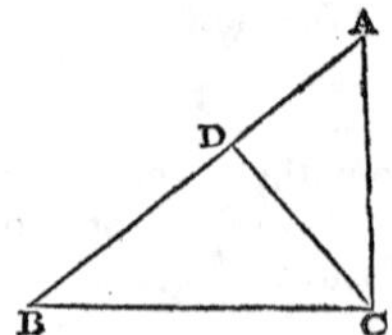

Put $p =$ perimeter, CD $= a$, AC $= x$, and BC $= y$; then AB $= p - (x + y)$.

But, by right-angled triangles (Euc. I. 47,) $AC^2 + BC^2 = AB^2$; whence $x^2 + y^2 = p^2 - 2p(x + y) + x^2 + 2xy + y^2$.

Or, by transposing the terms and dividing by 2,

$$p(x + y) - \tfrac{1}{2}p^2 = xy \; . \; . \; . \; . \; . \; . \; . \; (1).$$

And since, by similar triangles, AB : BC : : AC : CD, we shall also have, by multiplying the means and extremes,

$$AB \times CD = BC \times AC, \text{ or}$$

$$ap - a(x + y) = xy. \; . \; . \; . \; . \; . \; . \; . \; (2).$$

Whence, by comparing equation (1) with equation (2), there will arise $(a + p) \times (x + y) = ap + \frac{1}{2}p^2$.

Whence,

$$x + y = \frac{p(a + \frac{1}{2}p)}{a + p}, \text{ or}$$

$$y = \frac{p(a + \frac{1}{2}p)}{a + p} - x.$$

And if these values be now substituted for $x + y$ and y in equation (2), the result, when simplified and reduced, will give

$$(a + p)x^2 - p(a + \tfrac{1}{2}p)x = -\tfrac{1}{2}ap^2.$$

From which last equation, and the value of y above found, we shall have

$$x \text{ or } AC = \frac{p(a + \frac{1}{2}p)}{2(a + p)} \pm \frac{p}{2(a + p)} \sqrt{\left\{(a - \tfrac{1}{2}p)^2 - 2a^2\right\}};$$

and

$$y \text{ or } BC = \frac{p(a + \frac{1}{2}p)}{2(a + p)} \mp \frac{p}{2(a + p)} \sqrt{\left\{(a - \tfrac{1}{2}p)^2 - 2a^2\right\}}.$$

And if the sum of these two sides be taken from p, the result will give

$$AB = p - (x + y) = \frac{p^2}{2(a + p)}.$$

Which expressions are, therefore, respectively equal to the values of the three sides of the triangle.

PROBLEM 14.—Given the perpendicular, base, and sum of the sides of an obtuse angled plane triangle ABC, to determine the two sides of the triangle.

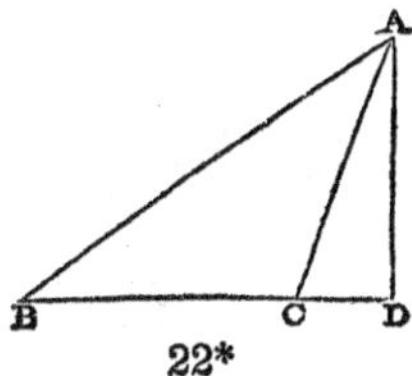

Let the perpendicular AD $= p$, the base BC $= b$, the sum of AB and AC $= s$, and their difference $= x$.

Then, since half the difference of any two quantities added to half their sum, gives the greater, and, when subtracted, the less, we shall have

$$\text{AB} = \tfrac{1}{2}(s + x), \text{ and AC} = \tfrac{1}{2}(s - x).$$

But, by Euc. I, 47, $\text{CD}^2 - \text{AC}^2 - \text{AD}^2$, or $\text{CD} = \sqrt{\left\{\tfrac{1}{4}(s - x)^2 - p^2\right\}}$; and, by B. II, 12, $\text{AB}^2 = \text{BC}^2 + \text{AC}^2 + 2\text{BC} \times \text{CD}$; whence

$$\tfrac{1}{4}(s + x)^2 = b^2 + \tfrac{1}{4}(s - x)^2 + 2b\sqrt{\left\{\tfrac{1}{4}(s - x)^2 - p^2\right\}}, \text{ or}$$

$$sx - b^2 = 2b\sqrt{\left\{\tfrac{1}{4}(s - x)^2 - p\right\}}.$$

And if each of the sides of this last equation be squared. there will arise, by transposition, and simplifying the result,

$$(s^2 - b^2)x^2 = b^2(s^2 - b^2) - 4b^2p^2, \text{ or}$$

$$x = b\sqrt{(1 - \frac{4p^2}{s^2 - b^2})}.$$

Whence, by addition and subtraction, we shall have

$$\text{AB} = \frac{s}{2} + \frac{b}{2}\sqrt{(1 - \frac{4p^2}{s^2 - b^2}}, \text{ and}$$

$$\text{AC} = \frac{s}{2} - \frac{b}{2}\sqrt{(1 - \frac{4p^2}{s^2 - b^2}.)}$$

Which are the sides of the triangle, as was required.

PROBLEM 15.—It is required to draw a right line BFE from one of the angles B of a given square BD, so that the part FE, intercepted by DE and DC, shall be of a given length.

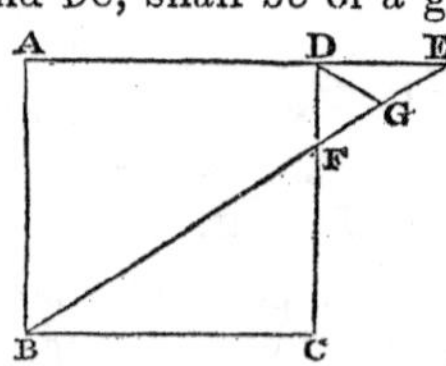

Bisect FE in G, and put AB or BC $= a$, FG or GE $= b$, and BG $= x$; then will BE $= x + b$ and BF $= x - b$.

But since, by right-angled triangles, $\text{AE}^2 = \text{BE}^2 - \text{AB}^2$, we shall have

$$\text{AE} = \sqrt{\{(x + b)^2 - a^2\}}.$$

And because the triangles BCF, EAB, are similar,

$$\text{BF} : \text{BC} :: \text{BE} : \text{AE}, \text{ or}$$

$$a(x + b) = (x - b)\sqrt{\{(x + b)^2 - a^2\}}.$$

Whence, by squaring each side of this equation, and arranging the terms in order, there will arise

$$x^4 - 2(a^2 + b^2)x^2 = b^2(2a^2 - b^2).$$

Which equation, being resolved after the manner of a quadratic, will give

$$x = \sqrt{\{a^2 + b^2 \pm a\sqrt{(a^2 + 4b^2)}\}}.$$

And consequently, by adding b to, or subtracting it from, this last expression, we shall have

$$\text{BE} = \sqrt{\{a^2 + b^2 \pm a\sqrt{(a^2 + 4b^2)}\}} + b, \text{ or}$$
$$\text{BF} = \sqrt{\{a^2 + b^2 \pm a\sqrt{(a^2 + 4b^2)}\}} - b.$$

Which values, by determining the point E, or F, will satisfy the problem.

Where it may be observed, that the point G lies in the circumference of a circle, described from the centre D, with the radius FG, or half the given line.

PROBLEM 16.—The perimeter of a right-angled triangle ABC, and the radius of its inscribed circle being given, to determine the triangle.

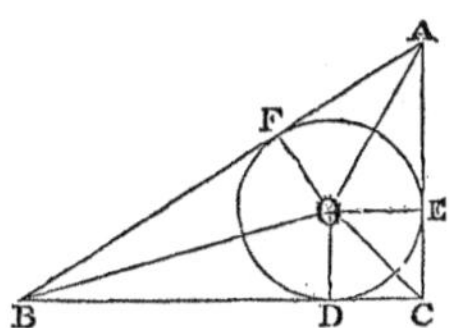

Let the perimeter of the triangle $= p$, the radius OD, or OE, of the inscribed circle $= r$, AE $= x$, and BD $= y$.

Then, since in the right-angled triangles AEO, AFO, OE, is equal to OF, and OA is common, AF will also be equal AE, or x.

And in like manner it may be shown, that BF is equal to BD, or y.

But by the question, and Euc. I, 47, we have

$$(x + r) + (y + r) + (x + y) = p, \text{ and}$$
$$(x + r)^2 + (y + r)^2 = (x + y)^2.$$

Or, by adding the terms of the first, and squaring those of the second,

$$x + y = \tfrac{1}{2}p - r, \text{ and}$$
$$r(x + y) = xy - r^2.$$

Hence, since, in the first of these equations, $y = (\frac{1}{2}p - r) - x$, if this value be substituted for y in the second, there will arise

$$x^2 - (\tfrac{1}{2}p - r)x = -\tfrac{1}{2}pr,$$

Which equation, being resolved in the usual manner, gives

$$x = \tfrac{1}{2}(\tfrac{1}{2}p - r) \pm \sqrt{\left\{\tfrac{1}{4}(\tfrac{1}{2}p - r)^2 - \tfrac{1}{2}pr\right\}},$$

and

$$y = \tfrac{1}{2}(\tfrac{1}{2}p - r) \mp \sqrt{\left\{\tfrac{1}{4}(\tfrac{1}{2}p - r)^2 - \tfrac{1}{2}pr\right\}}.$$

And consequently, if r be added to each of these last expressions, we shall have

$$\text{AC} = \tfrac{1}{2}(\tfrac{1}{2}p + r) \pm \sqrt{\left\{\tfrac{1}{4}(\tfrac{1}{2}p - r)^2 - \tfrac{1}{2}pr\right\}}, \text{ and}$$

$$\text{BC} = \tfrac{1}{2}(\tfrac{1}{2}p + r) \mp \sqrt{\left\{\tfrac{1}{4}(\tfrac{1}{2}p - r)^2 - \tfrac{1}{2}pr\right\}},$$

for the values of the perpendicular and base of the triangle, as was required.

PROBLEM 17.—From one of the extremities A, of the diameter of a given semicircle ADB, to draw a right line AE, so that the part DE, intercepted by the circumference and a perpendicular drawn from the other extremity, shall be of a given length.

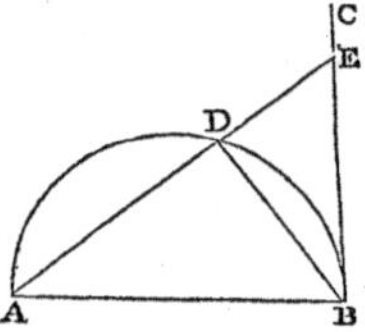

Let the diameter AB $= d$, DE $= a$, and AE $= x$; and join BD.

Then, because the angle ADB is a right angle, (Euc. III, 31,) the triangles ABE, ABD, are similar.

And consequently, by comparing their like sides, we shall have

$$\text{AE} : \text{AB} :: \text{AB} : \text{AD}, \text{ or}$$

$$x : d :: d : x - a.$$

Whence, multiplying the means and extremes of these proportionals, there will arise

$$x^2 - ax = d^2,$$

Which equation, being resolved after the usual manner, gives

$$x = \tfrac{1}{2}a + \sqrt{(\tfrac{1}{4}a^2 + d^2)}.$$

PROBLEM 18.—To describe the circle through two given points A, B, that shall touch a right line CD given in position.

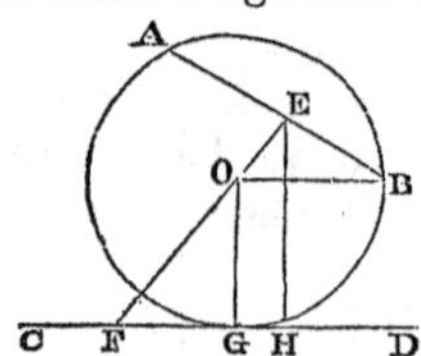

Join AB; and through O, the assumed centre of the required circle, draw FE perpendicular to AB; which will bisect it in E, (Euc. III, 3.)

Also, join OB; and draw EH, OG, perpendicular to CD; the latter of which will fall on the point of contact G, (Euc. III. 18.)

Hence, since A, E, B, H, F, are given points, put $EB = a$, $EF = b$, $EH = c$, and $EO = x$; which will give $OF = b - x$.

Then, because the triangle OEB is right-angled at E, we shall have

$$OB^2 = EO^2 + EB^2, \text{ or}$$
$$OB = \sqrt{(x^2 + a^2)}.$$

But, by similar triangles, FE : EH :: FO : OG or OB; or $b : c :: b - x : OB$; whence, also,

$$OB = \frac{c}{b}(b - x).$$

And consequently, if these two values of OB be put equal to each other, there will arise,

$$\sqrt{(x^2 + a^2)} = \frac{c}{b}(b - x).$$

Or, by squaring each side of this equation, and simplifying the result,

$$(b^2 - c^2)x^2 + 2bc^2x = b^2(c^2 - a^2).$$

Which last equation, when resolved in the usual manner, gives

$$x = -\frac{bc^2}{b^2 - c^2} + b\sqrt{\left\{\frac{c^4}{(b^2 - c^2)^2} + \frac{c^2 - a^2}{b^2 - c^2}\right\}},$$

for the distance of the centre O from the chord AB; where b must evidently be greater than c, and c greater than a.

PROBLEM 19.—The three lines AO, BO, CO, drawn from the angular points of a plane triangle ABC, to the centre of its inscribed circle, being given, to find the radius of the circle and the sides of the triangle.

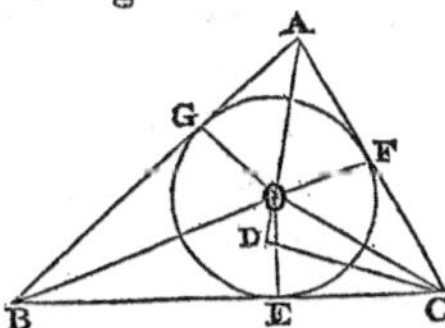

Let O be the centre of the circle, and, on AO produced, let fall the perpendiculars CD; and draw OE, OF, OG, to the points of contact, E, F, G.

Then, because the three angles of the triangle ABC are, together, equal to two right angles, (Euc. I, 32,) the sum of their halves OAC + OCA + OBE will be equal to one right angle.

But the sum of the two former of these, OAC + OCA, is equal to the external angle DOC; whence the sum of DOC + OBE, as also of DOC + OCD, is equal to a right angle: and consequently, OBE = OCD.

Let, therefore, AO $= a$, BO $= b$, CO $= c$, and the radius OE, OF, or OG $= x$.

Then, since the triangles BOE, COD, are similar, BO : OE :: CO : OD, or $b : x :: c :$ OD; which gives

$$\text{CD} = \frac{cx}{b}, \text{ and CD} = \surd\left(c^2 - \frac{c^2x^2}{b^2}\right), \text{ or } \frac{c}{b}\surd(b^2 - x^2).$$

Also, because the triangle AOC is obtuse angled at O, we shall have (Euc. II, 12,)

$$\text{AC}^2 = \text{AO}^2 + \text{CO}^2 + 2\text{AO} \times \text{OD}; \text{ or}$$

$$\text{AC} = \surd\left(a^2 + c^2 + \frac{2acx}{b}\right), \text{ or } \surd\left(\frac{b(a^2 + c^2) + 2acx}{b}\right).$$

But the triangles ACD, AOF, being likewise similar,

$$\text{AC} : \text{CD} :: \text{AO} : \text{OF}, \text{ or}$$

$$\surd\left(\frac{b(a^2 + c^2) + 2acx}{b}\right) : \frac{c}{b}\surd(b^2 - x^2) :: a : x.$$

Whence, multiplying the means and extremes, and squaring the result, there will arise

$$bx^2\{b(a^2 + c^2) + 2acx\} = a^2c^2(b^2 - x^2).$$

Or, by collecting the terms together, and dividing by the coefficient of the highest power of x,

$$x^3 + \left(\frac{ab}{2c} + \frac{ac}{2b} + \frac{bc}{2a}\right)x^2 = \frac{abc}{2}.$$

From which last equation x may be determined, and thence the side of the triangle.*

PROBLEM 20.—Given the three sides AB, BC, CD, of a trapezium ABCD, inscribed in a semicircle, to find the diameter or remaining side AD.

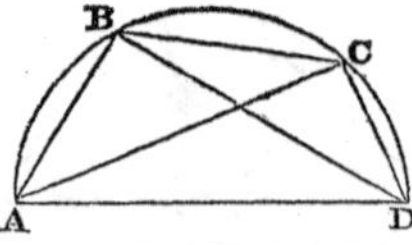

* This, and the following problem, cannot be constructed geometrically, or by means only of right lines and a circle, being what the ancients usually denominated solid problems, from the circumstance of their involving an equation of more than two dimensions; in which cases they generally employed the conic sections, or some of the higher orders of curves.

Let AB $= a$, BC $= b$, CD $= c$, and AD $= x$; then, by Euc. VI, Prop. D, AC $\times$ BD $=$ AD $\times$ BC $+$ AB $\times$ CD $= bx + ac$.

But ABD, ACD, being right angles, (Euc. III, 31,) we shall have

$$\text{AC} = \sqrt{(\text{AD}^2 - \text{DC}^2)}, \text{ or } \sqrt{(x^2 - c^2)}, \text{ and}$$
$$\text{BD} = \sqrt{(\text{AD}^2 - \text{AB}^2)}, \text{ or } \sqrt{(x^2 - a^2)}.$$

Whence, by substituting these two values in the former expressions, there will arise

$$\sqrt{(x^2 - c^2)} \times \sqrt{(x^2 - a^2)} = bx + ac.$$

Or, by squaring each side, and reducing the result,

$$x^3 - (a^2 + b^2 + c^2)\,x = 2abc.$$

From which last equation the value of a may be found, as in the last problem.*

MISCELLANEOUS PROBLEMS.

PROBLEM I.

To find the side of a square, inscribed in a given semicircle, whose diameter is d. Ans. $\frac{1}{5} d \sqrt{5}$.

PROBLEM II.

Having given the hypothenuse (13) of a right-angled triangle, and the difference between the other two sides (7), to find these sides.† Ans. 5 and 12.

PROBLEM III.

To find the side of an equilateral triangle, inscribed in a circle whose diameter is d; and that of another circumscribed about the same circle. Ans. $\frac{1}{2} d \sqrt{3}$, and $d \sqrt{3}$.

PROBLEM IV.

To find the side of a regular pentagon, inscribed in a circle, whose diameter is d. Ans. $\frac{1}{4} d \sqrt{(10 - 2\sqrt{5})}$.

PROBLEM V.

To find the sides of a rectangle, the perimeter of which shall be equal to that of a square, whose side is a, and its area half that of a square. Ans. $a + \frac{1}{2} a \sqrt{2}$ and $a - \frac{1}{2} a \sqrt{2}$.

---

* Newton, in his Universal Arithmetic, English edition, 1728, has resolved this problem in a variety of different ways, in order to show that some methods of proceeding, in cases of this kind, frequently lead to more elegant solutions than others; and that a ready knowledge of these can only be obtained by practice.

† Such of these equations as are proposed in numbers, should first be resolved generally, by means of the usual symbols, and then reduced to the answers above given, by substituting the numeral values of the letters in the results thus obtained.

PROBLEM VI.

Having given the side (10) of an equilateral triangle, to find the radii of its inscribed and circumscribing circles.

Ans. 2.8868 and 5.7736.

PROBLEM VII.

Having given the perimeter (12) of a rhombus, and the sum (8) of its two diagonals, to find the diagonals.

Ans. $4 + \sqrt{2}$ and $4 - \sqrt{2}$.

PROBLEM VIII.

Required the area of a right-angled triangle, whose hypothenuse is x^{3x}, and the base and perpendicular x^{2x} and x^{x}.

Ans. 1.029085.

PROBLEM IX.

Having given two contiguous sides (a, b) of a parallelogram, and one of its diagonals (d), to find the other diagonal.

Ans. $\sqrt{(2a^2 + 2b^2 - d^2)}$.

PROBLEM X.

Having given the perpendicular (300) of a plane triangle, the sum of the two sides (1150), and the difference of the segments of the base (495), to find the base and the sides.

Ans. 945, 375, and 780.

PROBLEM XI.

The length of three lines drawn from the three angles of a plane triangle, to the middle of the opposite sides, being 18, 24, and 30, respectively: it is required to find the sides.

Ans. 20, 28.844, and 34.176.

PROBLEM XII.

In a plane triangle, there is given the base (50), the area (796) and the difference of the sides (10), to find the sides and the perpendicular. Ans. 36, 46, and 33.261.

PROBLEM XIII.

Given the base (194) of a plane triangle, the line that bisects the vertical angle (66), and the diameter (200) of the circumscribing circle, to find the other two sides.

Ans. 81.36587 and 157.43865.

PROBLEM XIV.

The lengths of two lines that bisect the acute angles of a right-angled plane triangle being 40 and 50 respectively, it is required to determine the three sides of the triangle.

Ans. 35.80737, 47.40728, and 59.41143

PROBLEM XV.

Given the altitude (4), the base (8), and the sum of the sides (12), of a plane triangle to find the sides.

Ans. $6 + \frac{4}{5}\sqrt{5}$ and $6 - \frac{4}{5}\sqrt{5}$.

PROBLEM XVI.

Having given the base of a plane triangle (15), its area (45), and the ratio of its other two sides as 2 to 3, it is required to determine the lengths of these sides.

Ans. 7.7915 and 11.6872.

PROBLEM XVII.

Given the perpendicular (24), the line bisecting the base (40), and the line bisecting the vertical angle (25), to determine the triangle.

Ans. The base $\frac{250}{7}\sqrt{7}$.

From which the other two sides may be readily found.

PROBLEM XVIII.

Given the hypothenuse (10) of a right-angled triangle, and the difference of two lines drawn from its extremities to the centre of the inscribed circle (2), to determine the base and perpendicular. Ans. 8.08004 and 5.87447.

PROBLEM XIX.

Having given the lengths (a, b) of two chords, cutting each other at right angles, in a circle, and the distance (c) of their point of intersection from the centre, to determine the diameter of a circle. Ans. $\sqrt{\left\{\frac{1}{2}(a^2 + b^2) + 2c^2\right\}}$.

PROBLEM XX.

Two trees, standing on a horizontal plane, are 120 feet asunder; the height of the highest of which is 100 feet, and that of the shortest 80; whereabouts in the plane must a person place himself so that his distance from the top of each tree, and the distance of the tops themselves, shall be all equal to each other.

Ans. $20\sqrt{21}$ feet from the bottom of the shortest, and $40\sqrt{3}$ feet from the bottom of the other.

PROBLEM XXI.

Having given the sides of a trapezium, inscribed in a circle, equal to 6, 4, 5, and 3, respectively, to determine the diameter of the circle.

Ans. $\frac{1}{20}\sqrt{(130 \times 153)}$ or 7.051595.

PROBLEM XXII.

Supposing the town A to be 30 miles from B, B 25 miles from C, and C 20 miles from A; whereabouts must a house be erected that it shall be at an equal distance from each of them?

Ans. 15.118556 miles from each.

PROBLEM XXIII.

Given the area (100) of an equilateral triangle ABC, whose base BC falls on the diameter, and vertex A in the middle of the arc of a semicircle; required the diameter of the semicircle.

Ans. $20\sqrt[4]{3}$.

PROBLEM XXIV.

In a plane triangle, having given the perpendicular (p), and the radii (r, R) of its inscribed and circumscribing circles, to determine the triangle.

Ans. The base $\frac{2r\sqrt{(2pR - 4rR - r^2)}}{p - 2r}$.

PROBLEM XXV.

Having given the base of a plane triangle equal to $2a$, the perpendicular equal to a, and the sum of the cubes of its other two sides equal to three times the cube of the base; to determine the sides.

Ans. $a(2 + \frac{1}{3}\sqrt{6})$ and $a(2 - \frac{1}{3}\sqrt{6})$.

ADDITIONAL PROBLEMS FOR PRACTICE.

I. *Simple Equations involving only one unknown quantity*

1. Find that number to which 75 being added, the sum shall be quadruple the required number. Ans. 25.

2. A person has 1341 pieces of money, among which are $3\frac{1}{2}$ times as many of silver as of copper. How many has he of each? Ans. 298 copper, and 1043 silver.

3. A man pays his four labourers $40, giving the second twice, the third 3 times, and the fourth 4 times as much as the first. What did each receive? Ans. 4, 8, 12, and 16.

4. A legacy of 1080 dollars is to be divided between two persons, so that one is to receive $7 as often as the other $1. What are the shares? Ans. 135 and 945.

5. Divide 40 marbles between two boys in the proportion of 5 to 3. Ans. 25 and 15.

6. Two travellers, being 360 miles apart, set out at the same time towards each other, one travelling 8 miles an hour, the other 10. How far does each travel before they meet? Ans. 160 and 200 miles.

7. What two numbers are those whose difference is 36, and which are to each other as 7 to 4? Ans. 84 and 48.

8. Four persons gain $2040, of which B takes twice as much as A, C twice as much as A and B, and D as much as B and C. What are their shares? Ans. 120, 240, 720, and 960.

9. A man sells oxen, cows, and sheep, of each an equal number, for $700. He received for each ox $50, for each cow $17, and for each sheep $3. How many were there of each sort? Ans. 10.

10. A person buys 2 oranges, 4 lemons, and 10 apples, for 64 cents. He gives thrice as much for a lemon as for an apple, and as much for an orange as for a lemon and two apples. What did each cost? Ans. An apple 2, a lemon 6, and an orange 10 cts.

11. Find that number whose $\frac{1}{3}$ part exceeds its $\frac{1}{4}$ part by 12. Ans. 144.

12. What sum of money is that whose third, fourth, and fifth parts amount to $94? Ans. 120.

13. What number is that which being divided by 15, the dividend, divisor, and quotient shall amount to 79? Ans. 60.

14. A man lost $\frac{4}{9}$ of his estate, and had \$4500 left. What had he at first? Ans. 8100.

15. A person bestowed on two beggars 1*s*. 8*d*., giving the first $\frac{1}{8}$ and the second $\frac{1}{12}$ of what he had. How much had he? Ans. 8 shillings.

16. A and B talking of their ages, A says to B, your age is thrice and a third of my age, and the sum of both is 52. What are their ages? Ans. A's 12 and B's 40 years.

17. A man driving his geese to market, was met by another, who said, Good-morrow, master, with your hundred geese; said he, I have not a hundred, but if I had as many more, and half as many more, and two geese and a half, I should have a hundred. How many had he? Ans. 39.

18. What number is that whose third part, less 4, is equal to its fourth, plus 25? Ans. 348.

19. A father, at his death, left £1000 to be divided between his son and daughter, in such a manner that $\frac{1}{5}$ part of his share should exceed $\frac{1}{4}$ part of hers by £10. How must the money be divided?

Ans. The son, £577$\frac{7}{9}$. The daughter, £422$\frac{2}{9}$.

20. A gentleman would procure \$10 in quarters and dimes, and have just 64 pieces. How many of each kind must he get? Ans. 24 qrs. and 40 dimes.

21. The sum of two numbers is 40; if the greater be multiplied by 2, and the less by 3, the difference of the products will be 15. Required the numbers. Ans. 27 and 13.

22. A cask which held 146 gallons, was filled with a mixture of brandy, wine, and water. In it there were 15 gallons of wine more than there were of brandy, and as much water as both wine and brandy. What quantity was there of each?

Ans. 29, 44, and 73 gallons respectively.

23. A man hired a labourer on condition that for every day he wrought he should have 12*d*., and for every day he was idle he should forfeit 8*d*. After 390 days, neither was in debt. How many working days were there? Ans. 156.

24. A son asking his father how old he was, received the following reply: My age, says the father, 7 years ago was 4 times as great as yours at that time; but 7 years hence, if we live, my age will be only double of yours: it is required to find the age of each. Ans. 14 and 35.

25. A mercer having cut 19 yards from each of three equal pieces of silk, and 17 from another of the same length, found that the remnants together were 142 yards. What was the length of each piece? Ans. 54 yards.

26. A cistern can be emptied by three apertures; by the first in an hour, by the second in an hour and a half, and by the third in two hours. In what time will all three running together empty it? Ans. In $27\frac{9}{13}$ minutes.

27. A grocer has a quantity of tea, by which at 80 cents a pound, he will gain $2,50, but at 60 cents a pound, he will lose $7,50. What is the quantity and prime cost?
Ans. 50lb. at 75 cts.

28. How old is he, to the number of whose months, if you add its half and its eighth, and subtract a unit, there will be left the square of 21? Ans. 22 yrs. 8 mo.

29. Said A to B, double my money for me, and I will give you 6*d* out of the stock; with the remainder he applied to C, to double it and gave him also sixpence. He repeated the offer to D and then 6*d* was all he had left. What had he to begin with? Ans. 6*d*.

30. Find a number, which being added to itself, and the sum multiplied by the same, and from the product the same subtracted, and lastly, the remainder divided by the same, the quotient shall be 13. Ans. 7.

31. A draper bought three pieces of cloth, which together measured 159 yards. The second piece was 15 yards longer than the first, and the third 24 yards longer than the second. What was the length of each.
Ans. 35, 50 and 74 yards respectively.

32. A said he was 4 years older than C; B told them that his age exceeded the sum of theirs by 9 years; upon which D observed that his age was 45 and equal to the sum of the three. Required the several ages.
Ans. A, 11; B, 27; C, 7; D, 45.

33. A boy with a basket of peaches met three companions, and gave half his peaches to the first, who returned him 10; to the second he gave half of what remained, and received back 4. He then halved the remainder with the third, who returned one, and eighteen only remained in his basket. How many were there at first? Ans. 100.

34. One being asked his age, answered that if $\frac{1}{60}$ of the years he had lived were multiplied by $\frac{5}{8}$ of that number it would give his age. How old was he? Ans. 96.

35. A General having lost a batttle, found that he had only half his army plus 3600 men left, fit for action; one-eighth of his men plus 600 being wounded, and the rest, which were one-fifth of the whole army, either slain, taken prisoners, or missing. Of how many men did his army consist?
Ans. 24000.

36. If George lives till the quadruple of $\frac{2}{3}$ of his years, added

to half of them +50 amounts to as much above 100, as his years will actually be below 100, what will be his age?

Ans. 36.

37. A and B being at play, severally cut packs of cards, so as to take off more than they left. Now it happened that A cut off twice as many as B left, and B cut off seven times as many as A left. How were the cards cut by each?

Ans. A cut off 48, and B cut off 28 cards.

38. A courier, who travels 60 miles a day, had been dispatched five days, when a second is sent to overtake him, in order to do which he must travel 75 miles a day. In what time will he overtake the former?

Ans. 20 days.

39. What are those two numbers whose sum is 150 and their difference 38? Ans. 56 and 94.

40. A person being asked the time of day answered that the time past from noon was equal to $\frac{4}{5}$ of the time to midnight. What was the hour? Ans. 20 min. past 5.

41. Required two numbers in the proportion of 3 to 4 whose sum is $\frac{1}{12}$ of their product. Ans. 21 and 28.

42. What number is that to which if I add 12 and from $\frac{1}{12}$ of the sum subtract 12, the remainder shall be 12?

Ans. 276.

43. In a certain school there are 306 scholars of which $\frac{1}{6}$ study Algebra, and the numbers in the Latin, French, Writing and Reading classes are as 1, 3, 6, and 7. How many students in each?

Ans. 15 in Latin, 45 in French, 90 Read., and 105 Write.

44. There are three casks of wine whose contents are in proportion of 2, 3 and 5; from each of which if 10 gallons be drawn the whole quantity will be diminished in the proportion of 4 to 3. Required their contents severally.

Ans. 24, 36 and 60.

45. Four places are situated in the order of the four letters A, B, C, D. The distance from A to D is 34 miles, the distance from A to B : distance from C to D :: 2 : 3, and one-fourth of the distance from A to B added to half the distance from C to D, is three times the distance from B to C. What are the respective distances?

Ans. AB=12, BC=4, and CD=18 miles.

46. A trader maintained himself for 3 years at the expense of £50 a year; and in each of those years augmented that part of his stock which was not so expended by $\frac{1}{3}$ thereof. At the end of the third year his original stock was doubled. What was that stock? Ans. £740.

47. It is required to divide the number 104 into four such

parts, that the first being increased by 6, the second diminished by 5, the third multiplied by 4, and the fourth divided by 3, may all be equal. Ans. 14, 25, 5 and 60.

48. Find a number which being increased by 4, and the sum multiplied by 3, produces the same result as half the said number multiplied by 8, and the product diminished by 8. Ans. 20.

49. Two persons received equal sums of money; the first paid away £25, and the second £60, and then it appeared that the former had twice as much as the latter. What money did each receive? Ans. £95.

50. One being asked the time said it was between 10 and 11 and the hour and minute hands were together. Required the time. Ans. $54\frac{6}{11}$ minutes past 10.

51. A courier sets out from a certain place and travels at the rate of 15 miles in 4 hours; and 12 hours after, another sets out from the same place, and travels the same road at the rate of 9 miles in 2 hours. I demand how long and how far the first must travel before he is overtaken by the second? Ans. 72 hours and 270 miles.

52. It is required to divide the number 99 into five such parts, that the first may exceed the second by 3; be less than the third by 10; greater than the fourth by 9; and less than the fifth by 16. Ans. 17, 14, 27, 8, and 33.

53. A and B began to trade with equal sums of money. In the first year A gained 40 dollars, and B lost 40; but in the second A lost one-third of what he then had, and B gained a sum less by 40 dollars, than twice the sum that A had lost; when it appeared that B had twice as much money as A. What money did each begin with? Ans. 320 dollars.

54. A boy being caught stealing apples, was told by the owner that he should escape punishment if he would take a certain number of apples, and lay down at the first gate, half he had and half an apple over, and repeat this process with the remainder at the second gate, and also at the third, without dividing an apple at either, and then have one left. If he accomplished the task, how many apples did he take? Ans. 15.

55. Find two numbers in the proportion of 3 to 4, so that if 9 be added to each, the sums will be as 6 to 7. Ans. 9 and 12.

56. Upon measuring the corn produced in a field, being 105 bushels, it appears that it had yielded only one fourth part more than was sown. How much was sown? Ans. 84 bushels.

57. Given the sum of two numbers 12, and the difference of their squares 72, to find the numbers. Ans. 9 and 3.

58. If A can build 40 feet of wall in two days, B 13 feet in

one day, and C 50 feet in 3 days, how many days will all three be employed in building 596 feet? Ans. 12 days.

59. What number is that whose fourth part squared, is equal to its eighth part cubed? Ans. 32.

60. Three years ago A's age was half of B's, and 9 years hence it will be $\frac{5}{8}$ of it. Required the age of each.

Ans. 21, and 39.

61. A woman can get 25 apples for a penny, and twenty-five pears for three pence, if she lays out 1*s.* 8*d.* for 200 apples and pears together. How many of each will she have?

Ans. 50 apples, and 150 pears.

62. Find three numbers, such that the sum of the first and second shall be 15, of the first and third 16, and of the second and third 17. Ans. 7, 8, and 9.

63. A farmer would mix wheat at 4 shillings a bushel with rye at 2*s.* 6*d.* the bushel, so that the mixture may consist of 100 bushels at 3*s.* 2*d.* a bushel. How many bushels of each sort must be taken? Ans. $44\frac{4}{9}$ and $55\frac{5}{9}$.

64. A steamboat whose rate in still water is 10 miles *per* hour, descends a river whose velocity is 4 miles an hour, and returns in 10 hours. How far did she proceed? Ans. 42.

65. A gardener and his servant each dug a square piece of ground, whose side is as many feet long, as the labourer is years old. The gardener dug four times as much as his servant, and the sum of their ages is 45. What is the age of each?

Ans. 30 and 15 years.

66. A and B make a joint stock of £500, by which they gain £160, whereof A had for his share £32 more than B. What was each person's stock? Ans. A's 300, and B's 200.

67. Find two numbers in the proportion of 9 to 7, so that the square of their sum and the cube of their difference may be equal. Ans. 288, and 224.

68. Divide 100 into two parts, so that the difference of their squares may be 1000. Ans. 55, and 45.

69. A certain sum of money being put out at simple interest, amounts to £297 12*s.* in 8 months; and in 15 months to £306. What is the sum, and what the rate of interest?

Ans. £288 at 5 *per cent.*

70. A waterman can row down stream, on a certain river, keeping in the middle, 5 miles in 3 quarters of an hour; but it takes him a full hour and a half to return, though he keeps near shore, where the current is but half as strong as in the middle. What is the velocity of the river *per* hour?

Ans. $2\frac{2}{9}$ miles.

71. In a right angled triangle, given one leg 30, and the dif-

ference between the other leg and the hypothenuse 10; to find the two latter.

Ans. The hypothenuse 50, the other leg 40.

72. Given the base 12 and perpendicular 8, of any triangle, to find the side of the inscribed square. Ans. $4\frac{4}{5}$.

73. It is proposed to divide the number 10 into two parts, such that the greater multiplied by the less, may be to the greater, as 3 to 1. Ans. 7 and 3.

74. Find two numbers, such that if half the less be taken from the greater, 14 may remain; and if 1 be added to four times the less, the sum will be double the greater.

Ans. 9, and $18\frac{1}{2}$.

75. A person has two sorts of wine, one worth 20 pence a quart, and the other 12 pence; from which he would mix a quart to be worth 14 pence. How much of each must he take?

Ans. $\frac{1}{4}$ of the first, and $\frac{3}{4}$ of the second.

76. A labourer reaps 40 acres of barley and oats, and receives 4 shillings an acre for the oats. If the price *per* acre for the barley had been one shilling more, it would have been to the price *per* acre of the oats, as 3 to 2; and he receives in all £9 5*s*. How many acres were there of each sort?

Ans. 15 of oats, and 25 of barley.

77. Before noon, a clock which is too fast and points to afternoon time, is put back five hours and forty minutes; and it is observed that the time before shown, is to the true time, as 29 to 105. Required the true time.

Ans. 15 min. before 9 A. M.

78. A and B began to pay their debts. A's money was at first two-thirds of B's; but after A had paid £1 less than two-thirds of his money, and B £1 more than seven-eighths of his, it was found that B had only half as much as A had left. What sum had each at first?

Ans. A had £72, and B £108.

79. A regiment of 594 men is to be raised from three towns, A, B, and C. The contingents of A and B are in the proportion of three to five; and of B and C, in the proportion of eight to seven. Required the number raised by each.

Ans. 144 by A, 240 by B, and 210 by C.

80. A besieged garrison had such a quantity of bread, as would, if distributed to each man at 10 ounces a day, last 6 weeks; but having lost 1200 men in a sally, the governor was enabled to increase the allowance to 12 ounces *per* day, for 8 weeks. Required the number of men at first in the garrison? Ans. 3200.

81. A man gives the first beggar he meets $\frac{1}{6}$ of the pence he had and 4*d* more; to the second, $\frac{1}{6}$ of the remainder and

8*d.* more; and so on, increasing 4*d.* every time, till he had nothing left; and then all the beggars had equal shares. Required the number of beggars, and of pence at first.

Ans. 120*d.* and 5 beggars.

82. A gentleman gave in charity £46; a part in equal portions to 5 poor men, and the rest in equal portions to 7 poor women. Now a man and a woman had between them £8. What was given to the men, and what to the women?

Ans. The men received £25, the women £21.

II. *Simple Equations involving two or more unknown quantities.*

1. Find two numbers, such that if the first be multiplied by 4 and the second by 8, the sum of the products will be 184; but if the first be multiplied by 8 and the second by 4, the difference of the products will be 8. Ans. 10 and 18.

*2. What two numbers are those of which if the first be increased by 6 it will be three times the greater, and if the second be diminished by 4, it will be $\frac{1}{5}$ of the first?

Ans. 15 and 7.

*3. A and B together own 1320 acres of land, A sells $\frac{1}{3}$ of his share, and B $\frac{1}{5}$ of his, and the same number of acres remain to each. What were their shares? Ans. 720 and 600.

4. What fraction is that which, by subtracting 4 from both its terms, becomes $\frac{1}{8}$, and by adding 9 to both, becomes $\frac{2}{3}$?

Ans. $\frac{5}{12}$.

5. A party were to divide their expenses equally. Had there been 3 persons more and each paid 5 cents more, the bill would have been $3.75 more; but if there had been 10 less, and each had paid 7 cents less, it would have been $9.56 less. How many persons were there, and how much did each pay? Ans. 18 persons, paying 90 cts. each.

*6. Coffee and sugar were cheaper before the war than at present; coffee by 2 cts. and sugar by 4 cts. a pound, and yet the price of the former was double that of the latter. Now the prices are as 3 to 2. At what rate are they sold?

Ans. Coffee 18 and sugar 12 cts.

*7. What are those numbers, whose sum is 20, and the quotient obtained by dividing the greater by the less is $\frac{3}{2}$?

Ans. 12 and 8.

8. What fraction is that, whose numerator being doubled, and denominator increased by 7, the value becomes $\frac{2}{3}$; but the denominator being doubled, and the numerator increased by 2, the value becomes $\frac{3}{5}$? Ans. $\frac{4}{5}$.

9. There are two numbers, such, that $\frac{1}{2}$ the greater added to $\frac{1}{3}$ the less is 13; and if $\frac{1}{2}$ the less be taken from $\frac{1}{3}$ the greater, the remainder is nothing. What are the numbers?

Ans. 18 and 12.

10. There is a number expressed by two digits, which is double the sum of its digits, and 9 added to four times the number will invert the digits. What is the number? Ans. 18.

*11. A number consists of two places of figures, and their sum is 4 times the units figure; moreover, if 8 be added to $\frac{1}{3}$ of the number, the digits will be inverted. Required the number? Ans. 93.

12. There is a cistern into which water is admitted by three cocks, two of which are exactly of the same dimensions. When they are all open, five-twelfths of the cistern is filled in 4 hours: and if one of the equal cocks be stopped, seven-ninths of the cistern is filled in 10 hours and 40 minutes. In how many hours would each cock fill the cistern?

Ans. Each of the equal ones in 32 hours, and the other in 24.

*13. A and B speculate with different sums; A gains 150*l*, B loses 50*l*, and now A's stock is to B's as 3 to 2. But had A lost 50*l*. and B gained 100*l*, then A's stock would have been to B's as 5 to 9. What was the stock of each?

Ans. A's was 300*l*. and B's 350*l*.

*14. Two persons can perform a piece of work together in 21 days. After working in company 9 days, one leaves the other to complete the work alone, which he does in 20 days more. How many days would it take each to perform the whole alone? Ans. $52\frac{1}{2}$ and 35.

*15. Six persons comparing their gains found that A's and B's amounted together to \$30; C's and D's to \$140; E's and F's to 75; also that A had half as much as C, D four times as much as E, and F five times as much as B. What was the gain of each? Ans. A \$20, B 10, C 40, D 100, E 25, and F 50.

16. Two persons, A and B, had a mind to purchase a house rated at 1200 dollars; says A to B, if you give me $\frac{2}{3}$ of your money, I can purchase the house alone; but says B to A, if you will give me $\frac{3}{4}$ths of yours, I shall be able to purchase the house. How much money had each of them?

Ans. A had 800 and B 600.

17. A Vintner bought 6 dozen of port wine and 3 dozen of white, for 12*l*. 12 shillings; but the price of each afterwards falling a shilling *per* bottle, he had 20 bottles of port, and 3 dozen and 8 bottles of white more, for the same sum. What was the price of each at first?

Ans. the price of port was 2*s*. and of white 3*s*. *per* bottle.

18. It is required to find two numbers, such that $\frac{2}{5}$ of one added to $\frac{2}{3}$ of the other shall make 10: and if the first be multiplied by 3, and the second by 4, $\frac{1}{11}$ of the sum shall be 6. Ans. 10 and 9.

*19. A may-pole was broken by the wind in such a manner that 4 times the upper or broken part, added to 6 times the remaining part, was equal to 5 times the whole and 28 over: and the proportion of the former to the latter part was 9 to 16. Required the height at first. Ans. 100 ft.

20. The contents of two casks differ by 8 gallons. Each is worth as many shillings *per* gallon, as there are gallons in the other: and both together are worth £128. How many gallons in each? Ans. 40 and 32.

21. A lot of land was sold at the rate of $2.25 per foot, and came to $5400. Had it been 10 feet shorter, its value would have been $4725. What was its length and breadth?

Ans. 80 by 30 ft.

22. Find a number consisting of three digits, such that the middle digit shall be double the first, or left-hand digit, the sum of the digits 12, and if 792 be added to the number, the digits will be reversed. Ans. 129.

23. Find three numbers, such that the first with half the others, the second with one third of the others, and the third with one fourth of the others, shall always make 34.

Ans. 10, 22, and 26.

*24. A person having mixed a certain quantity of brandy and water, found that if he had mixed 6 gallons more of each, there would have been 7 gallons of brandy for every 6 of water; but if he had mixed 6 less of each there would have been 6 gallons of brandy for every 5 of water. How many gallons of each were mixed? Ans. 78 of brandy and 66 of water.

25. A person had two casks, the larger of which he filled with ale, and the smaller with cider. Ale being half a crown and cider 11*s.* *per* gallon, he paid £8 6*s.*; but had he filled the larger with cider, and the smaller with ale, he would have paid £11 5*s.* 6*d.* How many gallons did each hold?

Ans. The larger contained 18, and the smaller 11 gallons.

26. D and E each cut a separate pack of cards, in such a manner that the cards cut off by D, added to those left by E made 47; also the number left by both exceeded the number cut by both, by 50. How many cards did each cut?

Ans. 11 and 16.

27. What are the dimensions of a rectangular court, which if lengthened 7 feet, and made 4 feet broader, would contain 363 feet more; but if made 4 feet shorter and 3 feet narrower, would be diminished 208 feet? Ans. 40 long and 25 wide.

28. A Merchant finds that if he mixes sherry and brandy in quantities which are in the proportion of 2 to 1, he can sell the mixture at 78*s.* *per* dozen; but if the proportion be as 7 to 2, he must sell it at 79 shillings a dozen. Required the price of each liquor.

Ans. The price of sherry was 81*s.*, and of brandy 72*s.* *per* dozen.

*29. Two persons discoursing of their money—said A to B, give me 25 dollars, and I shall have as much as you; said B to A, give me 22 dollars, and I shall have as much again as you. What had each? Ans. A 116, B 166.

*30. Two pieces of muslin were bought, of equal fineness, but of unequal lengths; one for £4 10*s.*, the other for 5 guineas. Now if each had been 3 yards longer, they would have been to one another as 7 to 8. What was the length of each, and the price *per* yard? Ans. 18 and 21, at 5*s.*

*31. A farm, consisting of two kinds of land, is let at an annual rent of £390, the pasture being valued at 30*s.* *per* acre, and the arable at £3. Now the number of acres of arable is to half the excess of the arable above the pasture, as 5 to 1. Required the quantity of each.

Ans. 60 acres pasture, and 100 acres arable.

32. A sets out express from C towards D, and three hours afterwards B sets out from D towards C, travelling 2 miles an hour more than A. When they meet, it appears that the distances they have travelled are in the proportion of 13 to 15; but had A travelled five hours less, and B had gone 2 miles an hour more, they would have been in the proportion of 2 : 5. How many miles did each go *per* hour, and how many hours did they travel before they met?

Ans. A went 4, and B 6 miles an hour, and they travelled 10 hours after B set out.

33. A grocer bought a cask of wine and another of gin for $210, the wine at $1.50 a gallon, and the gin at $0.50 a gallon. He afterwards sold $\frac{2}{3}$ of his wine and $\frac{3}{7}$ of his gin for $150, by which he gained $15. How many gallons were there of each? Ans. 126 of wine, and 42 of gin.

*34. A vintner has two casks of wine, from each of which he draws 6 gallons, and finds the remainders in the proportion of 4 to 7. He then puts into the less 3 gallons, and into the greater 4 gallons, and the proportion is that of 7 to 12. How many gallons were there in each at first?

Ans. 38 and 62.

35. When wheat was 8 shillings a bushel, and rye 5 shillings, a man wished to fill his sack with a mixture of the two for the money he had with him. He found that if he bought

10 bushels of wheat, and laid out the rest of his money in rye, he would not fill his sack by two bushels ; and if he took 10 bushels of rye, and then filled his sack with wheat, he would have 5 shillings remaining. How many bushels would the sack hold, and how much money had he ?

Ans. 15 bushels, and 95 shillings.

*36. The weight of the first of two loaded wagons was to that of the second as 3 to 5. A portion of the load of the weightier wagon, however, having been taken out, and as much again being removed from the other, the weight of the first was to that of the second as 1 to 2, and both together weighed 21 tons. How much did each weigh at first, and how much was taken from each ?

Ans. One 9 tons, the other 15, and 1 ton taken from the latter, and 2 from the former.

37. A and B play at backgammon, and A bets three shillings to two on every game. After a certain number of games, it appears that A has won 3 shillings ; but had he bet 5*s*. to 2, and lost one game more out of the same number, he would have lost 30 shillings. How many games did they play ?

Ans. 34.

38. Some smugglers discovered a cave which would exactly hold their cargo, which consisted of 13 bales of cotton and 33 casks of wine. While they were unloading, a revenue cutter hove in sight, and they sailed away with 9 casks and 5 bales, leaving the cave two thirds full. How many bales or casks would it hold ? Ans. 24 bales, or 72 casks.

39. Find two numbers, the greater of which shall be to the less as their sum to 42, and as their difference to 6.

Ans. 32 and 24.

40. In one of the corners of a rectangular garden there is a fish-pond, whose area is one ninth of the whole garden ; the periphery of the garden exceeding that of the pond by 200 yards. Also, if the greater side be increased by 3 yards, and the other by 5 yards, the garden will be enlarged by 645 square yards. The pond is a rectangle, about the same diameter as the garden. What is the length of each side ?

Ans. 90 and 60 yards.

41. A and B engage together in play ; in the first game A wins as much as he had and four shillings more, and finds he has twice as much as B. In the second game B wins half as much as he had at first and one shilling more, and it appears that he has then three times as much as A. What sum had each at first ? Ans. A had 6*s*. and B 8*s*.

42. A farmer parting with his stock, sells to one person 9 horses and 7 cows for 300 dollars : and to another, at the

same prices, 6 horses and 13 cows for the same sum. What was the price of each?

Ans. The price of a cow was 12 dollars, and of a horse 24 dollars.

*43. A bill of £26 5*s*. was paid with half guineas and crowns, and twice the number of half guineas exceed 3 times the number of crowns by 17. How many were there of each?

Ans. 40 half guineas and 21 crowns.

44. Two labourers, A and B, received £5 17*s*. for their wages, A having been employed 15, and B 14 days; and A received for working four days, 11 shillings more than B did for three days. What were their daily wages?

Ans. A had 5*s*. and B 3*s*. a day.

*45. A boat has two sets of sails, the second worth $\frac{4}{7}$ths of the boat, and both sets together valued at £45. Now the worth of the boat with the second set, is to its worth with the first, as 11 to 12. Required the respective values.

Ans. The boat £35: the 1st set £25, and the 2d set £20.

*46. A garden, of which the annual rent is £50, is occasionally let with one of two houses. Now the rent of the garden and first house is to the rent of the second house, as 5 to 3, and the rent of the second house is to that of the first, as 3 to 4. At what yearly rent are the houses let?

Ans. The first at £200, and the other at £150.

47. There are 3 vessels which together hold 127 gallons; now 3 times the first with 4 times the second will fill the third; and four times the first with twice the second make 70 gallons. Required the content of each.

Ans. 8, 19, and 100 gallons respectively.

48. Find three numbers, such that $\frac{1}{2}$ of the first, $\frac{1}{3}$ of the second, and $\frac{1}{4}$ of the third shall make 62; $\frac{1}{3}$ of the first, $\frac{1}{4}$ of the second, and $\frac{1}{5}$ of the third, 47; and $\frac{1}{4}$ of the first, $\frac{1}{5}$ of the second, with $\frac{1}{6}$ of the third, 38. Ans. 24, 60, and 120.

49. Required those three numbers, whereof any one, added to half the other two, will make 51.

Ans. Impossible with unequal numbers. Each must be $25\frac{1}{2}$.

50. A, B, and C can together finish a piece of work in 9 days; A, B, and D, in 10 days; A, C, and D, in 11 days; and B, C, and D, in 12 days. In how long time can all four together finish it? Ans. $7\frac{1797}{2289}$ days.

The problems which are thus marked (*), may be neatly solved with only one unknown quantity; and it will be a useful exercise for the learner to attempt their solution by both methods.

III. *Questions producing Pure Equations of the second and higher degrees.*

1. What two numbers multiplied together give the product 245; and if the greater be divided by the less, the quotient is 5? Ans. 35 and 7.

2. Find two numbers whose sum is to the less, as 10 to 3, and the difference of whose squares is 640? Ans. 28 and 12.

3. There is a rectangular field containing 192 square rods, and the sum of the length and breadth is seven times their difference. Required its dimensions.

Ans. 16 long, and 12 wide.

4. What are the contents and value of that field whose sides are as 10 to 3;—which is worth as many dollars *per* acre as there are rods in the length, and will amount, at that rate, to 40 times as many dollars as there are rods in the width?

Ans. { 12 acres. $960 value.

5. A person has a quantity of sugar for sale. If he sells it for three times as many cents as there are pounds, he will get as much above $7, as he will get less than $7 by selling it for half as many cents as there are pounds. How much has he? Ans. 20 lbs.

6. Take a certain number and multiply it by 7, add 15 to the product, and multiply this sum by twice the number. Then divide by 90, and subtract from the quotient $\frac{1}{3}$ of the number, and 140 will remain. Required the number. Ans. 30.

7. What two numbers are as 2 to 3, the difference of whose cubes is 152? Ans. 4 and 6.

8. Two pieces of cloth which together amount to 20 yards were sold, each for as many dollars *per* yard, as it contained yards; and the sums received for the two were to each other as 9 to 4. How many yards were there in each piece?

Ans. 12 and 8 yds.

9. There is a wall containing 5400 cubic feet. The height is 5 times the thickness, and the length 8 times its height. What are the dimensions?

Ans. 3ft. thick, 15ft. high, and 120ft. long.

10. It is required to find that number to which 20 being added, and from which 10 being subtracted, the square of the sum added to twice the square of the remainder shall be 17475.

Ans. 75.

11. Divide 100 into two parts, so that their product may be 2100. Ans. 70 and 30.

Note. Let $x=$ the excess of the greater part above 50.

12. Find two numbers, one of which is triple the other, and the sum of their fourth powers is 1312. Ans. 2 and 6.

13. A parallelogram contains 1296 square yards; and if the length be divided by the breadth the quotient will be nine. Required the dimensions. Ans. 108 and 12.

14. A farmer sells a meadow at such a rate that the price of an acre was to the number of acres as 2 to 3. If he had received 270 dollars more for it, the price *per* acre would have been to the number of acres as 3 to 2. Required the number and price. Ans. 18 acres at $12.

15. Find those two numbers whose difference multiplied by the greater gives 160, but multiplied by the less, gives only 96. Ans. 20 and 12.

16. A number of boys set out to rob an orchard, each carrying as many bags as there were boys, and each bag capable of holding 6 times as many apples as there were bags. After filling their bags, they found the whole number of apples was 1536. How many boys were there? Ans. 4.

17. A departs from London toward Lincoln at the same time that B leaves Lincoln for London. When they met, A had travelled 20 miles more than B, having gone as far in $6\frac{2}{3}$ days as B had in all; and it appeared that B would not reach London under 15 days. What is the distance between those cities, and how far had each travelled?

Ans. 100 miles distance: A having gone 60, and B 40.

Note. Let $x=$ A's miles *per* day,
$y=$ the number of days for each,
and $z=$ B's miles *per* day.

18. It is required to find three numbers, such that the sum of the first and second multiplied by the third may be 63; the sum of the second and third multiplied by the first may be 28; and the sum of the first and third multiplied by the second may be 55. Ans. 2, 5, and 9.

19. Required those two numbers whose sum is 18, and the sum of their squares 170. Ans. 7 and 11.

Note. A neat solution may be had by substituting $m+n$ for x, and $m-n$ for y.

20. Find a number, such that twice its cube may exceed 106 as much as one fifth of the square root of its sixth power falls short of 70. Ans. 2 3/ 10.

21. It is required to find two numbers, whose sum, product, and the sum of whose squares shall be equal.

Let $x+y$ and $x-y$ be the two numbers,

Ans. It is impossible.

22. It is required to divide the number 18 into two such parts that their squares may be as 25 to 16. Ans. 10 and 8.

23. Two workman, A and B, were engaged to work for a certain number of days, at different rates. At the end of the time, A, who had played 4 of those days, received 75 shillings; but B, who had played 7 days, received only 48 shillings. Now had B played only 4 days, and A had played 7, they would have received exactly alike. For how many days were they engaged, and how many did each work?

Ans. Engaged for 19 days. A worked 15, and B 12, A at 5 shillings, and B at 4, *per* day.

24. What two numbers are those whose difference multiplied by the less produces 42, and by their sum 133?

Ans. 13 and 6.

25. In a mixture of rum and brandy, the difference between their quantities, is to the brandy, as 100 is to the number of gallons of rum; and the same difference is to the rum, as 4 to the number of gallons of brandy. How many gallons were there of each? Ans. 25 of rum, and 5 of brandy.

26. A Farmer has 2 cubical stacks of hay; the side of one is 3 yards longer than the side of the other, and the difference of their contents is 117 solid yards. Required the side of each.

Ans. 5 and 2 yards.

27. The captain of a privateer descrying a trading vessel 7 miles ahead, sailed 20 miles in direct pursuit of her, and then observing that the trader steered at right angles to her former course, changed his own course so as to overtake her without making another tack; the privateer running 10 knots, and the trader 8 knots *per* hour. How long did the chase continue?

Ans. $2\frac{1}{2}$ hours.

28. Given the product of two numbers, 12, and the sum of their cubes, 224, to find the numbers. Ans. 6 and 2

29. Find two numbers, so that the fifth power of one shall be to the cube of the other as 972 to 125. Ans. 6 and 10.

30. What three numbers are as $\frac{1}{2}$, $\frac{1}{3}$, and $\frac{1}{4}$, the sum of whose squares is 549. Ans 18, 12, and 9.

31. Given $\left.\begin{array}{l}\frac{x^2}{y}+\frac{y^2}{x}=18 \\ \text{and } x+y=12\end{array}\right\}$ to find the values of x and y.

Note. Put $z+v=x$, and $z-v=y$, $x=8$, or 4 and $y=4$ or 8.

32. Required those three numbers whose product divided by the first and second gives 200; by the first and third, 150, and by the second and third, 120. Ans. 10, 20, and 30.

33. What two numbers are those whose product is 320, and the difference of their cubes is to the cube of their difference as 61 to unity? Ans. 20 and 16.

34. Required four numbers in arithmetical progression,

which being increased by 2, 4, 8 and 15 respectively, the sums shall be in geometrical progression.

Ans. 6, 8, 10 and 12.

35. In a right angled triangle given the hypothenuse 60, and the radius of the inscribed circle 12 ; to find the other sides.

Ans. 36 and 48.

36. Of four numbers in geometrical progression there are given the sum of the two least =20, and the sum of the two greatest =45, to find the numbers. Ans. 8, 12, 18 and 27.

37. The sum of £700 was divided among four persons, whose shares were in geometrical progression ; and the difference between the greatest and least, was to the difference between the two means, as 37 to 12. What were the several shares ? Ans. 108, 144, 192, and 256 pounds.

38. Find a number consisting of two digits, such that if it be multiplied by the right hand digit, the product will be 159, the sum of the squares of the digits being 34 ; and if the sum of the digits be multiplied by the right hand digit, the product will be 24. Ans. 53.

39. Required two numbers whose sum is five-sixths of their product, and if the second be multiplied by the square of the first, and the first by the square of the second, the sum of those products shall be 30. Ans. 3 and 2.

Note.—The negative roots will also answer the conditions.

IV. *Affected Quadratic Equations involving one Unknown Quantity.*

1. The expenses of a party amount to $10 ; and if each pays 30 cents more than there are persons, the bill will be settled. How many are there ? Ans. 20.

2. A person took $6 with him to distribute equally among the paupers of his parish ; but as 5 of them were absent, the remainder received 10 cents apiece more than they otherwise would. How many paupers were there ? Ans. 20.

3. Divide 36 into three such parts, that the second may exceed the first by 4, and the sum of all their squares may be 464. Ans. 8, 12 and 16.

4. Divide 100 into two parts, so that the sum of their square roots shall be 14. Ans. 64 and 36.

5. Find two numbers whose difference is 10, and if 600 be divided by each, the difference of the quotients shall also be 10. Ans. 20 and 30.

6. Two droves of oxen differing 6 in number were each purchased for £672, those in one drove being valued at £2 a head more than those in the other. Of how many did

either drove consist, and at what rate *per* ox were they purchased?

Ans. The less drove of 42 at 16*l*., the other of 48 at 14*l*.

7. In a fleet of transports, the square root of half the number of ships expressed the number bound for the West Indies; $\frac{7}{8}$ of the fleet were for the Mediterranean, and the remaining 8, for the coast of Africa. Required the whole number. Ans. 128.

8. Two regiments, whose aggregate strength was 1305 men, stood respectively in solid square, the front of one presenting 3 men more than the front of the other. Of how many did each consist? Ans. 729 and 576.

9. A gentleman pays $180 more for his chaise than for his horse; and the price of the chaise is to that of the horse, as the latter is to 15. What is the price of each?

Ans. The chaise 240, and the horse 60.

10. If the square of a certain number be subtracted from 52, and the square root of the remainder be increased by 6, and that sum multiplied by 3, and the product divided by the number itself, the quotient will be 5. Required the number.

Ans. 6.

11. One lays out a certain sum of money in goods, which he sold again for 56 dollars, and gained as much per cent. as the goods cost him. What was the first cost? Ans. 40*l*.

12. What two numbers are those whose difference is 15, and half their product equals the cube of the less?

Ans. 3 and 18.

13. There are three numbers in continued geometrical progression; the sum of the first and second is 10, and the difference of the second and third is 24. What are the numbers? Ans. 2, 8, 32.

Note.—This gives two positive roots, only one of which answers the conditions.

14. In a rectangle, given the difference between the diagonal and the length, 2, and between the diagonal and the breadth, 9. Required the sides. Ans. 8 and 15.

15. The base of a right-angled triangle is 20, and is a geometrical mean between the hypothenuse and the perpendicular. What is the hypothenuse? Ans. 25,4+

16. There is a certain number consisting of two digits; if they be multiplied together, the square root of the product will equal the right hand digit, and if the square of their difference be divided by the left hand digit, the quotient will be one third of their sum. Required the number. Ans. 93.

17. A man playing at hazard won at the first throw as much money as he had in his pocket; at the second he won

the square root of what he then had, and 5 shillings more; at the third throw he won the square of all he then had, when he found he had in all, 112*l.* 16*s.* With how much did he begin to play? Ans. 18 shillings.

18. Find a number from the cube of which if you subtract 19, and multiply the remainder by that cube, the product shall be 216. Ans. 3.

19. Two partners, A and B gained 18*l.* by trade. A's money was in trade 12 months, and he received for his principal and gain 26*l.* Also B's money, which was 30*l.*, was in trade 16 months. What was A's stock? Ans. 20*l.*

20. The plate of a looking-glass is 18 inches by 12, and is to be framed with a frame of equal width, whose area is to be equal to that of the glass. Required the width of the frame. Ans. 3 inches.

21. A set out from C towards D, and travelled 7 miles a day. After he had gone 32 miles, B set out from D towards C, and went every day $\frac{1}{19}$th of the whole journey, and after travelling as many days as he went miles in one day, he met A. Required the distance from C to D. Ans. 152 miles.

Note.—The other value will also answer the conditions.

22. A and B hired a pasture into which A put 4 horses, and B as many as cost him 18 shillings a week. Afterwards B. put in two additional horses, and found that he must pay 20 shillings a week. At what rate was the pasture hired?

Ans. 30*s.* *per* week.

23. What number is that to which if 24 be added, the square root of the sum shall be 18 less than the original number? Ans. 25.

24. A wall was built round a rectangular court to a certain height. Now the length of one side of the court was 2 yards less than 8 times the height of the wall, and the length of the adjacent side was 5 yards less than 6 times the height of the wall; and the number of square yards in the court, was greater than the number in the wall by 178. Required the dimensions of the court, and the height of the wall.

Ans. The sides were 30 and 19, and the height 4 yards.

25. The sum of £190 was divided among three persons in geometrical progression, and the greatest share exceeded the least by £50. Required the shares.

Ans. 40, 60, and 90 pounds

26. Given the sum of two numbers, 20, and the sum of their cubes, 2240, to find the numbers. Ans. 12 and 8.

27. A traveller, bound to a place 14 miles distant, goes 26 miles the first day, 24 the second, and so on decreasing in

arithmetical progression. In how many days will he arrive at his journey's end ? Ans. 7 days.

28. Given $9x+\sqrt{(16x^2+36x^3)}=15x^2-4$, to find the value of x. Ans. $x=\frac{4}{3}$.

29. What are those two numbers whose product is 100, and the difference of their square roots 3 ? Ans. 25, and 4.

30. Given $\sqrt{x^5}-\frac{40}{\sqrt{x}}=3x$; to find the value of x.

Ans. $x=4$.

Affected Quadratic Equations involving more than one unknown quantity.

1. What two numbers are those whose product is 120, and if the greater be increased by 8, and the less by 5, the product of the two numbers thence arising shall be 300 ?

Ans. 10 and 12.

2. A merchant bought a piece of cloth for £8, and after cutting off 4 yards, sold the remainder for what the whole cost him, by which he gained 2 shillings a yard upon what was sold. How many yards were there, and what did they cost ?

Ans. 20 yds. at 8*s*.

3. Find two numbers, whose sum multiplied by the greater produces 130, and whose difference multiplied by the less gives 21. Ans. 10 and 3.

4. What number is that, which being divided by the product of its two digits, the quotient is 2 ; and if 27 be added to it the digits will be inverted ? Ans. 36.

5. There are three numbers, the difference of whose differences is 8 ; their sum is 41, and the sum of their squares 699. What are the numbers ? Ans. 7, 11, and 23.

6. There are three numbers, the difference of whose differences is 5 ; their sum is 44, and their continual product 1950. Required the numbers. Ans. 6, 13, and 25.

7. A farmer sold 80 bushels of barley and 100 bushels of oats for £65 ; but he sold 60 bushels more of oats for £20, than of barley for £10. What was the price of each ?

Ans. Barley 10*s* and oats 5.

8. The fore wheel of a carriage makes 5 revolutions more than the hind wheel in going 60 yards ; but if the periphery of each wheel be increased one yard, it will make only one revolution more than the hind wheel in the same space. Required the circumference of each.

Ans. The less 3 yds. the other 4.

9. A person bought two cubical stacks of hay for £41, each

of which cost as many shillings *per* solid yard, as there were yards in the side of the other; and the greater stood on more ground than the less by 9 square yards. What was the price of each? Ans. £25, and £16.

10. A farmer received £7 4*s*. for a certain quantity of wheat, and an equal sum, at a price less by 1*s*. 6*d*. *per* bushel, for a quantity of barley, which exceeded the quantity of wheat by 16 bushels. How many bushels were there of each?

Ans. 32 of wheat, and 48 of barley.

11. A student being questioned concerning his age and height, proposed the following equations; $x^2+xy=1155$, and $y^2-xy=2914$, from which it was required to find his age in years and his height in inches.

Ans. His age 15 yrs. his height 62 in.

12. It is required to find two such numbers that their sum being subtracted from the sum of their squares may leave 14; and if their product be added to their sum, it may make 14.

Ans. 4 and 2.

13. The hypothenuse of a right-angled triangle is 50 feet, and the side of the inscribed square is $17\frac{1}{7}$ feet. Required the base and perpendicular. Ans. 30 and 40 ft.

14. There are three towns, A, B and C; the road from B to A forming a right angle with that from B to C. Now a person travels a certain distance from B towards A, and then crosses by the nearest way, to the road leading from C to A, and finds himself 3 miles from A and 7 from C. Arriving at A, he finds he has gone farther by one-fourth of the distance from B to C, than he would have done, had he not left the direct road. Required the distance of B from A and C.

Ans. From C, 8 or 6 miles, and from A, 6 or 8.

15. A took cucumbers to market, and B took thrice as many eggs, and they found that if B gave all his eggs for the cucumbers, A would lose 10 pence, as they were then selling. A therefore reserves two-fifths of his cucumbers, by which B would lose six pence, at the same rate. But B selling the cucumbers at 6 pence apiece, gains upon the whole the price of six eggs. Required the number of eggs and cucumbers, and their price.

Ans. 30 eggs at a penny, and 10 cucumbers at 4 pence apiece.

16. Find two numbers whose difference is 279, and the difference of their cube roots 3. Ans. 343 and 64.

17. The product of five numbers in arithmetical progression is 945, and their sum is 25. Required the numbers.

Ans. 9, 7, 5, 3, and 1.

18. A gentleman divided £210 among three servants, in

geometrical progression; the first had £90 more than the last How much had each? Ans. 120, 60, and 30 pounds.

19. There are three numbers in geometrical progression, the greatest of which exceeds the least by 15. Also the difference of the squares of the greatest and least, is to the sum of the squares of all three, as 5 to 7. Required the numbers.
Ans. 5, 10, and 20.

20. There is a number consisting of three digits, the first of which is to the second as the second to the third; the number itself is to the sum of its digits, as 124 to 7; and if 594 be added to it, the digits will be inverted. What is the number?
Ans. 248.

21. There are three numbers in arithmetical progression, and the square of the first added to the product of the other two is 16; the square of the second added to the product of the other two is 14. What are the numbers?
Ans. 1, 3, and 5.

22. The sum of three numbers in geometrical progression is 3, and the product of the mean and the sum of the extremes is 30. Required the numbers. Ans. 1, 3, 9.

23. Given $x^2+xy=12$ and $xy-2y^2=1$ } to find the values of x and y
Ans. $x=3$, and $y=1$.

Note. Put $vy=x$. The other root produces a surd.

24. Find two numbers whose product is 300; and if 10 be added to the less, and 8 subtracted from the greater, the product of the sum and remainder shall still be 300.
Ans. 15 and 20.

25. Of four numbers in arithmetical progression, the sum of the squares of the means is 400; and the sum of the squares of the extremes, 464. Required the numbers.
Ans. 8, 12, 16, and 20.

26. The sum of two numbers added to the square root of their sum, equals 12, and the sum of their cubes is 189. What are the numbers? Ans. 5 and 4.

27. Given $\sqrt{x}+\sqrt{y}=5$ and $x-2\sqrt{xy}+y-\sqrt{x}+\sqrt{y}=0$ } to find the values of x and y. Ans. $x=9$, and $y=4$.

28. Given $x-y=4$ and $x^4-y^4=2320$ } to find x and y.
Ans. $x=7$ and, $y=3$.

Note. Put $m+n$ for x, and $m-n$ for y.

29. Given $x+y=7$ and $x^5+y^5=3157$ } to find x and y, the former being the less. Ans. $x=2$, and $y=5$.

Note. Substitute as before.

www.ingramcontent.com/pod-product-compliance
Lightning Source LLC
LaVergne TN
LVHW010218110826
845151LV00004B/1132

* 9 7 8 1 4 2 5 5 2 6 1 3 9 *